Achieving sustainable production of pig meat

Volume 2: Animal breeding and nutrition

It is widely recognised that agriculture is a significant contributor to global warming and climate change. Agriculture needs to reduce its environmental impact and adapt to current climate change whilst still feeding a growing population, i.e. become more 'climate-smart'. Burleigh Dodds Science Publishing is playing its part in achieving this by bringing together key research on making the production of the world's most important crops and livestock products more sustainable.

Based on extensive research, our publications specifically target the challenge of climate-smart agriculture. In this way we are using 'smart publishing' to help achieve climate-smart agriculture.

Burleigh Dodds Science Publishing is an independent and innovative publisher delivering high quality customer-focused agricultural science content in both print and online formats for the academic and research communities. Our aim is to build a foundation of knowledge on which researchers can build to meet the challenge of climate-smart agriculture.

For more information about Burleigh Dodds Science Publishing simply call us on +44 (0) 1223 839365, email info@bdspublishing.com or alternatively please visit our website at www.bdspublishing.com.

Related titles:
Achieving sustainable production of pig meat Volume 1: Safety, quality and sustainability
Print (ISBN 978-1-78676-088-3); Online (ISBN 978-1-78676-091-3, 978-1-78676-090-6)

Achieving sustainable production of pig meat Volume 3: Animal health and welfare
Print (ISBN 978-1-78676-096-8); Online (ISBN 978-1-78676-099-9, 978-1-78676-098-2)

Improving organic animal farming
Print (ISBN 978-1-78676-180-4); Online (ISBN 978-1-78676-182-8, 978-1-78676-183-5)

Chapters are available individually from our online bookshop: https://shop.bdspublishing.com

climate•SMART•publishing
IN AGRICULTURAL SCIENCE

BURLEIGH DODDS SERIES IN AGRICULTURAL SCIENCE

NUMBER 24

Achieving sustainable production of pig meat

Volume 2: Animal breeding and nutrition

Edited by Professor Julian Wiseman, University of Nottingham, UK

burleigh dodds
SCIENCE PUBLISHING

Published by Burleigh Dodds Science Publishing Limited
82 High Street, Sawston, Cambridge CB22 3HJ, UK
www.bdspublishing.com

Burleigh Dodds Science Publishing, 1518 Walnut Street, Suite 900, Philadelphia, PA 19102-3406, USA

First published 2017 by Burleigh Dodds Science Publishing Limited

Library of Congress Control Number: 2017938496

British Library Cataloguing in Publication Data
A catalogue record for this book is available from the British Library

ISBN 978-1-78676-092-0 (print)
ISBN 978-1-78676-094-4 (online)
ISBN 978-1-78676-095-1 (online)
ISSN 2059-6936 (print)
ISSN 2059-6944 (online)

Typeset by Deanta Global Publishing Services, Chennai, India
Printed by CPI Group (UK) Ltd, Croydon CR0 4YY

Contents

Series list

Title	Series number
Achieving sustainable cultivation of maize - Vol 1 From improved varieties to local applications Edited by: Dr Dave Watson, CGIAR Maize Research Program Manager, CIMMYT, Mexico	001
Achieving sustainable cultivation of maize - Vol 2 Cultivation techniques, pest and disease control Edited by: Dr Dave Watson, CGIAR Maize Research Program Manager, CIMMYT, Mexico	002
Achieving sustainable cultivation of rice - Vol 1 Breeding for higher yield and quality Edited by: Prof. Takuji Sasaki, Tokyo University of Agriculture, Japan	003
Achieving sustainable cultivation of rice - Vol 2 Cultivation, pest and disease management Edited by: Prof. Takuji Sasaki, Tokyo University of Agriculture, Japan	004
Achieving sustainable cultivation of wheat - Vol 1 Breeding, quality traits, pests and diseases Edited by: Prof. Peter Langridge, The University of Adelaide, Australia	005
Achieving sustainable cultivation of wheat - Vol 2 Cultivation techniques Edited by: Prof. Peter Langridge, The University of Adelaide, Australia	006
Achieving sustainable cultivation of tomatoes Edited by: Dr Autar Mattoo, USDA-ARS, USA & Prof. Avtar Handa, Purdue University, USA	007
Achieving sustainable production of milk - Vol 1 Milk composition, genetics and breeding Edited by: Dr Nico van Belzen, International Dairy Federation (IDF), Belgium	008
Achieving sustainable production of milk - Vol 2 Safety, quality and sustainability Edited by: Dr Nico van Belzen, International Dairy Federation (IDF), Belgium	009
Achieving sustainable production of milk - Vol 3 Dairy herd management and welfare Edited by: Prof. John Webster, University of Bristol, UK	010
Ensuring safety and quality in the production of beef - Vol 1 Safety Edited by: Prof. Gary Acuff, Texas A&M University, USA & Prof.James Dickson, Iowa State University, USA	011
Ensuring safety and quality in the production of beef - Vol 2 Quality Edited by: Prof. Michael Dikeman, Kansas State University, USA	012
Achieving sustainable production of poultry meat - Vol 1 Safety, quality and sustainability Edited by: Prof. Steven C. Ricke, University of Arkansas, USA	013
Achieving sustainable production of poultry meat - Vol 2 Breeding and nutrition Edited by: Prof. Todd Applegate, University of Georgia, USA	014
Achieving sustainable production of poultry meat - Vol 3 Health and welfare Edited by: Prof. Todd Applegate, University of Georgia, USA	015
Achieving sustainable production of eggs - Vol 1	016

Introduction

Pig meat is the most widely-consumed meat in the world, accounting for 40% of the world's overall meat consumption. The leading producers are China, the EU, USA, Brazil, Russia, Vietnam, Canada, Japan, the Philippines and Mexico. Consumption is growing, both in developed markets such as the United States and particularly in developing countries in Asia. Previous growth in pig production has been driven primarily by developments in breeding and the shift to larger, more intensive systems. These systems face a range of challenges in increasing production sustainably to meet rising demand. Pig production remains vulnerable to zoonotic and other diseases affecting pigs as well as the problem of antibiotic resistance. There is growing pressure to improve feed efficiency in the face of competition for raw materials and rising feed costs. At the same time, there is an increasing focus on reducing the environmental impact of animal production. Finally, consumers are increasingly concerned about animal welfare in intensive systems.

These challenges are addressed by *Achieving sustainable production of pig meat*. The three volumes are:

- Volume 1 Safety, quality and sustainability
- Volume 2 Animal breeding and nutrition
- Volume 3 Animal health and welfare

This volume, Volume 2, reviews the ways that developments in breeding and nutrition can contribute to more efficient and sustainable pig production. Part 1 looks at the continuing potential of and constraints in conventional breeding as well as new genetic breeding techniques to improve areas such as reproductive performance and feeding efficiency. Part 2 discusses nutritional strategies, from a better understanding of pig nutritional requirements to more targeted 'precision' nutrition linked to particular stages in the life cycle. The chapters show how improvements in nutrition can enhance reproductive performance and overall health (particularly gut health), promote more efficient feed conversion and more consistency in weight and meat quality at the point of slaughter.

Part 1 Genetics and breeding

Chapter 1 reviews advances and constraints in conventional breeding of pigs. Traditional methods of genetic improvement, both through selection programs and proper use of breed differences and heterosis, have been effective in improving production efficiency. This improvement has been realized primarily in reproductive performance, growth rate and carcass composition. Conventional approaches have resulted in significant increases in areas such as feed efficiency, daily weight gain, average carcass weight, number and sizes of litters. There remains, however, a need for improvement in traits associated with traits such as disease resistance, behavior and longevity. Whilst further improvement may depend upon development of molecular genetic tools, there are still opportunities to use conventional animal breeding technologies, albeit with the addition of some novel traits or measurements, to create genetic change.

As an example, as more is understood about the biology of growth and intake, alternative methods of measuring feed efficiency have been developed. Residual feed intake (RFI) (the difference between the observed feed intake and that predicted from the average requirements for growth and maintenance) has been recommended as a measure of efficiency. Selection for reduced RFI has decreased the amount of feed required for growth and also reduced backfat. Research at the University of Illinois, for example, has also suggested that inclusion of semen traits (volume, concentration, motility and presence of abnormal spermatozoa) in selection criteria will help further improve reproductive performance.

In contrast, efforts to select for genetic improvement of disease resistance have been limited. Conventional quantitative breeding methods have not been very successful because of low heritability for disease development, the time for symptoms to appear in infected animals and the fact that selection of animals that do not display the disease might result in higher tolerance but not reduced infectivity. In one study, scientists at the University of Nebraska used the selection lines (developed for reproductive performance or feed efficiency) to study genetic differences in resistance to Porcine Reproductive and Respiratory Syndrome (PRRS) and Porcine Circovirus Associated Disease (PCVAD) which can then be translated into breeding targets. This suggests that there are opportunities for genetic improvement in disease resistance through traditional genetic improvement techniques. The existence of breed differences, moderate heritabilities and evidence that selection may be effective are all encouraging. However, the substantial cost associated with such techniques remains. Much of this research also suggests that molecular techniques (discussed in the following chapter) will bear fruit and that improvement in disease resistance will, ultimately, result from a combination of quantitative and molecular techniques.

Chapter 2 builds on Chapter 1 by reviewing the use of molecular genetic information in genetic improvement programs for pigs. As discussed in Chapter 1, conventional genetic improvement of pigs has been achieved by selection of individuals (young males and females) for breeding based on estimates of breeding values (EBV) derived using phenotypes for important traits that have been recorded on the selection candidates themselves and/or their close relatives. For that purpose, sophisticated statistical methods based on mixed linear models and best linear unbiased prediction (BLUP) have been pioneered in animal breeding. These methods optimally utilize all available phenotypes on the individual itself and its relatives in order to obtain the most accurate estimate of the breeding value of the individual, which can then be used to rank individuals to identify animals that should be used to breed the next generation.

However, there are multiple challenges and limitations associated with these phenotype-based programs. Many of these can be overcome by using molecular information to help predict breeding values and inform selection decisions. Chapter 2 describes how molecular genetic information can be used to enhance selection programs in pigs, what is required to develop such information, and what strategies are available for the use of molecular information in breeding programs. Example cases, challenges and future developments are also reviewed.

In contrast to the 'black box' approach of phenotype-based prediction, DNA-based molecular genetic tests can provide information on the genetic basis of traits of interest and the genotype that individuals have at specific locations of the genome that may be related to the trait. If a genetic marker can be found that is in linkage disequilibrium (LD) with a quantitative trait locus (QTL) for a trait, individuals that have the marker genotype that is associated with the favorable genotype at the QTL are expected to have a higher breeding value. The procedure for determining whether a genetic marker can be used to identify animals that have better genetics for that trait is based on a statistical analysis of the association between the genotype of individuals at the genetic marker and their phenotype for the trait. Chapter 2 reviews how this works by discussing the principles of genome-wide association studies (GWAS) and their potential pitfalls. Nevertheless, despite the pitfalls, many QTL have been identified using genetic markers for many traits in livestock using different population designs.

Methods for detection of QTL or association analysis result in estimates of the effects of marker genotypes on a phenotype, often in the form of allele substitution effects. Based on the genotype of selection candidates at the identified markers, these estimates can be used to compute a molecular breeding value (MBV) of each individual. In addition to an MBV, each selection candidate will also have an EBV based on regular phenotype-based BLUP evaluation. As discussed in Chapter 1, the combination of MBV and EBV can then be used for selection.

Despite a large number of reports of significant QTL and genetic markers associated with important traits, the application of the resulting QTL or markers in marker-assisted selection (MAS) livestock breeding programs has been limited until recently. The main reasons for the limited use of MAS include the limited amount of genetic variation that the significant markers explain, the limited number of genetic markers available, the relatively high costs of genotyping, poor reproducibility of marker or QTL effects in populations of interest, in particular when discovery populations were experimental or of limited size, and estimation of marker or QTL effects on a within-family basis, which has made it more difficult to incorporate them into breeding programs. Research has shown, for example, that substantial (training) data sets of genotyped and phenotyped animals are required to obtain substantial accuracies of MBV.

Many of these limitations are, however, being overcome by recent advances in molecular technology such as genome sequencing, the identification of large numbers of SNPs across the genome, and cost-effective high-throughput genotyping of tens of thousands of such SNPs. The ability to effectively use this large number of SNPs for breeding value estimation, however, also required a paradigm shift in statistical models for estimation of SNP effects, which has taken us from MAS to new techniques such as genomic selection (GS). Since 2009, GS has been successfully implemented in dairy cattle breeding programs across the world. The use of GS has allowed a substantial decrease in the generation interval in dairy cattle because bulls can now be selected based on genomics before they reach reproductive age, rather than having to wait until they have received an EBV based on milk production performance of their daughters.

The implementation of GS in pigs has progressed more slowly because opportunities to reduce generation intervals are less prominent in swine breeding programs. However, research has shown that genomic selection is cost-effective in at least the larger swine breeding programs, especially with the use of low-density SNP genotyping and imputation, and, as the chapter shows, GS has now also been implemented in several swine breeding programs. As an example, trials have been implemented to investigate the genetic basis of resistance or susceptibility of pigs to infection with porcine reproductive and respiratory syndrome (PRRS) virus.

As Chapter 3 points out, the reproductive performance of sow herds has improved considerably over the last few decades. Farrowing rates and litter sizes are reaching levels that previously were unattainable. The chapter reviews the challenges in maintaining and further improving the reproductive efficiency of pigs. After first discussing measures of reproductive efficiency, the chapter highlights some of the factors affecting reproductive efficiency, including gilt development, litter size, control of the weaning to estrus interval, and factors affecting seasonal infertility such as heat stress.

As the chapter points out, gilt development is a key component for reproductive success and longevity in sow herds. The chapter discuses use of gilt development units (GDUs) to improve the identification of select gilts with superior reproductive potential and lifetime productivity, the importance of birth weights, the role of nutrition, as well as management of boar contact and timing of mating. The chapter goes on to discuss research on best practice in mating management and insemination techniques and the key overall role of stockmanship in influencing reproductive performance. It also assesses management strategies to sustain reproductive performance of sows in dry sow and other housing systems. Diminished conception rates and high pregnancy losses are also attributable to infections, and the chapter discusses ways of dealing with diseases such as porcine reproductive and respiratory syndrome virus (PRRSv), porcine circovirus (PCV-2), swine influenza and porcine epidemic diarrhea virus (PEDv) outbreaks.

Complementing Chapter 3, Chapter 4 reviews factors affecting the reproductive efficiency of boars. The main goal of breeding is efficient dissemination of genes of high-indexed boars. Efficient artificial insemination (AI) is essential in achieving this by diluting semen from high fertile breeding boars to allow insemination of many sows. The chapter provides an overview of factors affecting the reproductive efficiency of boars. It then reviews research and best practice in such areas as boar selection, semen management and quality assessment, implementation of AI and monitoring of its effectiveness. As an example, the chapter assesses ways of predicting porcine male fertility, including sub-fertile boars, by matching ejaculate records with field fertility data, to optimise boar selection.

Chapter 5 complements Chapter 2 by discussing genetic factors affecting feed efficiency, feeding behaviour and related traits in pigs. Feed resource efficiency contributes to sustainable production of pig meat, both economically and environmentally. This chapter begins by describing the relative merits of different measures of feed efficiency such as feed conversion (also called as food conversion) ratio (FCR) and residual feed intake (RFI).

The chapter also discusses the underlying genetics of feeding in pigs. It reviews research on the heritability of good RFI values in pigs. As the chapter points out (echoing Chapter 2), a key requirement for breeding is to establish genetic correlations between phenotypic, genetic and nutritional RFI measures. The chapter then reviews research on quantitative trait loci (QTL) mapping of traits associated with feed efficiency, the identification of candidate genes as well as other research describing biological pathways that might regulate RFI. The chapter also looks at genetic correlations between RFI and observed feeding behaviour. The chapter draws on studies of genetic correlations of RFI with pig production traits and Genome Wide Association Studies (GWAS). Understanding these relationships and the genetics underlying component traits of RFI helps in prioritizing of candidate genes for further investigation. It also helps assure that marker assisted selection based on candidate genes for RFI does not adversely affect daily weight gain. The chapter goes on to review the use of genomic selection (GS) in improved selection for feed efficiency, with GS estimated to increase the rate of genetic gain by up to 25% compared with traditional breeding. Finally, the chapter looks ahead to the future and the adoption of an integrative systems genomic approach which combines genomics, epigenomics, transcrtiptomics and metabolomics. This integrated approach will provide a more complete picture of the biological mechanisms underlying feed efficiency as well as more accurate genomic prediction for breeding.

Part 2 Animal nutrition

Part 2 begins with an overview of some of the key issues in improving pig nutrition. As Chapter 6 points out, our understanding of the nutritional requirements and metabolism of the pig has advanced significantly over the last 10-20 years. The magnitude of progress is evident when we compare the performance of pigs today compared to the late 1980s with whole of life average daily gain increasing by 60%. The chapter provides a wide-ranging and authoritative overview of some of the key ways this has been achieved and some of the key areas of research to ensure both continued improvement as well as more consistent feed efficiency and meat output across the herd. The need for continued research is all the more important because significant technologies like porcine somatotropin (pST) and ractopamine, that have been key contributors to this improved performance, are no longer viable technologies for use by the industry.

One key theme is nutrition at particular stages in the life cycle. Echoing Chapter 3, Chapter 6 first emphasises the importance of minimising sow replacement and maintaining sow health through optimal gilt development. As the chapter points out, the key to a long reproductive is to optimise the entry of the gilt into the breeding herd through meeting key weight targets, and then maintain an adequate level of nutrition throughout gestation. Nutritional programs need to support maintenance of body tissues and the growth of maternal and foetal tissues, as well as ensure a high level of feed intake during lactation to reduce losses of body reserves and maintain good ovarian function. The addition of spray-dried plasma products (SDPP) to diets has also revolutionized the feeding of newly-weaned pigs, enhancing performance through increased feed intake and feed efficiency in the immediate post-weaning period. Later in the life cycle, one method of inhibiting sexual development and aggressive behaviours in the late finisher phase is immunization against gonadotropin-releasing factor, referred to as immunocastration, which increases feed intake and average daily gain and decreases feed conversion rate.

A second key theme is ways to reduce variation at the point of sale to improve the profitability and efficiency of a pig enterprise. The chapter therefore looks at nutritional factors that can influence birthweight and variability in birthweight. While the greatest effects of lactation feeding will be on post-natal piglet performance and variation, research also shows, for example, that nutrition of the sow during lactation can also influence variation in litter birthweights. The chapter also looks at the role of supplementary milk on the performance of lightweight piglets before or after weaning, as well as studies on improving the performance of light-weight weaners through nutritional intervention.

A third key theme is the role on nutrition in 'gut health'. Gut health can be viewed as an outcome of the complex interactions occurring in the gastrointestinal tract (GIT) between nutrition, the mucosa and the microbiota of the GIT. The chapter reviews the wealth of research on the range of ingredients promoting gut heath such as: zinc oxide, probiotics (such as soluble non-starch polysaccharides (NSP) and resistant starch (RS)) as well as organic and inorganic acids in pig diets (including the encapsulation of acids for targeted delivery to different gut segments), phosphorylated mannans, amino acids (such as glutamine), yeast extract and peptides. The chapter also looks at research on ingredients such as dietary chromium in increasing feed efficiency and daily gain whilst improving aspects of meat quality such as decreased back fat thickness, as well as renewed research into safe and effective ways of using food waste as a potential source of feed.

A final key theme is the capacity to measure and control variation in the nutritional quality of feed ingredients prior to diet formulation. The chapter assesses the debate over the suitability of digestible energy (DE) and metabolisable energy (ME) versus net energy (NE) systems as the most appropriate measures of available energy in feed ingredients, particularly cereal grains. The chapter also describes advances in analysis of ingredients using either chemical or rapid techniques such as near infrared spectrophotometry (NIRS) as well as the progress in countries such as Australia in developing rapid, online assessment of digestible energy content in cereals and oilseed used in pig diets. Results have shown significant differences in energy intake which represent a major source of variation in pig growth which increases as the pigs grow. Technologies like NIR and other 'precision farming' technologies must be better utilised to ensure that the diets we are feeding our pigs actually meet their requirements.

The level of energy in a pig's diet influences the rate and efficiency of gain, the quality of the resulting carcass and even the quality of the pork produced from the carcass. The concentration of energy in the diet represents a critical balance among factors such as cost, the energy level of available ingredients and the level of growth performance desired. However, given the complexity of the subject, energy requirements and energy use in pigs remain poorly understood. Chapter 7 discusses the basics of energy metabolism, energy systems, energy sources, energy requirements and methods to improve the utilization of energy by the pig.

As the chapter notes, energy can be supplied in the diet by starch and other simple carbohydrates such as lactose, by more complex carbohydrates known as fibre, by individual amino acids within protein and by lipids. Each of these is utilized by the pig as an energy source in different ways. A key variable is the efficiency with which dietary components can be used by the pig to generate adenosine triphosphate (ATP) for the purpose of maintenance and protein accretion or to deposit lipid in the body. The digestibility of dietary components is also an important variable. Areas of research include the most effective way to incorporate highly fibrous – but often less expensive – ingredients in the diet of the pig as well as the use of dietary fats to enhance pig performance. The chapter also discusses ways of processing ingredients including pelleting to increase availability of energy or the addition of digestive enzymes to improve swine diets to improve the digestibility of energy. The chapter also reviews the relative merits of different ways of measuring energy such as net energy (NE), digestible energy (DE) and metabolizable energy (ME) as well as the potential role of modelling in measuring and optimising the delivery of energy through diet.

Chapter 8 discusses ways of meeting amino acid requirements in pig nutrition. Protein is the one of major components of the pig body, and dietary protein is the sole source of the essential amino acids required for protein synthesis for body maintenance, growth, and reproduction. Protein synthesis is limited when there is a deficiency of any amino acid, and it is therefore important to ensure that feed meets the amino acid requirements of pigs at their various growth stages. Chapter 8 reviews the principles and practical aspects of meeting the amino acid requirements of pigs, with information organised according to growth stage and physiological status. These growth stages include: gestating sows, lactating sows, nursing pigs, nursery pigs, growing and finishing pigs

The chapter shows how amino acid needs and their ratios for fetal and mammary gland growth affect dietary amino acid requirements which mainly occur during late gestation in sows. The chapter discusses amino acid needs and their ratios for milk production and mammary gland growth as well as the way amino acid contribution from maternal tissue mobilization during lactation affects dietary amino acid requirements for lactating sows. In feeding growing pigs, research shows that dietary amino acids should meet the requirements but also minimize excess provision to reduce amino acid catabolism. Use of supplemental amino acids allows the nutritionist formulating feed to reduce nitrogen excretion to the environment. With an increased number of supplemental amino acids available in swine feed, nutritionists can formulate feeds with ideal amino acid patterns based on their growth stages and physiological status which also significantly reduces the use of protein supplements.

Requirements for vitamins in modern intensive swine production are still based on genetically outdated lines of pigs and production conditions which have changed dramatically during the last 30 years. There is lack of scientific information on the vitamin requirement for the fast-growing lean meat type of pigs according to their physiological responses. There is evidence of lower levels of micronutrients in pork which may not only affect meat quality parameters such as oxidative stability, colour and water-binding but also affect consumer perception of the nutritional value of pork. At the same time, the criteria used to estimate vitamin requirements for pigs have moved from prevention of deficiencies to optimisation of performance for growth and reproduction. The estimation of requirements needs to include criteria related to other aspects potentially limiting for full expression of performances such as metabolic stress and disease resistance.

Chapter 9 reviews recent advances in understanding the role of vitamins and their importance for oxidative mechanisms, especially in relation to the development and competence of the immune system which are key contributors to optimal health status of pigs and to their ability to face pathogenic pressure during their life. The chapter outlines what current research shows about the potential contribution of both fat-soluble vitamins (e.g. A, D and E) and water-soluble vitamins (such as folates, vitamin B12 and the intermediary amino acid homocysteine) and offers new perspectives on the relationship between vitamins and antioxidation capacity. The chapter shows ways of balancing requirements for prevention of vitamin deficiencies and optimization of performance as well as disease protection.

As Chapter 10 points out, improvement of feed efficiency is crucial if pig production is meet the challenge of sustainability in terms of production costs and environmental impact. Chapter 10 describes advances in modelling approaches developed to predict the nutrient requirement of a single individual animal (growing pig or sow) in terms of protein/amino acids, energy, and minerals. The chapter reviews modelling approaches for growing pigs and reproducing sows such as the InraPorc sow model based on integrating on-farm data on reproductive performance, feeding practices, and housing conditions. The chapter explains recent advances in integration of individual variability among animals into models for pig feeding, including the use of stochastic modelling techniques and illustrates via a case study the potential for improving feed efficiency through the application of these models in precision feeding. Precision feeding is based on the dynamic adjustment (if possible day by day) of dietary nutrient supplies to requirements, at a group or at an individual level. In this approach, individual pigs are treated as such and each pig/group is to be modelled individually. The purpose is to improve feed efficiency whilst reducing feed cost and environmental impact. These themes are also picked up in Chapter 14

Echoing Chapter 6, Chapter 10 reviews ways of reducing variability at slaughter, including a herd modelling approach to evaluate different feeding strategies to control or reduce variability among pigs at slaughter. It also reviews the use of the InraPorc model to characterise the effect of different feeding strategies (e.g. amino acid content or feed allowance), on the mean and variation in growth rate and slaughter weight. As the chapter points out, new developments in sensors and data collection will allow access to more precise, detailed, real-time data on characteristics such as feed intake and body weight. When combined with historical data, it will be possible to predict the outcome of different feeding strategies with even greater accuracy to improve economic returns and environmental impact.

As Chapter 11 makes clear, exogenous feed enzymes have been in commercial use in swine diets for almost 30 years. This chapter focuses on the use of three classes of enzyme: NSP'ases, phytases and proteases. Lipolytic enzymes are also considered. The chapter reviews the evidence on their benefits and mechanisms of action as well as assays for measurement. As the chapter points out, it was the emergence of cost effective phytases that resulted in the now almost universal deployment of exogenous enzymes in swine diets. The commercialisation of microbial phytases has resulted in substantial reductions in rock phosphate utilisation and phosphorus (P) excretion into the environment. Phytase use to replace inorganic phosphate has significantly improved the sustainability of pork production through the reduction of raw material use coupled with reduced P excretion in manure. As the chapter discusses, novel uses for these evolved phytases have started to emerge with 'extra-phosphoric' effects of superdoses of phytase gaining considerable attention in the past few years. These extra-phosphoric effects of phytase are associated with the improvements in utilisation of nutrients other than P (such as zinc and nitrogen) and reduced maintenance costs to the animal, which may be attributed to the mitigation of the anti-nutritive effects of dietary phytate. Research shows that the value of feed enzymes is not only that performance on average is improved, or nutrient density reduced and performance maintained, but also the potential of exogenous enzymes to reduce the variation between the best and worst animals and herds. As more is learned of the factors which influence the response to exogenous enzymes, the greater the likelihood that their value and thus payback will increase.

Chapter 12 describes the use of growth promoters in pigs, specifically growth hormone (GH) and beta-adrenergic agonist (BA). Although banned in some parts of the world, such as the European Union, growth promoters have now been used in some countries for over 20 years with no adverse human safety issues, suggesting that they are relatively safe when used according to recommended guidelines. Despite the term 'growth promoter', these substances do not necessarily always result in increased growth rates, but they do tend to alter body composition and improve feed efficiency, as determined by the gain-to-feed ratio. After introducing each growth promoter, the chapter discusses current research on their individual effects on growth and feed efficiency, followed by sections on their mechanisms of action and effects on muscle fibre type and meat quality. The chapter also summarises recent studies on the use of growth promoters such as Ractopamine and Reporcin in combination and assesses their future use commercially.

Prebiotics and probiotics have attracted considerable interest as alternatives or replacements for growth promoting antibiotics and (or) some heavy metals in diets for pigs, particularly in the post-weaning period where the newly-weaned pig is subject to considerable challenges in its new environment. As Chapter 13 points out, the microbiota, i.e. the ecological community of commensal, symbiotic and pathogenic microorganisms, is of huge importance to the host, particularly in the gastrointestinal tract (GIT). The microbiota, particularly bacteria, plays a part in prevention of establishment of pathogenic bacteria, modulation of the immune system, detoxification, production of vitamins and short-chain fatty acids, and facilitation of digestion and absorption processes.

This chapter briefly reviews the microbiota of the gastrointestinal tract (GIT) of the young pig and the important roles it plays in the early stages of life, before introducing probiotics, prebiotics and synbiotics. Probiotics are live beneficial microorganisms (bacteria) that may exert a beneficial effect on the host and can aid in the establishment or composition of the GIT microbiota. Prebiotics are non-digestible food ingredients and are thought to benefit the host by providing substrates for beneficial microbiota, giving them a competitive advantage. The term synbiotics is used when probiotics and prebiotics are combined.

The most commonly used probiotic organisms are bacillus, yeast and lactic-acid producing bacteria (LAB such as *Lactobacillus* spp.). As research shows, the mode of action varies between different probiotics, but typically involves competitive exclusion through mucosal adhesion, competition of nutrients in the GIT, decrease in luminal pH, production of bacteriocins/defensin, and (or) modulation of the mucosal and systematic immune responses and strengthening of the intestinal barrier. The chapter reviews research on the benefits from supplementation of probiotics in pigs. These include improved absorption and digestion, stimulation of GIT immunity and increased resistance to infectious diseases of the GIT, which all together may improve production performance. To illustrate the benefits of probiotics, the chapter includes a case study of a probiotic product (Peribios™) (which contains a strain of *Enterococcus*) used commercially for sows and suckling piglets to modulate production around parturition. The probiotic was found to be beneficial in increasing the number of liveborn piglets, the number of weaned piglets per litter, and an increase in piglets weaned per sow per year.

Prebiotics that have been identified for use in pigs are generally carbohydrates containing a different molecular structure, namely dietary carbohydrates such as fibres and resistant starch, and non-digestible oligosaccharides (NDOs). They also include fructooligosaccharide (FOS), oligofructose (OF) and inulin, galactooligosaccharides (GOS), transgalacto-oligosaccharides (TOS), and lactulose. Prebiotics exert their effect in two ways: they may be fermented by beneficial bacteria such as *Lactobacilli* and *Bifidobacteria,* which would give the beneficial bacteria a competitive advantage by giving them the opportunity to work more efficiently; and, secondly, they appear to interfere with the attachment of pathogenic bacteria to the GIT wall, which would alter the microbiota. As the chapter shows, it has been suggested that the use of synbiotics would be a more beneficial strategy for preventing and controlling diseases during early life and weaning, compared with supplementing with either by itself. The prebiotics would provide a readily available substrate for the probiotics, allowing enhanced growth and enhancing their survival rate, colonisation and effects.

As the chapter concludes, given the complexity of the GIT, the range of probiotics and prebiotics and many other variables, it has been hard to compare studies or achieve consistent results. Fundamentally, there remains a lack of detailed knowledge on the influence of the complex interactions inside the GIT ecosystem as well as the definite mode(s) of action of each probiotic strain or prebiotic material. Advances in this fundamental understanding will make it easier to predict and control performance of individual probiotics and prebiotics.

The final chapter in the book builds on Chapter 10 by looking at developments in precision feeding which involves feeding techniques that provide individual animals with diets tailored daily to production objectives (e.g. maximum growth). A method of estimating energy and nutrient requirements by integrating current relevant knowledge on pig metabolism has been developed and incorporated into a mathematical model. This chapter reviews recent research projects which indicate that feeding pigs a diet tailored daily to their individual requirements is essential to maximize nutrient efficiency and ensure the sustainability of the pig industry by reducing the excretion of nutrients and nutrient constituents and lowering feeding costs. This new nutritional approach represents a paradigm shift in pig feeding, because the optimal dietary nutrient level is no longer considered a static population attribute, but rather a dynamic process that evolves independently for each animal. Precision feeding is a highly promising avenue for improving resource-use efficiency in comparison with conventional group phase-feeding programs. The chapter includes a case study on the use of precision feeding in practice.

Part 1

Genetics and breeding

Chapter 1

Advances and constraints in conventional breeding of pigs

David S. Buchanan, North Dakota State University, USA

1 Introduction

Genetic improvement of the efficiency of pork production has been the focus of swine geneticists at least as far back as the 1930s. The USDA Regional Swine Breeding Laboratory was established at Ames, Iowa, in 1937 under the leadership of Dr. W.A. Craft. That laboratory, along with animal geneticists at several US land-grant universities, established the ways that principles of genetic improvement would be utilized for the improvement of efficiency of pork production for next several decades (Craft, 1958). In 1974, geneticists, extension personnel, test station managers and interested pork industry leaders started the National Swine Improvement Federation which continues to the present in its efforts to provide tools for genetic improvement to pork producers.

Starting in the 1950s and accelerating during the 1970s and 1980s, various companies expanded their influence upon genetic improvement of pork production throughout the world. Scientists in those companies, with considerable cooperation with scientists in both federal and state agencies in the United States and other countries, continue their efforts and the rate of genetic improvement is rapid.

Much has been accomplished but much remains. The amount of fat in pig carcasses dramatically dropped from the 1940s to the 1970s. The standard for pigs/sow/year has transitioned from 20 in the 1970s to more than 30 at the present (Anonymous, 2011). These, and other, improvements in performance have already led to striking change in sustainability of pork production. From 1959 to 2009 average daily gain increased 33%, dressed carcass weight increased 36%, litter size increased 30%, the number of litters/sow/year increased

http://dx.doi.org/10.19103/AS.2017.0013.02

36% and the kilograms of feed per kilogram of hot dressed carcass weight decreased 34% (Boyd, 2012). These improvements, along with dramatic increase in corn and soybean yield, work together to result in a more than 400% improvement in hot dressed carcass weight produced per hectare (Boyd, 2012). This improvement in the overall efficiency of pork production has been paralleled by similar changes in beef production (Capper, 2011) and milk production (Capper, 2009).

2 Historical overview

After the end of Second World War, pork production was largely on small operations that were multipurpose enterprises. Even as late as 1997, more than 75% of 647 000 US swine farms had fewer than 100 heads (Pork Checkoff Quick Facts). By 2012, the number of swine operations had decreased to just more than 63 000, the average inventory was nearly 1000 and more than 3000 operations had more than 5000 pigs in the inventory. These changes were being mirrored in many countries of the world as commercial entities in Europe and North America were expanding throughout those regions as well as making inroads into Asia.

Along with the changing demographics of the worldwide pork industry, this was also a time period when record keeping and measurement of performance was also coming to the forefront. Advances in computing and in animal breeding technology enabled the assembling of extremely large sets of data, along with pedigree records, so that the basic tools for genetic improvement were being put into place by individual producers and by the large pork companies which were taking a larger and larger role in genetic improvement of pigs. The traits of economic importance having to do with reproduction, efficient growth and body composition were well studied and recommended performance programmes were put into place (National Swine Improvement Federation, 2003).

The estimated heritabilities (Table 1) of these various performance traits suggest that improvement through selection is possible. Early selection experiments that examined selection for improvements in reproduction and performance showed mixed results (Dickerson et al., 1954; Krider et al., 1946; Dickerson and Grimes, 1947; Stothart and Fredeen, 1950 as reviewed by Fredeen, 1958).

The apparent difficulty associated with improvement of litter size and feed conversion was especially troubling. Both traits are economically important and are also very important in terms of the sustainability of swine production. Production efficiency can be greatly improved if an improvement of pigs/sow/year or a reduction in feed costs, relative to pork output, can be realized. The relatively low estimated heritability of litter size makes the difficulty in selecting for improvement less surprising, but the estimated heritability of feed conversion suggests that improvements should be possible.

Improvement of reproductive efficiency has been a central part of the research effort at the University of Nebraska (Zimmerman and Cunningham, 1975; Young et al., 1978; Cunningham et al., 1979; Johnson et al., 1984, 1999; Hsu and Johnson, 2014). The first step in that research was to establish that ovulation rate could be changed through selection even though, initially, a corresponding change in litter size was not obtained. Further efforts established that breaking down litter size into its components (e.g. ovulation rate,

Table 1 Estimates of heritability and phenotypic and genetic correlations for various traits in swine

Trait	h^2	AP	OR	NBA	S	R	NW	LBW	L21W
Age at puberty (AP)	0.37		−0.06	0.07	−0.25	0.14	0.09	−0.10	−0.15
Ovulation rate (OR)	0.32	0.05		0.24	−0.38		0.01	0.24	0.03
Number born alive (NBA)	0.11	−0.03	0.13		−0.14		0.81	0.55	0.55
Survival to weaning (S)	0.07		−0.11	−0.22			0.15	−0.07	0.65
Rebreeding interval (R)	0.23								
Number weaned (NW)	0.08	−0.01	0.03	0.79	0.55			0.67	0.81
Litter birth wt (LBW)	0.24	−0.03	0.07	0.82	0.09		0.71		0.65
21-day litter wt (L21W)	0.14	−0.04	0.02	0.46	0.65		0.80	0.61	

		ADG	AGE	BF	FE	RFI	LEA	DP	LEN	PL
Average daily gain (ADG)	0.29		−0.93	0.22	−0.37	0.10	−0.10	0	0.10	−0.15
Days to 230 pounds (AGE)	0.25	−0.90		−0.20	0.65		0.05	0	−0.10	0.10
Backfat thickness (BF)	0.49	0.20	−0.18		0.31	0.44	−0.35	0.15	−0.28	−0.85
Feed efficiency (FE)	0.30	−0.65	0.60	0.25		0.76	−0.35	−0.43	0	−0.07
Residual feed intake (RFI)	0.24									
Loin eye area (LEA)	0.47	−0.06	0.03	−0.28	−0.20			0.50	−0.18	0.65
Dressing % (DP)	0.30	−0.15	0.10	0.20	0.10		0.32		−0.32	−0.10
Carcass length (LEN)	0.56	0.08	−0.06	−0.21	−0.04		−0.12	−0.21		0.18
% lean (PL)	0.48	−0.11	0.10	−0.71	−0.25		0.62	0	0.10	

[a]Phenotypic correlations below diagonal, genetic correlations above diagonal.

Adapted from Lamberson and Cleveland (1988) and Young (1990).

embryo survival, and foetus survival) as well as considering age at puberty and pig weight at birth should allow improvement of litter size through selection.

The subject of feed efficiency has been well-reviewed recently (Patience et al., 2015). They described the importance of feed efficiency to overall production efficiency by pointing out that feed accounts for more than 60% of the total cost of pork production. Selection for improvement of feed efficiency has, historically, been difficult because of the difficulty of obtaining good measurement of feed intake. This has placed the emphasis on selection for traits which are correlated with efficiency. Selection to increase growth rate has been successfully accomplished in several experiments (Bullock et al., 1991; Kuhlers and Jungst, 1990, 1991a,b; Rahnefeld, 1971a,b, 1973; Rahnefeld and Garnett, 1976; Garnett and Rahnefeld, 1976; Woltmann et al., 1992, 1995; Clutter et al., 1995). However, such selection may be accompanied by increased fatness and increased feed intake (Woltmann et al., 1995), so that the improvement in efficient production of lean tissue was minimal. Inclusion of backfat thickness, in an index with growth rate, has successfully improved both traits (Cleveland et al., 1982, 1983a,b; Fredeen and Mikami, 1986a,b,c,d,e). Such an index has also been effective in increasing protein growth and improving the efficiency of production of edible lean (Cleveland et al., 1983a).

Breed differences in performance have been well documented (see review by Buchanan and Stalder, 2010). There are differences in reproduction, maternal ability, efficiency and body composition among breeds commonly used in North American and European pork production. These differences, along with the benefits of heterosis (Johnson, 1980), can be used to improve the efficiency, and therefore the sustainability, of pork production. Genetic stocks that include novel performance for body composition (e.g. the Pietrain) or reproduction (e.g. Meishan and other Chinese breeds) have been incorporated into some breeding programmes, although these also bring some disadvantages such as the incidence of porcine stress syndrome in the Pietrain and very poor growth and body composition of many of the Chinese breeds.

The primary benefits of heterosis (Table 2) in swine derives from its effect on liveability and reproduction. While breed effects change over time, the benefits of heterosis should

Table 2 Estimates of heterosis (%) for several swine traits

	Individual heterosis	Maternal heterosis	Paternal heterosis
Ovulation rate		0.3	
Testis weight			24.6
Semen volume			9.6
Sperm number			26.7
Sperm concentration			5.9
Sperm motility			2.8
Time to first mating			23.5
% boars mating each time			156.1
Conception rate		3.8	3.4
1st service conception rate			17.1
Litter size born	1.0	4.7	−1.1
Birth weight	3.1	1.5	−1.4
Litter size at 21 days	8.0	8.7	
21-day weight	3.1	3.7	
Litter size at 42 days	10.1	7.7	−2.5
42-day weight	4.8	8.2	−1.2
Average daily gain	9.4	0	1.2
Age at 230 pounds	6.5	1.2	
Feed efficiency	2.3	0	−1.2
Length	0	0.2	
Backfat thickness	2.5	4.4	1.3
Loin eye area	1.8	0.4	−1.4

Adapted from Johnson (1980), Buchanan (1987).

remain relatively static. Heterotic advantages contribute greatly to improving the overall efficiency of production and, as a result, improving sustainability.

3 New approaches to genetic improvement: feed efficiency and disease resistance

3.1 Feed efficiency

As more is understood about the biology of growth and intake, alternative methods of measuring feed efficiency have been developed. Residual feed intake (RFI) (the difference between the observed feed intake and the feed intake predicted from the average requirements for growth and maintenance) has been recommended as a measure of efficiency (Cai et al., 2007, 2011). In their research at Iowa State University they found that selection for reduced RFI decreased the amount of feed required for growth and also reduced backfat. It was also determined that selection for lower RFI resulted in pigs that ate faster and spent less time eating (Young et. al., 2011). A correlated decline in insulin-like growth factor-I was also observed (Bunter et al., 2010). This suggests that insulin-like growth factor-I may be an early indicator of feed efficiency. Selection for decreased RFI also resulted in a small increase in sow reproductive performance (Young et al., 2016). A summary (Dekkers, 2009) of several of the findings from the Iowa State selection for RFI revealed the following: the low RFI line was more efficient under both *ad libitum* and restricted feeding, required less feed to maintain weight, and had indications of lower protein turnover and less energy expenditure in muscle.

Pigs at a Danish central testing station were studied to examine RFI, and its relationship to other measures of performance in Duroc, Landrace and Yorkshire pigs (Do et al., 2013). Selection for reduced RFI could be successful without an adverse effect on average daily gain or backfat thickness. The authors recommended that RFI be considered as a replacement for conventional measures of feed conversion ratio.

Selection for reduced RFI in France has resulted in improvement in feed conversion and little change in fat thickness (Gilbert et al., 2007). Further research for development of more efficient sows (Gilbert et al., 2015). It was also found that selection for decreased RFI may have a favourable impact on nitrogen and phosphorus excretion (Saintilan et al., 2013).

Further improvements in the effectiveness of selection for feed efficiency may arise as the technologies associated with electronic identification and feeding systems that allow individual measurement of feed intake improve (David et al., 2015).

3.2 Disease resistance

Disease has long been a concern to swine producers. The costs associated with swine disease are extremely large (Rothschild, 1998). Despite this, until recently, efforts to select for genetic improvement of disease resistance have been minimal. During the early years of performance testing, a lot of effort had to be placed on encouraging producers to measure performance and to use that information in making selection decisions. Even by the time that measurement of performance became routine, the maintenance of records of disease incidence, which could be used to identify sires with less disease outbreak among their offspring, was sparse. In order to have much selection differential it would be desirable to actually infect selection candidates with economically important diseases but the cost associated with care and disposal of the infected animals is prohibitive (Rothschild, 1998). Individual seedstock producers would have difficulty maintaining either the facilities or the animals for sustaining such efforts.

A first question might be whether there are breed differences for disease resistance. Such differences would suggest that there is genetic variation which might be exploited. Pigs representing German Landrace, Pietrain, Hampshire and Large White were exposed to *Actinobacillus pleuropneumoniae* and were scored for respiratory health with consideration of clinical, sonographic and radiographic examination (Hoeltig et al., 2009). Hampshire pigs were predominantly found in the lowest scoring quartile while German Landrace and Pietrain were predominantly found in the highest scoring quartile. This suggests differences among breeds for response to this bacterium. Duroc, Hampshire and Meishan pigs were inoculated with porcine reproductive and respiratory syndrome virus (PRRSV) between 22 and 38 days of age (Halbur et al., 2009). Hampshire pigs had more severe lung lesion scores while Meishan pigs had lower PRRSV antigen in the lungs. In addition, many Meishan pigs had myocarditis and encephalitis. Duroc pigs had lower normalized serum antibody titres to PRRSV. Porcine circovirus type 2 was the focus of research evaluating Duroc, Landrace and Large White pigs (Opriessnig et al., 2006). In this study, with limited numbers, only the Landrace pigs showed signs of severe lymphoid lesions. These studies show promise for exploitation of genetic variation for disease resistance.

A report from a herd of Sinclair swine that had been affected by malignant melanoma revealed heritabilities of 0.27 and 0.26 for number of tumours at birth and at six weeks, respectively (Gomez-Raya et al., 2009). Research conducted at Iowa State University (Dunkelberger et al., 2015, 2016) suggested that selection for performance under coinfection from PRRSV and PCV2b should be effective. They also identified unique genomic regions that could make it possible to effectively select for better response to vaccination. In addition, they found evidence that pigs selected for reduced RFI would be more robust to a challenge from PRRSV.

Challenges in modelling infectious disease were reviewed (Brooks and Pollack, 2014). The reviewers identified disease control by selective breeding as one of the challenges. Conventional quantitative breeding methods have not been very successful because of low heritability for disease development, the time for symptoms to appear in infected animals and the selection of animals that do not display the disease may result in either lower susceptibility or higher tolerance but not reduced infectivity.

Scientists at the University of Nebraska used the selection lines (developed for reproductive performance or feed efficiency) to study genetic differences in resistance to porcine reproductive and respiratory syndrome (PRRS) and porcine circovirus associated disease (PCVAD) (Johnson, 2007). They observed an interaction between line (Index line selected for litter size vs Hampshire-Duroc line selected for feed efficiency) and PRRSV infection status. The Hampshire-Duroc pigs grew faster than Index when not infected but of pigs that were infected, the Index line pigs grew faster. This suggested genetic variation in the mechanisms that affect immune response to PRRS. In a study that examined PCVAD incidence, they observed little direct genetic variation ($h^2 = 0.01$) but a higher maternal heritability ($h^2 = 0.11$). This would indicate that genes in the dam may have an effect on the expression of PCVAD in their offspring. Intrauterine environment may play a role since birth weights were lower in PCVAD-positive pigs, even though actual expression of PCVAD did not arise until three months later. These efforts suggest that there is genetic variation for resistance to PRRS and PCVAD.

On balance, it appears that there are opportunities for genetic improvement in disease resistance through traditional genetic improvement techniques. The existence of breed differences, moderate heritabilities and evidence that selection may be effective are all encouraging. However, the substantial cost associated with such techniques remains. Much of this research also suggests that molecular techniques will bear fruit and that improvement in disease resistance will, ultimately, result from a combination of quantitative and molecular techniques. Among the complexes of traits discussed in this presentation, disease resistance seems to be the complex that will be most dependent upon inclusion of molecular biology techniques. There may be indicator traits which are suggestive of disease resistance but genomic information concerning individual genes that influence disease resistance is likely to be enormously beneficial.

4 New approaches to genetic improvement: reproduction, longevity and behaviour

4.1 Reproduction

Selection for improved sow reproduction has been demonstrated to be effective, although predicted improvement is slow (Hsu and Johnson, 2014). There have, however, been some recent suggestions for ways to expand genetic improvement in reproduction through selection. Research at Illinois has suggested the inclusion of semen traits (volume, concentration, motility and presence of abnormal spermatozoa) in selection criteria (Gonzalez-Pena et al., 2015). Selection for total number born or for number born alive may have the detrimental effect of increasing mortality rate (Johnson et al., 1999) so that measurement of litter size between birth and weaning has been recommended (Su et al., 2008). Research at North Carolina State University indicated selection for number of pigs at ten days could be effective for increasing reproductive performance without a detrimental effect on pig mortality (Putz et al., 2015).

Continued genetic improvement of reproduction may continue to be possible through conventional quantitative genetic evaluation systems, although there may also be considerable opportunity for the use of techniques associated with molecular biology.

4.2 Longevity and behaviour

Genetic improvement programmes have placed considerable emphasis on standard measures of reproduction, performance and carcass merit. Behaviour and longevity have not received much attention until more recently. Stalder (2008) reviewed the situation pertaining to longevity and recommended that the existing genetic variability for longevity should be exploited through genetic evaluation that is somewhat similar to what is done with stayability in several beef cattle breed genetic evaluation programmes. Part of the variation to be exploited may be through crossbreeding programmes. Crossbred advantages in stayability and accumulated born alive through four parities were observed in field data from swine farms in the United States (Engblom et al., 2006). The heritabilities for stayability traits, through four parities, were observed to be low (Engblom et al., 2015) suggesting that genetic improvement may be difficult. However, the number of pigs born alive, within parity, could be used as an indicator for longevity.

Aggressive behaviour between pigs in a growing-finishing programme can be problematic. An attempt to address this through a combination of a direct breeding value and a social breeding value, obtained through evaluation of behaviour and counting skin lesions, revealed that considering social interactions could be effective to increase growth while modulating aggressiveness (Canario et al., 2012).

Pigs need to be robust in a variety of environments. A challenging environment can be identified if there is a decline in reproductive performance during certain weeks of the year. A study involving nearly one million records from 13 countries used the measured environmental challenge to enhance estimation of breeding value (Herrero-Medrano et al., 2015) and found that inclusion of such a measurement may result in enhanced ability to maintain high production in challenging environments.

To summarize some of these novel approaches to genetic improvement, Merks et al. (2012) recommended consideration of vitality from birth to slaughter; uniformity at different levels of production; robustness, welfare and health; and reduction of carbon footprint.

Genetic improvement of these complex traits may continue to be achieved through traditional genetic evaluation methodologies, although it may require a larger complex of standard reporting of information, especially for improving longevity which will require information about when the sows are culled from the herd. It may also require measures of reproductive performance and pig size at times that have not been traditionally a part of routinely obtained data on the pigs.

5 Conclusions

Traditional methods of genetic improvement, both through selection programmes and proper use of breed differences and heterosis, have been effective in improving production efficiency. This improvement has been realized primarily on reproductive performance, growth rate and carcass composition. There remain needs for improvement in traits associated with efficiency, disease resistance, behaviour and longevity. In recent years, various conventional genetic improvement tools have been developed for these traits but the demonstrated effectiveness of these tools is still incomplete. It seems apparent that further improvement may depend upon development of molecular genetic tools. However, there do appear to still be opportunities to use conventional animal breeding technologies, albeit with the addition of some novel traits or measurements, to create genetic change.

6 Where to look for further information

Bidanel, J. P. 2011. Biology and genetics of reproduction. In *The Genetics of the Pig*, 2nd Ed. (Eds: M. F. Rothschild and A. Ruvinsky). CAB International, pp. 218–41.

Buchanan, D. S. and K. Stalder. 2011. Breeds of pigs. In *Genetics of the Pig* ,2nd Ed. (Eds: M. F. Rothschild and A. Ruvinsky). CAB International, pp. 445–72.

Ciobanu, D. C., S. M. Lonergan and E. A. Huff-Lonergan. 2011. Genetics of meat quality and carcass traits. In *The Genetics of the Pig*, 2nd Ed. (Eds: M. F. Rothschild and A. Ruvinsky). CAB International, pp. 355–89.

Clutter, A. C. 2011. Genetics of performance traits. In *The Genetics of the Pig*, 2nd Ed. (Eds: M. F. Rothschild and A. Ruvinsky). CAB International, pp. 325–54.

7 References

Anonymous. 2011. Understanding the physiology of 30+ pigs/sow/year. *National Hog Farmer*, 18 October 2011.

Bidanel, J. P. 2011. Biology and genetics of reproduction. In *The Genetics of the Pig*, 2nd Ed. (Eds: M. F. Rothschild and A. Ruvinsky). CAB International, pp. 218–41.

Boyd, G. and R. Cady. 2012. A 50-Year Comparison of the Carbon Footprint of the US Swine Herd: 1959–2009. A report submitted to the Environmental Programs of the National Pork Board.

Brooks-Pollock, E., M. C. M. de Jong, M. J. Keeling, D. Klinkenberg and J. L. N. Wood. 2015. Eight challenges in modelling infectious livestock diseases. *Epidemics* 10:1–5.

Buchanan, D. S. 1987. The crossbred sire: Experimental results for swine. *J. Anim. Sci.* 65(1):117–27.

Buchanan, D. S. and K. Stalder. 2011. Breeds of pigs. In *Genetics of the Pig* 2nd Ed. (Eds: M. F. Rothschild and A. Ruvinsky). CAB International, pp. 445–472.

Bullock, K. D., D. L. Kuhlers and S. B. Jungst. 1991. Effects of mass selection for increased weight at two ages on growth rate and carcass composition of Duroc-Landrace pigs. *J. Anim. Sci.* 69(4):1409–19.

Bunter, K. L., W. Cai, D. J. Johnston and J. C. M. Dekkers. 2010. Selection to reduce residual feed intake in pigs produces a correlated response in juvenile insulin-like growth factor-I concentration. *J. Anim. Sci.* 88:1973–81.

Cai, W., D. S. Casey and J. C. M. Dekkers. 2007. Selection response and genetic parameters for residual feed intake in Yorkshire swine. *J. Anim. Sci.* 86(2):287–98.

Cai, W., M. S. Kaiser and J. C. M. Dekkers. 2011. Genetic analysis of longitudinal measurements of performance traits in selection lines for residual feed intake in Yorkshire swine. *J. Anim. Sci.* 89(5):1270–90.

Canario, L., S. P. Turner, R. Roehe, N. Lundeheim, R. B. D'Eath, A. B. Lawrence, E. Knol, R. Bergsma and L. Rydhmer. 2012. Genetic associations between behavioral traits and direct-social effects of growth rate in pigs. *J. Anim. Sci.* 90:4706–15.

Capper, J. L. 2011. The environmental impact of beef production in the United States: 1977 compared with 2007. *J. Anim. Sci.* 89(12):4249–61.

Capper, J. L., R. A. Cady and D. E. Bauman. 2009. The environmental impact of dairy production: 1944 compared with 2007. *J. Anim. Sci.* 87(6):2169–7. https://doi.org/10.2527/jas.2009-1781

Ciobanu, D. C., S. M. Lonergan and E. A. Huff-Lonergan. 2011. Genetics of meat quality and carcass traits. In *The Genetics of the Pig*, 2nd Ed. (Eds: M. F. Rothschild and A. Ruvinsky). CAB International, pp. 355–89.

Cleveland, E. R., P. J. Cunningham and E. R. Peo. 1982. Selection for Lean Growth in Swine. *J. Anim. Sci.* 54(4):719–27.

Cleveland, E. R., R. K. Johnson and R. W. Mandigo. 1983a. Index selection and feed intake restriction in swine. 1. Effect on rate and composition of growth. *J. Anim. Sci.* 56(3):560–9.

Cleveland, E. R., R. K. Johnson, R. W. Mandigo and E. R. Peo. 1983b. Index selection and feed intake restriction in swine. 2. Effect on energy utilization. *J. Anim. Sci.* 56(3):570–8.

Clutter, A. C. 2011 Genetics of performance traits. In *The Genetics of the Pig*, 2nd Ed. (Eds: M. F. Rothschild and A. Ruvinsky). CAB International, pp. 325–354.

Clutter, A. C., L. J. Spicer, M. D. Woltmann, R. W. Grimes, J. M. Hammond and D. S. Buchanan. 1995. Plasma growth hormone, insulin-like growth factor I, and insulin-like growth factor binding proteins in pigs with divergent genetic merit for postweaning average daily gain. *J. Anim. Sci.* 73(6): 1776–83.

Clutter, A. C., R. Jiang, R., J. P. McCann and D. S. Buchanan. 1998. Plasma cholecystokinin-8 in pigs with divergent genetic potential for feed intake and growth. *Domest. Anim. Endocrinol.* 15(1):9–21.

Craft, W. A. 1958. Fifty years of progress in swine breeding. *J. Anim. Sci.* 17(4):960–80.

Cunningham, P. J., M. E. England, L. D. Young and D. R. Zimmerman. 1979. Selection for ovulation rate in swine: Correlation response in litter size and weight. *J. Anim. Sci.* 48:509–16.

David, I., J. Ruesche, L. Drouilhet, H. Garreau and H. Gilbert. 2015. Genetic modeling of feed intake. *J. Anim. Sci.* 93:965–77.

Dekkers, J. 2009. A geneticist's perspective on improvement of feed efficiency. Proceedings of the 2009 Annual Meeting of the National Swine Improvement Federation.

Dickerson, G. E. and Grimes, J. C. 1947. Effectiveness of selection for efficiency of gain in Duroc swine. *J. Anim. Sci.* 6:265–87.

Dickerson, G. E., C. T. Blunn, A. B. Chapman, R. M. Kottman, J. L. Krider, E. J. Warwick and J. A. Whatley. 1954. Evaluation of selection in developing inbred lines of swine. Res. Bull. Mo Agric. Exp. Sta. No. 551.

Do, D. N., A. B. Strathe, J. Jensen, T. Mark and H. N. Kadarmideen. 2013. Genetic parameters for different measures of feed efficiency and related traits in boars of three breeds. *J. Anim. Sci.* 91:4069–79.

Dunkelberger, J. R., N. J. Boddicker, N. V. L. Serao, J. M. Young, R. R. R. Rowland and J. C. M. Dekkers. 2015. Response of pigs divergently selected for residual feed intake to experimental infection with the PRRS virus. *Livest. Sci.* 177:132–41.

Dunkelberger, J. R., N. V. L. Sarao, M. A. Kerrigan, J. K. Lunney, R. R. R. Rowland and J. C. M. Dekkers. 2016. Genetic parameters and genomic regions associated with piglet response to vaccination for porcine reproductive and respiratory syndrome (PRRS) virus and coinfection with PRRS virus and porcine circovirus type 2b (PCV2b). *J. Anim. Sci.* 94(suppl. 2):52–3. https://doi.org/10.2527/msasas2016-112

Engblom, L., J. A. Calderon Dıaz, M. Nikkila, K. Gray, P. Harms, J. Fix, S. Tsuruta, J. Mabry and K. Stalder. 2015. Genetic analysis of sow longevity and sow lifetime reproductive traits using censored data. *J. Anim. Breed. Genet.* 133:138–44.

Engblom, L., K. Stalder, M. Nikkila and J. Mabry. 2006. Genetic improvement of sow longevity. http://benchmark.farms.com/PopUp_Genetic_Improvement_Sow_Longevity.html (accessed 29 November 2016).

Fredeen, H. T. 1958. Selection and swine improvement. *Anim. Breed. Abstracts* 26(3):229–41.

Fredeen, H. T. and H. Mikami. 1986a. Mass selection in a pig population: Experimental design and responses to direct selection for rapid growth and minimum fat. *J. Anim. Sci.* 62(6):1492–508.

Fredeen, H. T. and H. Mikami. 1986b. Mass selection in a pig population: Realized heritabilities. *J. Anim. Sci.* 62(6):1509–22.

Fredeen, H. T. and H. Mikami. 1986c. Mass selection in a pig population: Correlated responses in reproductive performance. *J. Anim. Sci.* 62(6):1523–32.

Fredeen, H. T. and H. Mikami. 1986d. Mass selection in a pig population: Correlated responses in preweaning growth. *J. Anim. Sci.* 62(6):1533–45.

Fredeen, H. T. and H. Mikami. 1986e. Mass selection in a pig population: Correlated changes in carcass merit. *J. Anim. Sci.* 62(6):1546–54.

Garnett, I. and G. W. Rahnefeld. 1976. Mass selection for postweaning growth in swine: 5 correlated response of reproductive traits and preweaning growth. *Can. J. Anim. Sci.* 56(4):791–801.

Gilbert, H., J. P. Bidanel, Y. Billon, H. Lagant, P. Guillouet, P. Sellier, J. Noblet and S. Hermesch. 2012. Correlated responses in sow appetite, residual feed intake, Body composition and reproduction after divergent selection for residual feed intake in the growing pig. *J. Anim. Sci.* 90:1097–108.

Gilbert, H., J. P. Bidanel, J. Gruand, J. C. Caritez, Y. Billon, P. Guillouet, H. Lagant, J. Noblet and P. Sellier. 2007. Genetic parameters for residual feed intake in growing pigs, with emphasis on genetic relationships with carcass and meat quality traits. *J. Anim. Sci.* 85:3182–8.

Gomez-Raya, L., M. S. Amoss, Y. Da, C. W. Beattie, O. Ash and W. M. Rauw. 2009. Role of selection and inbreeding on the incidence of cutaneous malignant melanoma in Sinclair swine. *J. Anim. Brd. Gen.* 126:242–9.

González-Peña, D. R., V. Knox, M. D. MacNeil and S. L. Rodriguez-Zas. 2015. Genetic gain and economic values of selection strategies including semen traits in three- and four-way crossbreeding systems for swine production. *J. Anim. Sci.* 93:879–91. https://doi.org/10.2527/jas.2014-8035

Halbur, P. G., M. F. Rothschild, B. J. Thacker, X. J. Meng, P. S. Paul and J. D. Bruna. 1998. Differences in susceptibility of Duroc, Hampshire, and Meishan pigs to infection with a high virulence strain (VR2385) of porcine reproductive and respiratory syndrome virus (PRRSV). *J. Anim. Brd. Gen.* 115:181–9.

Herrero-Medrano, J. M., P. K. Mathur, J. ten Napel, H. Rashidi, P. Alexandri, E. F. Knol and H. A. Mulder. 2015. Estimation of genetic parameters and breeding values across challenged environments to select for robust pigs. *J. Anim. Sci.* 93:1494–502.

Hoeltig, D., I. Hennig-Pauka, K. Thies, T. Rehm, M. Beyerback, Katrin Strutzbert-Minder, G. F. Gerlach, K. H. Waldmann and FUGATO-consortium IRAS. 2009. A novel respiratory health score supports a role of acute lung damage and pig breed in the course of an Actinobacillus pleuropneumoniae infection. *BMC Vet. Res.* 5:14–21.

Hsu, W. L. and R. K. Johnson. 2014. Analysis of 28 generations of selection for reproduction, growth and carcass traits in swine. *J. Anim. Sci.* 92:4806–22.

Johnson, R. K. 1980. Heterosis and Breed Effects in Swine. NC Reg. Pub 262.

Johnson, R. K. 2007. Genetics of disease (PRRS and PCVAD) resistance. Proceedings of the 2007 Annual Meeting of the National Swine Improvement Federation.

Johnson, R. K., M. K. Nielsen and D. S. Casey. 1999. Responses in ovulation rate, embryonal survival, and litter traits in swine to 14 generations of selection to increase litter size. *J. Anim. Sci.* 77:541–57.

Johnson, R. K., D. R. Zimmerman and R. J. Kittok. 1984. Selection for components of reproduction in swine. *Livest. Prod. Sci.* 11:541–58.

Krider, J. L., B. W. Fairbanks, W. E. Carroll and E. Roberts. 1946. Effectiveness of selecting for rapid and slow growth rate in Hampshire swine. *J. Anim. Sci.* 5:3–15.

Kuhlers, D. L. and S. B. Jungst. 1990. Mass selection for increased 70-day weight in a closed line of Landrace pigs. *J. Anim. Sci.* 68(8): 2271–8.

Kuhlers, D. L. and S. B. Jungst. 1991a. Mass selection for increased 200-day weight in a closed line of Duroc pigs. *J. Anim. Sci.* 69(2):507–16.

Kuhlers, D. L. and S. B. Jungst. 1991b. Mass selection for increased 200-day weight in a closed line of Landrace pigs. *J. Anim. Sci.* 69(3):977–84.

Lamberson, W. R. and E. R. Cleveland. 1988. Genetic parameters and their use in swine breeding. Swine Genetics Fact Sheet Number 3, NSIF-FS3.

Merks, J. W. M., P. K. Mathur and E. F. Knol. 2012. New phenotypes for new breeding goals in pigs. *Animal* 6(4):535–43.

National Swine Improvement Federation. 2003. Guidelines for Uniform Swine Improvement Programs. http://www.nsif.com/guidel/guidelines.htm. Accessed 28 November 2016.

National Swine Improvement Federation. 2003. http://www.nsif.com/guidel/guidelines.htm.

Opriessnig, T., M. Fenaux, P. Thomas, M. J. Hoogland, M. F. Rothschild, X. J. Meng and P. G. Halbur. 2006. Evidence of breed-dependent differences in susceptibility to porcine circovirus Type–2-associated disease and lesions. *Vet. Pathol.* 43:281–93.

Patience, J. F., M. C. Rossoni-Serao and N. A. Gutierrez. 2015. A review of feed efficiency in swine: Biology and application. *J. Anim. Sci. Biotechnol.* 6:33–42.

Pork Checkoff Quick Facts. 2013. http://www.pork.org/pork-quick-facts/home/stats/structure-and-productivity/number-of-u-s-hog-operations-by-year/Accessed 28 November 2016.

Putz, A. M., F. Tiezzi, C. Maltecca, K. A. Gray and M. T. Knauer 2015. Variance component estimates for alternative litter size traits in swine. *J. Anim. Sci.* 93:5153–63.

Rahnefeld, G. W. 1971a. Mass selection for postweaning growth in swine: 1. The value of a pedigreed control population. *Can. J. Anim. Sci.* 51(2):481–6.

Rahnefeld, G. W. 1971b. Mass selection for postweaning growth in swine: 2. Response to selection. *Can. J. Anim. Sci.* 51(2):497–502.

Rahnefeld, G. W. 1973. Mass selection for postweaning growth in swine: 3. Correlated response in weaning weight and feed efficiency to recurrent selection for postweaning average daily gain in swine. *Can. J. Anim. Sci.* 53(2):173–8.

Rahnefeld, G. W. and I. Garnett. 1976. Mass selection for postweaning growth in swine: 4. Selection response and control population stability. *Can. J. Anim. Sci.* 56(4):783–90.
Rothschild, M. F. 1998. Selection for disease resistance in the pig. Proceedings of the 1998 Annual Meeting of the National Swine Improvement Federation.
Saintilan, R., I. Merour, L. Brossard, T. Tribout, J. Y. Dourmad, P. Sellier, J. Bidanel, J. van Milgen and H. Gilbert. 2013. Genetics of residual feed intake in growing pigs: Relationships with production traits, and nitrogen and phosphorus excretion traits. *J. Anim. Sci.* 91:2542–54.
Stalder, K. J., T. Serenius, M. Nikkila and R. F. Fitzgerald. 2008. Genetic evaluation of longevity traits. Proceedings of the 2008 Annual Meeting of the National Swine Improvement Federation.
Stothart, J. G. and Fredeen H. T. 1950. Selection on the basis of performance in swine. *Proc. Can. Soc. Anim. Prod.* 1950:49–51.
Su, G., D. Sorenson and M. S. Lund. 2008. Variance and covariance components for liability of piglet survival during different periods. *Animal* 2(2):184–9.
Woltmann, M. D., A. C. Clutter and D. S. Buchanan. 1995. Effect of divergent selection for postweaning average daily gain on front end soundness of market weight pigs. *J. Anim. Sci.* 73(7):1940–7.
Woltmann, M. D., A. C. Clutter, D. S. Buchanan and H. G. Dolezal. 1992. Growth and carcass characteristics of pigs selected for fast or slow gain in relation to feed intake and efficiency. *J. Anim. Sci.* 70(4):1049–59.
Young, J. M., R. Bergsma, E. F. Knol, J. F. Patience and J. C. M. Dekkers. 2016. Effect of selection for residual feed intake on sow reproductive performance and lactation efficiency. *J. Anim. Sci.* 94(10):4120–32.
Young, J., R. Bergsma, E. F. Knol and J. C. M. Dekkers 2010. Effect of selection for residual feed intake on sow reproductive performance and lactation efficiency. 9th World Congress on Genetics Applied to Livestock Production.
Young, J. M., W. Cai. and J. C. M. Dekkers. 2011. Effect of selection for residual feed intake on feeding behavior and daily feed intake patters in Yorkshire swine. *J. Anim. Sci.* 89:639–47.
Young, L. D., R. A. Pumfrey, P. J. Cunningham and D. R. Zimmerman. 1978. Heritabilities and genetic and phenotypic correlations for prebreeding traits, reproductive traits and principal components. *J. Anim. Sci.* 46:937–49.
Young, L. G. (Ed.). 1990. *Genetics of Swine*. North Central Regional Research Report NC-103. Meat Animal Research Center, Nebraska.
Zimmerman, D. R. and P. J. Cunningham. 1975. Selection for ovulation rate in swine: Population, procedures and ovulation response. *J. Anim. Sci.* 40:61–9.

Chapter 2

The use of molecular genetic information in genetic improvement programmes for pigs

Jack C. M. Dekkers, Iowa State University, USA

1 Introduction

Over the past decades, genetic improvement has been a very important contributor to improved productivity of pigs, as outlined in the previous chapter. Most of that improvement has been achieved by selection of individuals (young males and females) for breeding based on estimated breeding values (EBVs) derived using phenotypes for important traits that have been recorded on the selection candidates themselves and/or their close relatives (phenotype-based prediction, see top of Fig. 1). For that purpose, sophisticated statistical methods based on mixed linear models and best linear unbiased prediction (BLUP) have been pioneered in animal breeding. These methods optimally utilize all available phenotypes on the individual itself and its relatives in order to obtain the most accurate EBV of the individual, which can then be used to rank individuals to identify animals that should be used to breed the next generation.

The rate of genetic improvement that can be achieved per year using the conventional phenotype-based approach is theoretically equal to $i*r*\sigma_g/L$ (Falconer and MacKay, 1996),

http://dx.doi.org/10.19103/AS.2017.0013.03

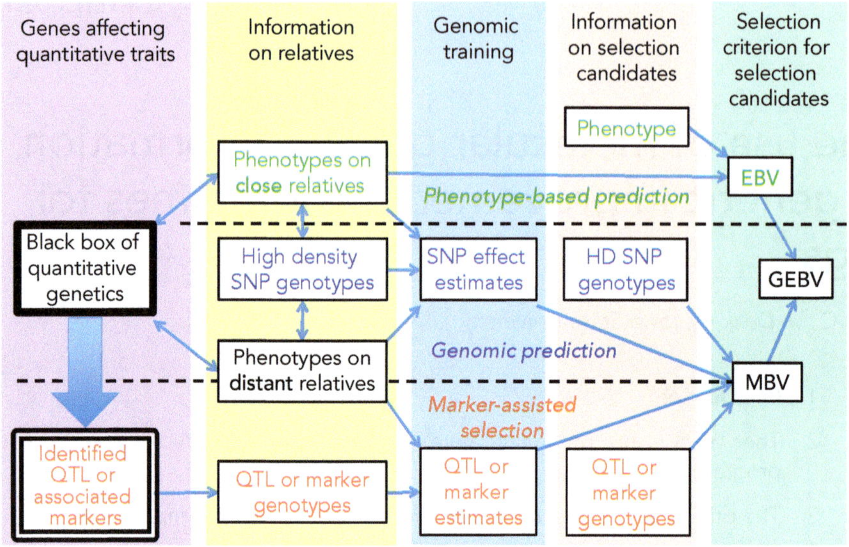

Figure 1 Information needed or available for traditional phenotype-based prediction, genomic prediction and marker-assisted selection.

where *i* is the selection intensity, which increases as the top % of selection candidates that are selected to be used for breeding decreases; *r* is the accuracy of selection, which is equal to the correlation between the estimated and the true breeding values of selection candidates; σ_g is the genetic standard deviation of the population for the trait, which is equal to the standard deviation of the true breeding values; and *L* is the generation interval, which is equal to the average age of parents when their progeny are born. Thus, a breeder can increase the rate of improvement by increasing selection intensity, by increasing accuracy of selection and/or by reducing the generation interval, that is, by selecting and using individuals for breeding at a younger age. For this purpose, extensive phenotype recording programmes have been developed, primarily in the purebred sector of the industry, where most selection and genetic improvement takes place (the nucleus herd in the pyramid breeding structure shown in Fig. 2).

Although these approaches have been very successful, the previous chapter also outlined the multiple challenges and limitations that are associated with these phenotype-based programmes. Most of these relate to the need to have phenotypes on, ideally, the selection candidates themselves, and if that is not possible, at least on some of their close relatives, in order to obtain a high accuracy (*r*), as well as to obtain those phenotypes early in the life of the selection candidates, such that early selection decisions can be made and the generation interval (*L*) can be reduced. Many of these limitations can be overcome, at least to some extent, by using molecular information to help predict breeding values and inform selection decisions. Thus, the objective of this chapter is to describe how molecular genetic information can be used to enhance selection programmes in pigs, what is required to develop such information, and what strategies are available for the use of molecular information in breeding programmes. Several examples and challenges and future prospects are also reviewed.

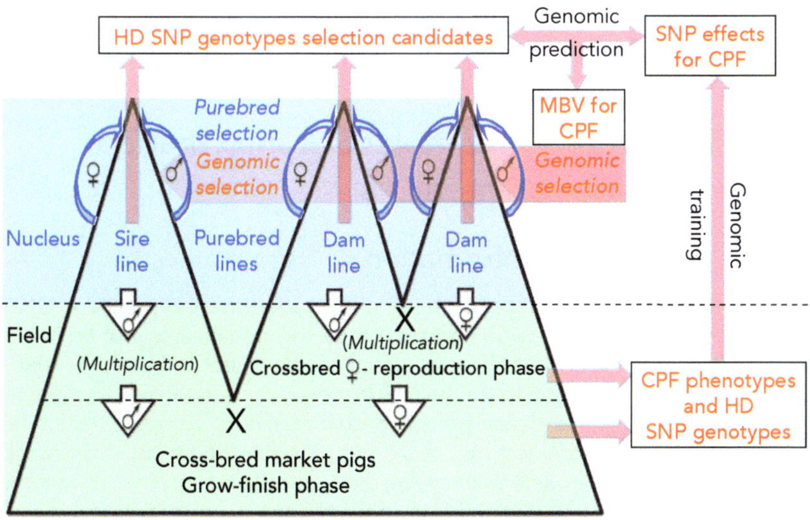

Figure 2 Pyramid breeding structure and the use of phenotypes and high density (HD) SNP genotypes collected on crossbred animals to effect genomic selection based on molecular breeding values (MBVs) of purebred animals for Crossbred Performance in the Field (CPF).

2 The black box of quantitative genetics for phenotype-based breeding programmes

The purpose of breeding value estimation is to obtain an estimate of the collective effects of all alleles that an individual has at genes that affect a particular trait. Most traits of interest in pork production are so-called quantitative traits, which means that they are affected by a large number of genes (hundreds to thousands), as well as environment. The observed phenotype is the sum of the collective effects of all these alleles (G) and the effect of all environmental factors that affect a trait (E): P = G + E (Falconer and Mackay, 1996). The aim of EBV is to obtain an estimate of G. In phenotype-based prediction of breeding values, this is achieved by using phenotypes of the individual itself and its close relatives in BLUP breeding value estimation. This prediction method weighs phenotypic information that is available on relatives depending on their pedigree-based relationships with the selection candidate. Genetic relationships quantify the proportion of alleles that the selection candidate and relative are expected to share (Falconer and MacKay, 1996). This is, however, done without knowledge of the genotype that individuals have at the individual genes that affect the trait, let alone knowledge of where those genes are, what their specific effects on phenotype are and even without knowledge of the number of such genes. Thus, from the perspective of the underlying genetic basis of the trait, phenotype-based selection approaches very much treat the genetic architecture of the trait as a black box (Fig. 1). The only information that is available is the extent to which the trait is affected by genetics (heritability) and what proportion of genes relatives are expected to share and, therefore, to what extent relatives are expected to have similar phenotypes. For multiple-trait selection, the extent to which traits share genes (quantified by the genetic correlation between traits (Falconer and MacKay, 1996)) is also utilized but, again, without knowledge of how many or which genes they share.

3 The principle of using molecular information for genetic improvement

3.1 Types of molecular information

In contrast to the black box approach of phenotype-based prediction, DNA-based molecular genetic tests can provide information on the genetic basis of traits of interest and the genotype that individuals have at specific locations of the genome that may be related to the trait. With regard to the trait of interest, an important distinction has to be made between so-called Mendelian traits and quantitative or complex traits. Mendelian traits are traits that are controlled by one or just a couple of genes. In livestock, these include genetic defects (e.g. deleterious recessives genes) and external appearance traits (e.g. coat colour and horned/polled). Many of these have been catalogued in the Online Mendelian Inheritance in Animals (http://omia.angis.org.au/), which is the equivalent of the Online Mendelian Inheritance of Man catalogue for humans (http://www.omim.org/).

Quantitative or complex traits are affected by potentially a large number of genes across the genome, along with environment. This includes most of the traits that are of economic importance in livestock (e.g. growth rate, milk yield and litter size). Although most quantitative traits are measured on a continuous scale (e.g. kg) or on a semi-continuous scale (e.g. counts), many production diseases are also quantitative or complex because they are affected by many genes, although they may be recorded on a categorical scale (healthy or sick, dead or alive). Examples of such traits in humans include obesity and the risk of getting cancer. In livestock, these complex disease traits include mastitis in dairy cattle (udder infections) and susceptibility or resistance to many other diseases. Causative loci for a quantitative trait are often referred to as quantitative trait loci (QTLs).

For both Mendelian and quantitative traits, two types of DNA-level tests can be developed:

- Genetic tests for the causative genes or mutations
- Genetic tests for so-called genetic markers at loci that have no direct effect on the trait themselves, but that have a statistical association with phenotype for the trait because they are closely linked in the genome to the causative genes or mutations.

Because of the difficulty of finding the causative genes and the current abundance and feasibility of genotyping individuals for large numbers of genetic markers across the genome (see Archibald chapter), most genetic tests that are used in livestock are for linked genetic markers. In order for such genetic markers to be useful, however, they not only need to be closely linked to a causative gene but they must also be in so-called linkage disequilibrium (LD) with the causative gene (Falconer and MacKay, 1996). A marker locus is in LD with a causative locus if knowledge of the genotype that an individual has at the marker locus provides information on the genotype that it has at the causative gene. If genotypes are available at both loci, LD can be quantified by the correlation between genotypes at the two loci, coded as 0, 1 and 2 based on the number of copies of a

Table 1 Illustration of linkage disequilibrium (r^2) between a genetic marker and a QTL and the association between marker genotype and phenotype in a population based on a sample of seven individuals

	Genotype		Genotype code			
Individual	Marker	QTL	Marker	QTL	Phenotype	Average phenotype
1	mm	qq	0	0	11	
2	mm	qq	0	0	13	$\mu_{mm} = 12$
3	Mm	qq	1	0	10	
4	Mm	Qq	1	1	14	
5	Mm	Qq	1	1	15	$\mu_{Mm} = 13$
6	MM	Qq	2	1	14	
7	MM	QQ	2	2	18	$\mu_{MM} = 16$
			$r^2 = 0.65$[1]			

[1] LD is strictly defined based on the correlation between allele codes (0/1) on individual chromosomes received from parents (one from the mother and one from the father) (Hill and Robertson, 1968), rather than as correlations between genotype codes, but these are expected to be equal for larger sample sizes and when mating is at random in relation to genotype.

particular allele that an individual has at the locus (see Table 1). Because the coding of alleles is arbitrary, the sign of the correlation is not informative and, hence, LD is often quantified as the square of the correlation between genotypes (r^2).

LD between loci can be generated by a number of processes, including the use of a limited number of individuals for breeding (inbreeding or drift), crossing, migration, selection and mutation. However, LD between loci is eroded each generation by recombination or cross-overs, which results in reshuffling of alleles at linked loci when an individual generates gametes that are passed on to its progeny (see Fig. 3). The rate of recombination between loci depends on their distance on the genome, with recombination being less frequent between loci that are closely linked. Thus, the amount of LD that is present between loci in a population in a given generation is expected to be higher at loci that are close together, which is what is observed in practice (Fig. 4). Note, however, that loci that are close together are not guaranteed to be in high LD because both the creation and erosion of LD are random processes. In addition, the amount of LD between the marker locus and the causative gene can change over time within a population, especially when they are further apart. For the same reasons, the LD between a given pair of loci can also differ between populations. Thus, a genetic marker that is informative for a causative gene in one population may not be informative for that gene in another breed or line.

3.2 Identification of useful genetic markers for quantitative traits

If a genetic marker can be found that is in LD with a QTL for a trait, individuals that have the marker genotype that is associated with the favourable genotype at the QTL are expected to have a higher breeding value and are preferred for breeding (assuming no other information is available). Thus, in the example of Table 1, if allele Q at the QTL has a favourable effect on phenotype, individuals with genotype MM at the marker are expected to have a higher breeding value. The challenge, however, is that we do not know

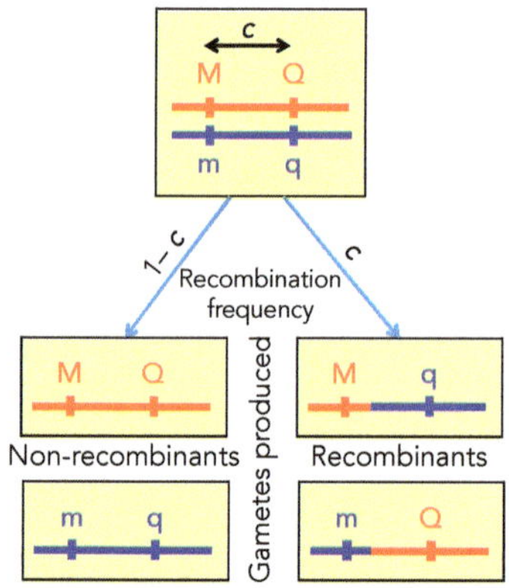

Figure 3 Illustration of recombination between linked loci in the process of producing gametes, which results in reshuffling of alleles between loci. The rate of recombination is c, which depends on the distance between loci.

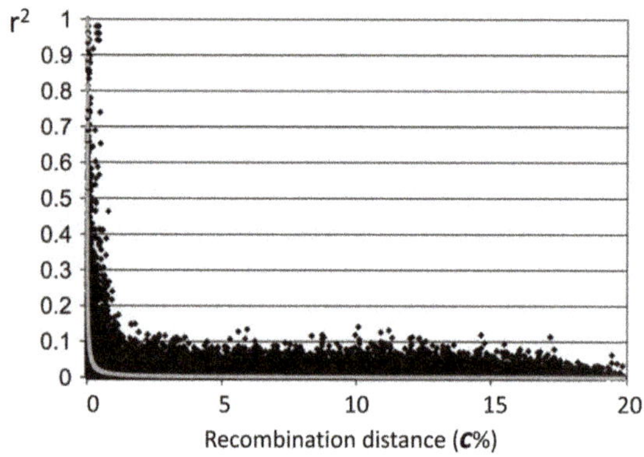

Figure 4 Observed linkage disequilibrium (r^2) between pairs of SNP loci across a chromosome in a population as a function of recombination distance between the loci. Each dot represents the linkage disequilibrium between a pair of loci plotted against their distance. The line represents the expected average linkage disequilibrium.

where the QTLs are and, thus, we do not have the genotypes at the QTL to determine which markers are in high LD with the QTL by simple correlation of genotypes. However, in addition to genotyping individuals for genetic markers across the genome, we can also record the phenotype of these individuals. If the marker is in LD with a QTL that affects the phenotype, we expect individuals that have alternate genotypes at the genetic marker to have a different genotype at the QTL and, therefore, to have a different phenotype. Thus, the procedure for determining whether a genetic marker can be used to identify animals that have better genetics for that trait is based on a statistical analysis of the association between the genotype of individuals at the genetic marker and their phenotype for the trait. As illustrated in Table 1, if the average growth rate of pigs that have the MM genotype at a genetic marker is significantly greater than the average growth rate of pigs that have the mm genotype at this marker, this indicates that this genetic marker is associated with growth rate. This is probably because the genetic marker is in LD with a QTL in this population. Because LD tends to be highest for loci that are close together, this statistical association also implies that the QTL likely is located near this genetic marker. Thus, in addition to identifying whether a genetic marker is associated with phenotype and which genotype at the marker is favourable, this marker–phenotype association procedure can also be used to identify the approximate location of QTL (so-called QTL mapping).

The association between marker genotype, QTL genotype and phenotype is further illustrated in Fig. 5. This figure shows that the phenotype of an individual with favourable genotype QQ at the QTL is not guaranteed to have a phenotype that is higher than that of individuals with genotype Qq or qq (Fig. 5a), because a particular QTL only explains a limited proportion of the genetic variance for a trait and, in addition, observed phenotypes are affected by environmental effects, which also differ between individuals. However, on average, the phenotype of individuals that have the favourable QQ genotype is higher than that of individuals with the Qq and qq genotypes. If the marker is in complete LD with the QTL (Fig. 5a), marker genotype is equivalent to QTL genotype and individuals with alternate marker genotypes are expected to have equivalent differences in average phenotypes as individuals with alternate QTL genotypes. Note that marker genotypes can be observed, in contrast to QTL genotypes. If, however, the marker is only in partial LD with the QTL (Fig. 5b), differences in average phenotypes between individuals with alternate marker genotypes are expected to be reduced relative to the QTL effects. Thus, with partial LD, the marker only explains a fraction of the variance contributed by the QTL. This fraction actually is equal to the LD between the marker and QTL (i.e. r^2).

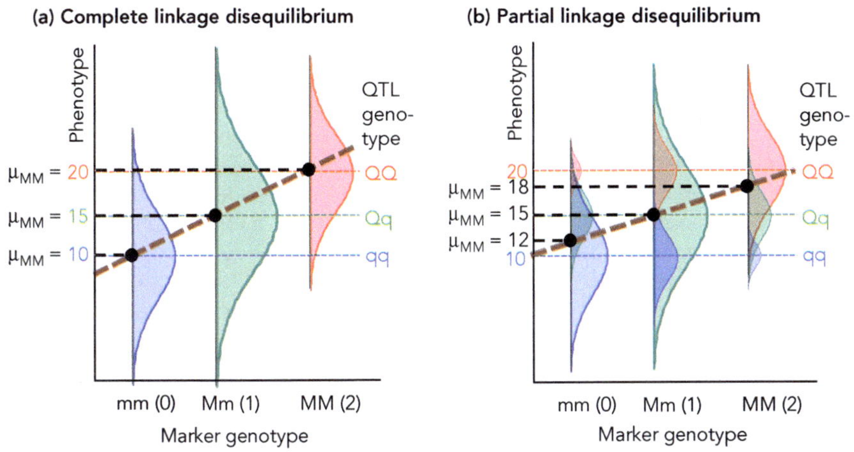

Figure 5 Relationship between genotype at a marker, at a QTL, and observed phenotype, depending on the extent of linkage disequilibrium between the marker and QTL. The Q allele at the QTL increases average phenotype by 5 units, which is also the effect observed for the M allele at the marker if the marker and QTL are in complete LD (a). Normal distributions around the mean represent the effects of other genes and environment. However, when the marker is not in complete LD with the QTL (b), differences in average phenotype between alternate marker genotypes are less than 5. The diagonal line represents the regression of phenotype on marker genotype (coded 0/1/2), with a slope of 5 with complete LD and 3 with partial LD.

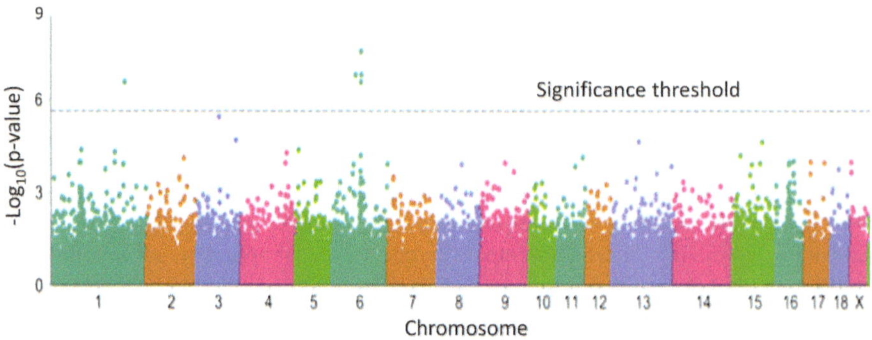

Figure 6 Example Manhattan plot resulting from a genome-wise association study. The X-axis represents individual SNPs on each chromosome and the Y-axis the dy. han 5. The dotted horizontal line represents the regression of phenotype at an SNP with phenotype for the trait.

Procedures to identify genetic markers that are associated with a trait are typically implemented within the context of a statistical analysis of the phenotype of the trait by fitting the genotype at the marker locus as an effect or factor in a statistical model and testing for its significance. This results in a *P*-value for genotype at the marker, as well as estimates of the effect of each genotype on trait phenotype. Because parents pass on alleles rather than genotypes, marker genotypes are often fitted as a coded covariate (0,1,2). This results in estimates of allele substitution effects (Falconer and MacKay, 1996), which is the average effect of having one extra copy of the allele on the phenotype of the progeny of that individual. These are represented as the slope of the regression line of phenotype on genotype code in Fig. 5a and b. Note that, due to incomplete LD, this slope is smaller for Fig. 5b than 5a.

This single-marker association analysis procedure can be repeated for each of many genetic markers that a population is genotyped for across the genome, which is referred to as a genome-wide association study (GWAS). Results of a GWAS are typically represented by a so-called Manhattan plot of the log-base-10 of the *P*-value (such that low *P*-values show higher peaks) obtained at each genetic marker (see Fig. 6 for an example). Manhattan plots often show multiple sharp peaks that represent neighbouring SNPs that are all in high LD with a putative QTL.

Although the principle of a GWAS is fairly simple, it is fraught with statistical pitfalls. These include:

- The *large number of statistical tests* that are conducted (one for each genetic marker genotyped, which can run in the tens or hundreds of thousands). This requires careful multiple test correction to control the number of false-positive results. As a result, thresholds on *P*-values for declaring significance are very stringent (typically $\alpha < 0.000001$) and large data sets (at least 1000 individuals with genotypes and phenotypes) are needed to even detect moderate to large effects.
- The so-called *curse of dimensionality or the 'Beavis effect'* of significance testing when power is limited (because of the stringent thresholds). This results in the true effects of significant markers often to be substantially smaller than estimated in the original data and in poor repeatability of results in independent data sets.

- *Population stratification*, which can result in genetic markers that are significant to be quite far removed or even on another chromosome than the QTL. The reason for this is that statistical association does not imply causality. For example, Holstein cattle that have the red coat colour gene on average have lower milk yield than black-and-white Holstein cattle. However, this does not mean that the red coat colour gene causes lower milk production or even that it is closely linked to a gene that causes lower milk production. Instead, the red Holstein population is almost a separate breed from the black-and-white Holstein population and has not been as intensely selected for milk yield. Thus the red Holstein population tends to have a lower frequency of favourable alleles at QTL for milk yield across the genome and any marker locus for which the red and black-and-white populations differ in allele frequencies will pick these effects up as a statistical association, unless 'breed' is fitted as an extra effect in the model. Thus, population stratification can result in spurious associations and poor repeatability of results in other populations or over time, unless it is accounted for in the statistical model of analysis. Population stratification can also be overcome by conducting the QTL mapping study in a carefully designed study such as a breed cross, or within families, but such studies also have multiple limitations (see Dekkers, 2004).

These pitfalls also contribute to the so-called 'missing heritability' that has been popularized in human genetics, that is, that the genetic markers that are identified in even large GWAS studies only explain a limited proportion of the genetic variance that is known to exist for a trait. In addition to missing many small effects due to limited power and overestimation of significant effects, other suspects in 'the case of the missing heritability' include incomplete LD between markers and QTL, as well as the rare allele hypothesis for QTL, which states that the frequency of either the favourable or the unfavourable allele may be low at many QTLs. This not only makes such QTLs difficult to detect but also makes it difficult to have a genetic marker that is in high LD with it, especially because genetic markers that are included on SNP panels are typically selected to have moderate minor allele frequencies, which is referred to as ascertainment bias of SNP panels.

Nevertheless, despite these pitfalls, many QTLs have been identified using genetic markers for many traits in livestock using different population designs (see e.g. Lamont et al., 2013). Most of these QTLs are catalogued in the pig QTL database (Zhang et al., 2005). And, while some of these QTLs have been replicated across studies, many have not and, together, they only explain a limited proportion of the genetic variance for reasons outlined above. In addition, the position of many QTL is not very precise, which makes replication and use of these results in other populations problematic.

4 The use of molecular information in selection: genetic tests

4.1 Genetic tests for Mendelian traits

The aim of the use of genetic tests for gene(s) that affect undesirable Mendelian traits is to reduce the frequency of the undesirable allele(s) in the population by avoiding the use of animals that carry one or more undesirable alleles for breeding. Alternatively, if the

Mendelian trait is a recessive, the production of homozygous recessive progeny can be avoided by not mating carriers to each other.

With regard to selection against the undesirable allele(s), several approaches can be considered:

- *Complete selection* against undesirable allele(s): reduces the frequency of the undesirable allele(s) as quickly as possible by not using animals that carry one or two copies of the undesirable allele(s). This strategy can remove an undesirable allele from the population within one or two generations, depending on the frequency of the allele and the proportion selected. However, if the number of selection candidates that are available and the number of animals that must be selected for breeding are fixed, such selection will substantially reduce the selection emphasis that can be placed on other traits of interest. As a result, overall economic response to selection may be reduced.
- *Index selection* against unfavourable allele(s): to optimize overall economic response, genetic tests should be properly weighted against other traits in an overall economic objective by making the Mendelian trait part of the breeding goal. Such index selection would result in only individuals that carry one or more undesirable alleles and also have less than superior EBV for other traits of interest to be removed from consideration for breeding. Over time, this index selection strategy would still reduce the frequency of undesirable alleles in the population but not as fast as the *complete selection* strategy. This also means, however, that selection candidates need to be genotyped for the trait for more generations, whereas genotyping can be discontinued once the undesirable allele has been removed from the population. This extra cost also needs to be taken into account. In addition, there may be a marketing advantage to having undesirable alleles removed from the population.
- *Pre-selection* against undesirable allele(s): because genetic tests can be conducted as soon as DNA can be collected from an individual, selection against the undesirable allele(s) can take place prior to the normal selection step. Then, if the number of individuals that are raised up to the time of selection is fixed but the number of animals that are produced is not, extra animals could be produced to provide room for early selection against unfavourable allele(s) based on genetic tests. In particular, if such selection takes place within litters, that is, littermates that carry the unfavourable allele are removed, this may have limited impact on selection response for other traits. The cost of producing these additional animals must, however, be taken into account.

4.2 Genetic tests for quantitative traits

4.2.1 Marker-assisted selection (MAS)

Methods for detection of QTL or association analysis result in estimates of the effects of marker genotypes on phenotype, often in the form of allele substitution effects. Based on the genotype of selection candidates at the identified markers, these estimates can be used to compute a molecular breeding value (MBV) of each individual, as illustrated at the bottom of Fig. 1 and in Table 2 (Lande and Thompson, 1990; Dekkers and Hospital, 2002). Note that estimation of the effect of an identified QTL or of a marker that is in LD

Table 2 Example of the calculation of molecular breeding values (MBVs) of five individuals for three additive loci (QTLs or SNPs) with additive effects (effect of allele A vs. B) of +10, +5 and –10 for locus 1, 2 and 3, respectively, and of selection on MBV, phenotype-based EBV, combined genomic EBV

	Locus 1		Locus 2		Locus 3		Selection criteria[1]		
Individual	Genotype	Value	Genotype	Value	Genotype	Value	MBV	EBV	GEBV[2]
1	AA	10	AA	5	AA	–10	5	13	**12**
2	AB	5	AA	5	BB	10	**20**	–5	10
3	AB	0	BB	–5	AB	0	–5	–7	–8
4	AB	0	BB	–5	AA	–10	–15	9	–4
5	BB	–10	AA	5	AB	0	–5	**17**	8

[1] The top individual based on each selection criterion is indicated in bold.
[2] An index of MBV and EBV with weights equal to $^2/_3$ and $^2/_3$ based on the scenario where the QTLs jointly explain 50% of the genetic variance and EBVs are based on own phenotype only for a trait with heritability 0.5 (Lande and Thompson, 1991).

with the QTL still requires estimation of its effect on phenotype. However, in contrast to requiring phenotypes on close relatives, QTL or marker effect estimates can be based on phenotypes of more distant relatives or even unrelated individuals (depending on consistency of the LD).

In addition to an MBV, each selection candidate will also have an EBV based on regular phenotype-based BLUP evaluation (top of Fig. 1). The combination of MBV and EBV can then be used for selection in different manners:

- *Select on the MBV alone*
- *Independent culling level selection*, with selection of individuals that meet both a threshold on MBV and a threshold on EBV
- *Index selection* on a combination of MBV and EBV, called a genomic EBV (GEBV)
- *Pre-selection* on MBV followed by selection on EBV

Selection on MBV, EBV, and an index of MBV and EBV (GEBV) is illustrated in Table 2, while index selection versus independent culling level selection is illustrated in Fig. 7.

Selection on MBV alone ignores information that is available on all other genes (polygenes) that affect the trait and is expected to result in the lowest response to selection, unless all genes that affect the trait are included in the MBV. If both EBV and MBV are available on selection candidates, index selection is expected to result in greater response to selection than independent culling selection. The reason for this is similar to why two-trait selection using independent culling levels is expected to give lower multiple-trait response than index selection. Independent culling selection and pre-selection do not allow some individuals that have a low MBV but an EBV that is high enough to more than compensate for the low MBV (individuals in area b in Fig. 7) to be selected. In some situations, there can, however, be an advantage to pre-selection on MBV if phenotypic records to compute an accurate EBV are available only at a later stage, since an MBV can

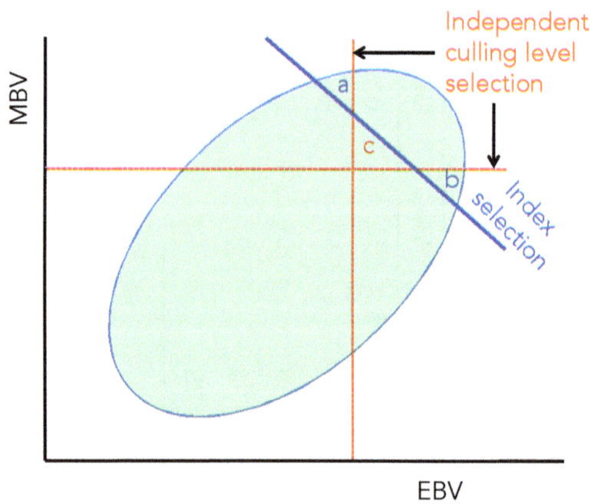

Figure 7 Illustration of independent culling level selection and index selection on MBV and EBV. Index selection uses equal emphasis on MBV and EBV. With index selection, individuals in areas a and b are selected in favour of individuals in area c, which are selected instead with independent culling level selection.

be obtained as soon as individuals can be genotyped. Similar to pre-selection against undesirable Mendelian trait alleles, this pre-selection would allow a reduction in the number of individuals to be raised to selection age, thereby saving costs, or an increase in response by increasing the number of selection candidates that are produced for the first selection step based on MBV.

In cases where selection on EBV is delayed past reproductive age because phenotypic records to obtain an EBV with sufficient accuracy are only available at a later age and, therefore, the accuracy of EBV at reproductive age is low, availability of an MBV could allow all selection decisions to be made before animals reach reproductive age, thereby reducing the generation interval. Optimal selection would then be on an index of MBV and the less accurate EBV. Depending on the accuracy of the MBV, this could result in the accuracy of the index to lower than the accuracy of the EBV at a later age. However, if this reduction in accuracy is less than the drop in generation interval, response to selection would increase, as response per year is a function of r/L. In addition, selection at a younger age could allow for an increase in the number of selection candidates, and thereby an increase in selection intensity and a reduction in the cost of keeping selection candidates. Thus, in addition to increasing accuracy of selection at a given age, an important advantage of molecular information is that it allows for an increase in the accuracy of selection at, in particular, a younger age, which may allow for a reduction in the generation interval.

4.2.2 Genomic selection (GS)

Despite a large number of reports with significant QTL and genetic markers associated with important traits, the application of the resulting QTL or markers in livestock breeding programmes has been limited until recently. The main reasons for the limited use of marker-assisted selection (MAS) have been described in Dekkers (2004) and Dekkers and van der Werf (2007). These reasons include:

- the limited amount of genetic variation that markers explain
- the limited number of genetic markers available
- relatively high genotyping costs
- poor reproducibility of marker or QTL effects in populations of interest, in particular when discovery populations were experimental or of limited size
- estimation of marker or QTL effects on a within-family basis, which makes it more difficult to incorporate them into breeding programmes

Many of these limitations have, however, been partially overcome by recent advances in molecular technology, as described in the Archibald chapter, in particular genome sequencing, the identification of large numbers of SNPs across the genome and cost-effective high-throughput genotyping of tens of thousands of such SNPs.

The ability to effectively use this large number of SNPs for breeding value estimation, however, also required a paradigm shift in statistical models for estimation of SNP effects, which has taken us from MAS to genomic selection (GS) (Meuwissen et al., 2001). In GS (often and more appropriately referred to as genomic prediction), the association of each SNP with phenotype is estimated using sophisticated statistical and quantitative genetics models without pre-screening markers based on significance. The MBV of an individual then is the sum of the estimated effect at each SNP summed across all SNPs, that is, extending the example of Table 2 for three markers to thousands of markers across the genome. Similar to MAS, genomic prediction allows phenotypes of more distant relatives to be used to estimate marker effects, as illustrated in the middle portion of Fig. 1. However, in contrast to the two-step approach that is employed in traditional MAS, which uses only markers that have been found to be significant for EBV, genomic prediction allows the effects of SNPs with small effects to be included in the MBV by using all markers across the genome. Further, by fitting the effect of each SNP as a random effect, the estimate of the association of each SNP with phenotype is shrunken towards zero, depending on the amount of data that is available to estimate the effect and the prior distribution that is assumed for these effects (Meuwissen et al., 2001). This addresses overestimation of effects caused by the Beavis effect. Note, however, that because the nature of the estimated effects is unknown, genomic prediction still relies on the black box of quantitative genetics, but with additional DNA-level information in the form of SNP genotypes.

Over the past 15 years, many models that assume different distributions as the precondition for SNP effects have been developed and employed for genomic prediction, using so-called Bayesian variable selection methods (Meuwissen et al., 2001; Gianola et al., 2008; Habier et al., 2010; de los Campos et al., 2013). However, in most applications, a model in which genetic variance is assumed to be equally distributed across all SNPs across the genome has often proved as accurate in determining MBV as more complex models (Hayes et al., 2009). This model is often referred to as the genomic BLUP (GBLUP) model. It has been shown to be equivalent to the standard animal model BLUP procedure that is extensively used to estimate breeding values in livestock. The difference is the use of a genetic relationship matrix that is based on genome-wide SNP genotypes, rather than pedigree, to quantify how closely individuals are related (Stranden and Garrick, 2009; Koivula et al., 2012).

One way to conceptualize the benefit of GBLUP over pedigree-based BLUP is that SNP genotypes allow relationships of close relatives to be better characterized. For example, based on pedigree, full-sibs are expected to share 50% of their alleles because each progeny receives a random one of the two alleles that a parent carries at each locus. The pedigree-based relationship of 0.5 between full-sibs is based on the 50% probability that one full-sib received the same allele at a particular locus from a parent as another full-sib. However, by chance, some full-sib pairs can share more or less than 50% of their alleles, and this is captured by the genomic relationship. Using these more accurate relationships can lead to more accurate EBV from GBLUP. In addition, if the recorded pedigree only goes back a limited number of generations, many individuals that do not appear to be related based on pedigree may actually be related because they share a common ancestor prior to the recorded pedigree. These distant relationships are also captured by genomic relationships. Thus, the use of genomic relationships in GBLUP results in individuals having many more distant relatives that can be used to inform prediction. The GBLUP method also greatly facilitates the implementation of genomic prediction, because of its technical similarity to pedigree-based BLUP.

However, unless most animals have been genotyped and the markers capture all genetic variation, the MBV obtained from genomic prediction still must be combined with a phenotype-based EBV, as illustrated in Fig. 1. This holds for both MBV derived using GBLUP, as well as for Bayesian methods, and this 'blending' is typically achieved using a selection index approach (Hayes, 2009). To address this, Misztal et al. (2009) developed a so-called single-step GBLUP method that allows data on both genotyped and non-genotyped individuals to be included and directly computes the 'blended' GEBV. In this method, the relationship matrix that is used is based on a combination of pedigree and genomic information. More recently, a single-step method that allows Bayesian variable selection models to be applied has also been developed (Fernando et al., 2015). These Bayesian models give better prediction accuracies if the trait is affected by some genes with larger effects. Many of the quantitative traits of interest, however, appear to be highly polygenic, with no or only a few SNPs or QTLs with major effects, allowing the GBLUP method to be effective.

5 Phenotyping and genotyping requirements for genomic selection (GS) or marker-assisted selection (MAS)

The use of molecular information in principle removes many of the limitations of traditional selection based on EBV derived from phenotype, which requires phenotypic records on the selection candidates themselves and/or their close relatives. Instead, the phenotype of any individual that has been genotyped for an SNP can in principle contribute to estimation of the effect of that SNP and, therefore, to estimation of the MBV of any selection candidate. Alternatively, based on the GBLUP model, a phenotyped individual will have many more individuals with which it has a quantified genomic relationship, although many may be low, allowing that individual's phenotype to contribute to the estimation of the breeding value for many more selection candidates than with pedigree-based BLUP. Thus, although MAS and GS still require phenotypic records, each phenotypic record can be leveraged to contribute to EBV for many more individuals.

Nevertheless, research has shown that substantial (training) data sets of genotyped and phenotyped animals are required to obtain substantial accuracies of MBV (Goddard and Hayes, 2009). For example, to obtain an MBV accuracy of 0.7 for a trait with heritability 0.25, a training data set of ~5000 genotyped and phenotyped individuals is needed. In addition, it has been shown that the accuracy of MBV is lower if the selection candidates do not have close relatives that are in the training data set (Habier et al., 2007; Wolc et al., 2010). This is because markers capture relationships and co-inheritance (co-segregation) of linked markers and QTL within families, although they may be in incomplete LD across families. Thus, for routine application, continued phenotype recording is required but with GS, the requirement that these phenotypes are on close relatives of selection candidates is relaxed. This also means that the recording of phenotypes can lag behind selection, allowing selection decisions to be made before the selection candidates have own phenotypes or phenotypes on close relatives. Genomic predictions have, however, not been found to be accurate across breeds. Thus, in order to implement GS within a population, breed or line, routine phenotype recording must be established or maintained within that line.

5.1 Use of low-density genotyping and imputation

The large numbers of individuals that must be genotyped for genomic training, combined with the requirement to genotype selection candidates in order to obtain genomic predictions on them, substantially increase the cost of a breeding programme, unless other structural changes can be made to reduce other costs of the programme. The cost of genotyping can, however, be reduced substantially by using low-density SNP panels to genotype selection candidates, combined with imputation of ungenotyped SNPs, as proposed by Habier et al. (2009). The latter study showed that selection candidates could be accurately imputed up to higher density if the parents of selection candidates or their sires and maternal grandsires were high-density SNP genotyped, with selection candidates themselves genotyped for as few as 400 SNPs across the genome. The same strategy can also be used when establishing the training data set, which can be constructed based on high-density genotypes on all male breeders and low-density genotypes on all female breeders and other individuals that have phenotypes. This data structure allows effective imputation of all individuals up to high density. Genotype imputation has now been implemented in most genomic prediction applications in livestock (e.g. Erbe et al., 2012)

5.2 Implementation of genomic selection

Since 2009, GS has been successfully implemented in the US dairy cattle breeding programme and since then in many other dairy cattle breeding programmes across the world (Hayes et al., 2009). In contrast to other livestock species, the use of GS has allowed a substantial decrease in the generation interval in dairy cattle because bulls can now be selected based on genomics before they reach reproductive age (<1.5 years), rather than having to wait until they have received an EBV based on milk production performance of their daughters (at >5 years of age of the bull). In addition, a very large training set could be developed in a relative cost-effective manner by capitalizing on decades of milk production data collected through milk recording programmes and extensive progeny testing of bulls, with DNA being available on all bulls that have been progeny-tested over the past decades in the form of frozen semen. Thus, by genotyping these progeny-tested bulls and using the average performance of their daughters as a very accurate and heritable phenotype, a powerful data set of several thousand bulls was developed to estimate SNP effects and for genomic prediction in the United States shortly after the release of the first high-density SNP chip for cattle (VanRaden et al., 2009). These factors, combined with the high value of individual dairy bulls, resulted in very rapid adoption of GS in the dairy industry, not only in the United States but in most countries with a well-developed dairy industry.

The implementation of GS in pigs has progressed more slowly because opportunities to reduce generation intervals are less prominent in swine breeding programmes. In addition, swine breeding programmes typically utilize multiple purebred parental lines to produce F1 crossbred sows, which are then bred to a terminal male pure line to produce market pigs, capitalizing on breed complementarity and heterosis, as illustrated in Fig. 2. Thus, full implementation of GS requires implementation in multiple lines, compared to the breeding programmes in dairy cattle, which do not extensively utilize crossbreeding. In addition, whereas competing breeding programmes in dairy cattle are genetically well connected through the use of the common bulls for breeding across the United States and even the globe, individual breeding programmes in swine are typically closed and, thus, training populations need to be developed within each separate population. Finally, compared to dairy cattle, the number of selection candidates is larger in swine breeding programmes and the relative value of each selection candidate is lower. Nevertheless, research has shown that implementation of GS is cost-effective in at least the larger swine breeding programmes (Tribout et al., 2012), especially with the use of low-density SNP genotyping and imputation, and GS has now also been implemented in several swine breeding programmes (Cleveland et al., 2013; Van Eenennaam et al., 2014). For the most part, this implementation has been without major restructuring of the breeding programme. Optimal implementation of GS may, however, require substantial restructuring of existing breeding programmes. Such strategies and additional issues associated with the use of GS in breeding programmes were reviewed in Dekkers (2010). An example of opportunities to restructure the breeding programme in egg laying chickens is in Wolc et al. (2015).

5.3 Implementation of genomic selection for 'difficult' traits

To date, the main implementation of GS in pigs has been for traits that are routinely recorded in existing breeding programmes. In swine breeding programmes, opportunities to use GS for such traits have been explored for both growth performance in terminal sire lines (e.g. Tribout et al., 2012) and reproductive performance in maternal lines (Lillehamer et al., 2011). The use of GS for meat quality traits that cannot be recorded on selection candidates, since they require harvesting the animal, was recently reviewed by Younes et al. (2015). Opportunities for the use of GS to improve efficiency, reduce emissions and improve adaptation have been described by Hayes et al. (2013).

Most applications of genomic prediction in pig breeding to date have been for traits that are recorded on animals in the selection nucleus. However, for several important traits, for example, disease traits, obtaining data for training is an issue because populations that are under selection are typically maintained in disease-free environments. Using disease-related traits as an example, several possibilities exist to collect phenotypes to inform selection decisions related to improving such traits:

- collect phenotypes on disease indicator traits within the nucleus
- collect disease-related phenotypes in production farms in the field
- collect disease-related phenotypes in special experimental facilities

As described previously, the phenotypes that are collected can be used to derive the following types of information to inform selection decisions in the nucleus:

- phenotype-based EBV on selection candidates
- MBV based on genomic prediction
- MBV based on identified major genes.

Disease indicator traits that can be collected on nucleus animals without jeopardizing their health status could be based on specific immune response tests and much research is currently ongoing to develop such indicator traits (Flori et al., 2011). Their value depends on the magnitude of the genetic correlation they have with disease-related traits in the field. A focus is on developing indicator traits for general immune response, rather than for immune response that is specific to a particular disease.

Collecting disease-related phenotypes on production farms is logistically complicated because the phenotypes must be recorded on individually identified animals that can be connected back to animals in the nucleus either through pedigree or genomics. An important benefit of such phenotypes is that they are obtained in the target environment for genetic improvement (i.e. commercial production farms in the field) and on the target animals, which in the case of pigs typically are crossbred animals, compared to the purebred animals that are selected in the nucleus farms (Fig. 5). Thus, even for traits that are routinely recorded in the nucleus, such as growth rate, backfat and litter size, collecting these traits on crossbred animals in the field potentially has great benefits. Because of genotype by environment interactions and genetic differences between purebred and crossbred performance, phenotypes collected on crossbreds in the field often have a genetic correlation with phenotypes on the same trait collected on purebreds in a nucleus environment that is substantially less than one. These low genetic correlations indicate that the genes or the effects of genes that control the same trait can be substantially different when that trait is recorded on crossbreds in the field versus purebreds in the nucleus. Since crossbred performance in the field is the target for genetic improvement, there is great value in incorporating this information in selection decisions.

Strategies to incorporate commercial crossbred data in traditional pedigree-based selection have been developed (Wei and van der Werf, 1997). These, however, require knowledge of pedigree that connects the crossbred data back to selection candidates. Also, because crossbred phenotypes can only be obtained on family members of the purebred selection candidates, increases in the accuracy of EBV from incorporating crossbred data are limited. Thus, genomic prediction provides an attractive alternative strategy and can substantially increase accuracy of EBV for crossbred performance in the field of purebred selection candidates in the nucleus, as outlined by Dekkers (2007) and illustrated in Fig. 5. One strategy to implement GS for crossbred performance in the field is to phenotype and genotype commercial crossbred animals, use this as training data to estimate the effect of markers on crossbred performance and then use these estimates to develop MBV of purebred selection candidates, as illustrated in Fig. 5. With training data of sufficient size, Dekkers (2007) showed that this can result in substantial increases in response to crossbred performance in the field. Note that this strategy does not require crossbreds to be pedigreed but continued phenotyping and genotyping of crossbreds will be needed to ensure that the training data remains relatively close to the selection candidates in terms of genetic relationships. A recent study by Lillehammer et al. (2015), however, showed that this will be beneficial only for traits with substantial economic importance and does require selection on crossbreds that are not too distantly related to the selection candidates.

Data collected in experimental disease trials can also be used for genomic prediction of resistance or susceptibility to disease of purebred nucleus animals. Such trials have been implemented to investigate the genetic basis of resistance or susceptibility of pigs to infection with porcine reproductive and respiratory syndrome (PRRS) virus in the PRRS host genomics consortium (Lunney et al., 2009). In these trials, more than 2500 commercial crossbred nursery piglets from multiple breeding programmes have been experimentally infected with specific strains of the PRRS virus and genotyped using the 80K SNP panel. GWAS of the resulting data identified one major gene on chromosome 4 for host response to PRRS infection (Boddicker et al., 2012). This gene was shown to be segregating in all commercial genetic products that were evaluated and shown to have an effect on host response to PRRS in all lines. Thus, a genetic test for this gene can be used to select pigs within the nucleus lines to improve response to PRRS infection. However, although it is considered a major gene, this gene still only explains 10–15% of the genetic differences in host response to infection. Attempts to use genomic prediction to capture the rest of the genetic variance had mixed results, depending on the relationship between the training and validation data (Boddicker et al., 2014). Thus, while experimental trials can identify genes of moderate to large effect, which can be used for selection based on an MBV, if most of the host response is due to many genes of small effects, genomic prediction will be needed, which requires routine updating of training data specific to the population under selection.

6 Other benefits of molecular information for swine breeding programmes

In addition to allowing MAS or GS, information on genetic markers can also be used for other purposes within a breeding programme, including parentage assignment or validation and control of inbreeding.

If genotypes on potential parents are available, genotyping individuals for a limited number (~100) of genetic markers across the genome is sufficient to assign parentage (Harlizius et al., 2011). This allows pedigree information to be generated when this is unknown or parentage to be verified in cases of pedigree errors. Depending on their rate, correction of pedigree errors can result in substantial increases in accuracy of even pedigree-based EBV.

One of the limitations of pedigree-based EBV is that the same set of phenotypes (i.e. on close relatives) contributes to the EBV of selection candidates that are from the same family, in particular when the selection candidates have no own phenotypes or phenotypes on progeny, in which case the EBV are based on phenotypes of full-sibs, half-sibs, parents and other ancestors. This results in family members to rank very similarly based on their EBV, and thus, when the individuals with the highest EBV are selected, it is likely that the selected individuals consist of family groups. This is referred to as having a high probability of co-selection of relatives, which ultimately results in higher levels of inbreeding in the population. Accumulation of inbreeding is undesirable because it results in reduced genetic variation (alleles become fixed), inbreeding depression and a greater incidence of homozygous deleterious recessive genotypes. This co-selection of relatives can, of course, be overcome by putting limits on the number of individuals that are selected from each family, but not at a cost of reducing genetic gain, at least in the short term. Thus, obtaining information that can be used to differentiate family members is crucial, both for increasing response to selection and for reducing rates of inbreeding. In traditional breeding programmes, the only information that can be used to differentiate family members based on their EBV is the own phenotypes and/or progeny phenotypes obtained. Using such phenotypes in pedigree-based genetic evaluations will differentiate the EBV of family members from each other, because each family member will have a different 'own phenotype' and/or 'progeny phenotypes' contributing to their EBV, and these phenotypes reflect the collective effect of the specific sample of alleles that the individual received from its parents. Collecting own or progeny phenotypes is, however, problematic for many traits and can also lead to longer generation intervals. Genomic information is another important source that allows family members to be differentiated; because each progeny receives a different sample of alleles from its parents, progeny will have different breeding values and these differences can be captured by marker genotypes (Daetwyler et al., 2007). Thus, MBV show substantial differences between family members because they relate to the actual set of alleles that each family member received from its parents and, therefore, reduce the probability of co-selection of relatives and, thereby, rates of inbreeding. Furthermore, because molecular information can be obtained at a young age, increases in inbreeding are expected to be lower with MAS and GS, without requiring the increases in generation intervals that are associated with phenotype-based information to reduce co-selection of relatives.

In addition to reducing probabilities of co-selection of relatives and, thereby future increases in inbreeding, genomic information can also be used to assign matings that avoid high levels of inbreeding of the resulting progeny. This can, in principle, also be done based on pedigree, that is, by avoiding mating close relatives, but is more effective when based on molecular information (e.g. Sun et al., 2013). For example, SNP genotypes of selected males and females can be used to assign matings that result in the greatest heterozygosity in the progeny.

7 Summary

The use of molecular genetic information has promised to enhance genetic improvement of livestock since the 1970s, but it is not until recently that this has substantially materialized through the use of large-scale high-density SNP genotyping and genomic prediction. The use of genomics is now instrumental to genetic improvement of livestock, leading to substantial increases in rates of genetic improvement and more effective tools to maintain genetic diversity. Recent advances in the use of genomics also have opened opportunities to enhance selection for disease resistance and performance under suboptimal health situations in the field. Genomics can also be used to more effectively implement genetic improvement in developing countries, whereas traditional genetic improvement programmes require extensive phenotypic data recording programmes on animals with known pedigree, GS programmes do not require knowledge of pedigree, which is often a problem in developing and even developed countries. Although genomic prediction still requires extensive recording of phenotypic data, such data can be collected on animals in the field without pedigree.

8 Future prospects and challenges

As the costs of whole-genome sequencing decreases, the possibility to have whole-genome sequence information on all selection candidates (either actual or imputed) will become a reality. In that case, the causative mutations at genes that result in differences in phenotype will be part of the genomic data that will be analysed. In principle, the statistical methods that have been developed for genomic prediction can incorporate the large number of sequence variants that will be discovered. It has also been shown that, if genomic training is only based on causative variants, the accuracy of the resulting MBV is substantially higher and more persistent across generations and populations (Kizilkaya et al., 2012). This would substantially reduce the need for routine phenotype recording. The challenge, however, is to identify the causative mutations among the larger number of genetic variants that are present in whole-genome sequence data. When all genetic variants are included, availability of whole-genome sequence data may not result in a substantial increase in accuracy of prediction, because even Bayesian variable selection methods will not be able to differentiate between true causative effects and other genetic variants that are in LD with them (Hayes et al., 2013; van Binsbergen et al., 2015). This reduces accuracy and persistence across generations and populations.

Several methods are currently being pursued to help prediction models put more emphasis on causative variants. These include incorporating other biological or bioinformatics information.

9 Where to look for further information

The use of genomics to understand the genetic basis of traits of importance in livestock and to enhance programmes for genetic improvement of livestock is a very fast moving field of research, development and application. Details are provided in the numerous research articles that are published in this field. A comprehensive overview of the use of GS in livestock is provided by a recent book on the topic by Weller (2016).

10 References

Andersson, L. 2001. Genetic dissection of phenotypic diversity in farm animals. *Nat. Rev. Genet.* 2:130–8.

Beavis, W. D. 1998. QTL analysis: Power, precision, and accuracy. In Paterson, A. H. (Ed.), *Molecular Dissection of Complex Traits*. New York: CRC Press, pp. 145–62.

Bijma, P. and van Arendonk, J. A. M. 1998. Maximizing genetic gain for the sire line of a crossbreeding scheme utilizing both purebred and crossbred information. *J. Anim. Sci.* 66:529–42.

Boddicker, N., Waide, E. H., Rowland, R. R., Lunney, J. K., Garrick, D. J., Reecy, J. M. and Dekkers, J. C. M. 2012. Evidence for a major QTL associated with host response to PPRS virus challenge. *J. Anim. Sci.* 90:1733–46

Boddicker, N., Bjorquist, A., Rowland, R. R. R., Lunney, J. K., Reecy, J. M. and Dekkers, J. C. M. 2014. Genome-wide association and genomic prediction for host response to PRRSV infection. *Genet. Select. Evol.* 46:18.

Cleveland, M. A. and Hickey, J. M. 2013. Practical implementation of cost-effective genomic selection in commercial pig breeding using imputation. *J. Anim. Sci.* 91:3583–92.

Daetwyler, H. D., Villanueva, B., Bijma, P. and Woolliams, J. A. 2007. Inbreeding in genome-wide selection. *J. Anim. Breed. Genet.* 124:369–76.

Daetwyler, H. D., Calus, M. P. L., Pong-Wong, R., de Los Campos, G. and Hickey, J. M. 2013. Genomic prediction in animals and plants: Simulation of data, validation, reporting, and benchmarking. *Genetics* 193:347–65.

de los Campos, G., Hickey, J. M., Pong-Wong, R., Daetwyler, H. D. and Calus, M. P. L. 2013. Whole genome regression and prediction methods applied to plant and animal breeding. *Genetics* 193:327–45.

Dekkers, J. C. M. and van der Werf, J. H. J. 2007. Strategies, limitations and opportunities for marker-assisted selection in livestock. In Guimarães, E. P., Ruane, J., Scherf, B. D., Sonnino, A. and Dargie, J. D. (Eds), *Marker-assisted Selection: Current Status and Future Perspectives in Crops, Livestock, Forestry and Fish*. Rome, Italy: Food and Agriculture Organization of the United Nations.

Dekkers, J. C. M. 2004. Commercial application of marker- and gene-assisted selection in livestock: Strategies and lessons. *J. Anim. Sci.* 82:E313–28, http://jas.fass.org/content/82/13_suppl/E313.long.

Dekkers, J. C. M. 2007. Marker-assisted selection for commercial crossbred performance. *J. Animal Sci.* 85:2104–14.

Dekkers, J. C. M. 2010. Use of high-density marker genotyping for genetic improvement of livestock by genomic selection. *CAB Reviews: Perspectives in Agriculture, Veterinary Science, Nutrition and Natural Resources* 5(037):1–13 .

Dekkers, J. C. M., 2012. Application of genomics tools to animal breeding. *Curr. Genomics* 13:207–12.
Dekkers, J. C. M. and Hospital, F. 2002. The use of molecular genetics in improvement of agricultural populations. *Nature Rev. Genet.* 3:22–32.
Do, D. N., Janss, L. L. G., Jensen, J. and Kadarmideen, H. N. 2015. SNP annotation-based whole genomic prediction and selection: An application to feed efficiency and its component traits in pigs. *J. Anim. Sci.* 93(5):2056–63.
Erbe, M., Hayes, B. J., Matukumalli, L. K., Goswami, S., Bowman, P. J. and et al. 2012. Improving accuracy of genomic predictions within and between dairy cattle breeds with imputed high-density single nucleotide polymorphism panels. *J. Dairy Sci.* 95:4114–29.
Falconer, D. S. and MacKay, T. F. C. 1996. *Introduction to Quantitative Genetics*. Harlow, UK: Longman. xii, p. 438.
Flori, L., Gao, Y., Laloë, D., Lemonnier, G., Leplat, J.-J., Teillaud, A. and et al. (2011). Immunity traits in pigs: Substantial genetic variation and limited covariation. *PLoS One* 6(7):e22717. doi:10.1371/journal.pone.0022717.
Goddard, M. E. and Hayes, B. J. 2009. Mapping genes for complex traits in domestic animals and their use in breeding programmes. *Nat. Rev. Genet.* 10:381–91.
Goddard, M. E., Hayes, B. J. and Meuwissen, T. H. E. 2010. Genomic selection in livestock populations. *Genet. Res. (Camb).* 92:413–21.
Habier, D., Fernando, R. L. and Dekkers, J. C. M. 2007. The impact of relationship information genome-assisted breeding values. *Genetics* 177:2389–97.
Habier, D., Fernando, R. L. and Dekkers, J. C. M. 2009. Genomic selection using low-density marker panels. *Genetics* 182:343–3.
Habier, D., Fernando, R. L., Kizilkaya, K. and Garrick, D. J. Extension of the bayesian alphabet for genomic selection. *BMC Bioinformatics*. 2011;12:186.
Harlizius, B., Lopes, M. S., Duijvesteijn, N., van de Goor, L. H., van Haeringen, W. A., Panneman, H., Guimarães, S. E., Merks, J. W. and Knol, E. F. 2011. A single nucleotide polymorphism set for paternal identification to reduce the costs of trait recording in commercial pig breeding. *J Animal Sci.* 89:1661–8.
Hayes, B. J., Bowman, P. J., Chamberlain, A. J. and Goddard, M. E. 2009. Invited review: Genomic selection in dairy cattle: Progress and challenges. *J. Dairy Sci.* 92:433–43.
Hayes, B. J., Lewin, H. A. and Goddard, M. E. 2013. The future of livestock breeding: Genomic selection for efficiency, reduced emissions intensity, and adaptation. *Trends Genet.* 29:206–14.
Hayes, B. J., Macleod, I. M., Daetwyler, H. D., Bowman, P. J., Chamberlain, A. J., Vander Jagt, C. J. and et al. 2014. Genomic prediction from whole genome sequence in livestock: The 1000 bull genomes project. In: Proceedings of the 10th World Congress of Genetics Applied to Livestock Production: 17–22 August 2014; Vancouver. https:// asas. org/docs/default-source/wcgalp-proceedings-oral/183_paper_10441_manuscript_1644_0. pdf? sfvrsn=2.
Hayes, B. J., Visscher, P. M. and Goddard, M. E. 2009. Increased accuracy of artificial selection by using the realized relationship matrix. *Genet. Res. (Camb).* 91:47–60.
Hu, Z.-L., Dracheva, S., Jang, W., Maglott, D., Bastiaansen, J., Rothschild, M. F. and Reecy, J. M. 2005. A QTL Resource and Comparison Tool for Pigs: PigQTLDB. *Mamm. Genome.* 16(10):792–800. http://www.animalgenome.org/cgi-bin/QTLdb/SS/index.
Kizilkaya, K., Fernando, R. L. and Garrick, D. J. 2010. Genomic prediction of simulated multibreed and purebred performance using observed fifty thousand single nucleotide polymorphism genotypes. *J. Animal Sci.* 88: 544–51.
Koivula, M., Stranden, I., Su, G. and Mantysaari, E. A. 2012. Different methods to calculate genomic predictions – comparisons of BLUP at the single nucleotide polymorphism level (SNP-BLUP), BLUP at the individual level (G-BLUP), and the one-step approach (H-BLUP). *J. Dairy Sci.* 95:4065–73.
Lamont, S. J., Dekkers, J. C. M. and Zhou, H. 2013. Immunogenetics and mapping immunological functions. In Schat, K. A., Kaspers, B. and Kaiser, P. (Eds), *Avian Immunology 2nd Ed.* London, San Diego: Elsevier.
Lande, R. and Thompson, R. 1990. Efficiency of marker-assisted selection in the improvement of quantitative traits. *Genetics* 124:743–56.

Lillehammer, M., Meuwissen, T. H. and Sonesson, A. K. 2011. Genomic selection for maternal traits in pigs. *J. Anim. Sci.* 89:3908–16.

Lillehammer, M., Sonesson, A. K. and Meuwissen, T. H. E. 2015. Use of field data in pig genomic selection schemes: a simulation study. *Animal.* 10(6):1025–32. DOI:10.1017/S1751731115002669.

Lunney, J. K., Steibel, J. P., Reecy, J. M., Fritz, E., Rothschild, M. F., Kerrigan, M., Trible, B. and Rowland, R. R. 2011. Probing genetic control of swine responses to PRRSV infection: Current progress of the PRRS host genetics consortium. *BMC Proc.* 5 Suppl 4:S30.

Lynch, M. and Walsh, B. 1998. *Genetics and Analysis of Quantitative Traits.* Sunderland, MA: Sinauer Associates, Inc., p. 980.

Meuwissen, T. and Goddard, M. 2010. Accurate prediction of genetic values for complex traits by whole genome resequencing. *Genetics* 185:623–31.

Meuwissen, T., Hayes, B. J. and Goddard, M. E. 2013. Accelerating improvement of livestock with genomic selection. *Annu. Rev. Anim. Biosci.* 1:221–37.

Meuwissen, T. H. E., Hayes, B. J. and Goddard, M. E. 2001. Prediction of total genetic value using genome-wide dense marker maps. *Genetics* 157:1819–29.

Meuwissen, T. H. E. and Goddard, M. E. 2010. Accurate prediction of genetic values for complex traits by whole-genome resequencing. *Genetics* 185:623–31.

Misztal, I., Legarra, A. and Aguilar, I. 2009. Computing procedures for genetic evaluation including phenotypic, full pedigree, and genomic information. *J. Dairy Sci.* 92:4648–55.

Morota, G., Abdollahi-Arpanahi, R., Kranis A. and Gianola, D. 2014. Genome-enabled prediction of quantitative traits in chickens using genomic annotation. *BMC Genomics* 15 : 1109 .

Online Mendelian Inheritance in Animals (OMIA) http://omia.angis.org.au/home/.

Ramos, A. M., Crooijmans, R. P. M. A., Affara, N. A., Amaral, A. J., Alan, L., Beever, J. E., Bendixen, C., Churcher, C., Clark, R., Dehais, P., Hansen, M. S., Hedegaard, J., Hu, Z., Kerstens, H. H., Law, A. S., Milan, D., Nonneman, D. J., Rohrer, G. A., Rothschild, M. F. and Smith, T. P. L. 2009. Design of a high density SNP genotyping assay in the pig using SNPs identified and characterized by next generation sequencing technology. *PLoS One* 4:e6524.

Strandén, I. and Garrick, D. J. 2009. Technical note: Derivation of equivalent computing algorithms for genomic predictions and reliabilities of animal merit. *J. Dairy Sci.* 92:2971–5 .

Sun, C., VanRaden, P., O'Connell, J., Weigel, K. and Gianola, D. 2013. Mating programmes including genomic relationships and dominance effects. *J. Dairy Sci.* 96(12):8014–23.

Tribout, T., Larzul, C. and Phocas, F. 2012. Efficiency of genomic selection in a purebred pig male line. *J. Anim. Sci.* 90:4164–76.

Tribout, T., Larzul, C. and Phocas, F. 2013. Economic aspects of implementing genomic evaluations in a pig sire line breeding scheme. *Genet. Sel. Evol.* 45: 40.

van Binsbergen, R., Calus, M. P. L., Bink, M. C. A. M., van Eeuwijk, F. A., Schrooten, C. and Veerkamp, R. F. 2015. Genomic prediction using imputed whole-genome sequence data in Holstein Friesian cattle. *Genet. Sel. Evol.* 47:1 .

Van Eenennaam, A. L., Weigel, K. A., Young, A. E., Cleveland, M. A. and Dekkers, J. C. M. 2014. Applied animal genomics: Results from the field. *Ann. Rev. Anim. Biosci.* 2:105–39.

VanRaden, P. M., Van Tassell, C. P., Wiggans, G. R., Sonstegard, T. S., Schnabel, R. D. and et al. 2009. Invited review: Reliability of genomic predictions for North American Holstein bulls. *J. Dairy Sci.* 92:16–24.

Wang, C., Habier, D., Peiris, B. L., Wolc, A., Kranis, A., Watson, K. A., Avendano, S., Garrick, D. J., Fernando, R. L., Lamont, S. J. and Dekkers, J. C. M. 2013. Accuracy of genomic prediction using an evenly spaced, low-density single nucleotide polymorphism panel in broiler chickens. *Poultry Sci.* 92:1712–23.

Wei, M. and van der Werf, J. H. 1994. Maximizing genetic response in crossbreds using both purebred and crossbred information. *Anim. Prod.* 58:401–13.

Weller, J. I. 2016. *Genomic Selection in Animals.* John Wiley and Sons Inc., New York, USA. DOI:10.1002/9781119213628.

Wolc, A., Zhao, H. H., Arango, J., Settar, P., Fulton, J. E., O'Sullivan, N. P., Preisinger, R., Stricker, C., Habier, D., Fernando, R., Garrick, D. J., Lamont, S. J. and Dekkers, J. C. M. 2015. Response and inbreeding from a genomic selection experiment in layer chickens. *Genet Sel. Evol.* 47:59 .

Wolc, A., Arango, J., Settar, P., Fulton, J. E., O'Sullivan, N. P., Preisinger, R., Habier, D., Fernando, R., Garrick, D. J. and Dekkers, J. C. M. 2011. Persistence of accuracy of genomic estimated breeding values over generations in layer chickens. *Genet. Sel. Evol.* 43:23. DOI:10.1186/1297-9686-43-23.
Younes, M., Plastow, G. and Wang, Z. 2015. Genomic Selection, a New Era for Pork Quality Improvement. *Springer Sci. Rev.* 3:27–37.

Chapter 3

Factors affecting the reproductive efficiency of pigs

Glen W. Almond and Emily Mahan-Riggs, North Carolina State University, USA

1 Introduction

1.1 Overview

Over the last few decades, pig producers dramatically improved the reproductive performance of their sow herds. Enhanced reproductive efficiency is achievable; however, biological improvements are almost meaningless unless there also are increased economic returns for the producer. Solving problems associated with reproductive failure in sow herds typically is a challenge. It is critical to meet breeding targets on a routine basis, achieve the appropriate farrowing rates and litter sizes, and minimize the number of non-productive sow days. Reproductive failure interferes with the consistent production of pigs, which ultimately results in the suboptimal number of pigs for market.

Unfortunately, attempts to improve reproductive performance are hampered by suboptimal management, small profit margins, legislative mandates, public demands, disease problems and a variety of other factors. The range of factors that affect reproductive efficiency of sows is exhaustive, and some are beyond the scope of this chapter. In fact, entire books (Gordon, 1997; Bonet et al., 2013) and scientific conferences (Rodriguez-Martinez et al., 2009) provide information on reproduction in pig. This chapter describes some of the current factors affecting reproductive efficiency in modern pig production.

http://dx.doi.org/10.19103/AS.2017.0013.04

1.2 Measures of productivity

Since the advent of computerized record systems, it definitely is easier to assess the reproductive success or failure of individual sows, groups of sows, sow herds and entire systems of sow production. Despite the distinct advantage of the record systems, it still is necessary for farm personnel to record the information for subsequent entry into computer programmes. This creates three potential opportunities for problems, namely, the initial gathering and recording of on-farm data, the data entry and the data analysis. Interpretation of the records, including the definition of each production parameter, poses another challenge for producers and veterinarians; however, a complete list of definitions published previously assist with the assessment of productivity (Almond et al., 2006).

The metrics to assess sow and farm productivity often are highly variable. Some producers and veterinarians consider litter size (pigs born alive) and pigs weaned per sow per year as important criteria, while others prefer pigs produced per sow per lifetime or the number of non-productive days (NPD) as key indicators of sow performance. Regardless of the metric to evaluate reproductive performance, the methodology must be translated into profitability. Most metrics are interrelated and the alteration of one factor influences the outcome of another factor. For example, pigs weaned per sow per year is affected by pigs weaned per litter, which in turn, is influenced by pigs born alive per litter and pre-weaning mortality. Similarly, NPD are influenced by pre-breeding intervals, post-breeding pregnancy losses and reasons for removal or culling of sows from herds. The flow charts to illustrate these interrelationships initially were provided in the 1990s (Dial, 1990) and continue to be used in current production systems and farms. The flow charts and metrics are useful to identify problematic areas within a farm or among several farms; however, it is the role of the owner, farm manager, production manager and/or veterinarian to actually institute the appropriate changes to rectify the problems.

2 Gilt development, reproductive efficiency and litter size

2.1 Gilt development and reproductive efficiency

Albeit variable, it is common to observe sow replacement rates approaching or surpassing 50% per year. The high replacement rates dictate appropriate rearing, selection and introduction of gilts into the breeding herd. The high replacement rates also raise concerns over sow longevity. At one time, gilts were raised in a fashion similar to market pigs without concern about their reproductive performance and long-term returns to the herd; however, times changed considerably and the pork industry recognizes the need for optimal performance of replacement gilts. If possible, it is beneficial to raise gilts in a gilt development unit (GDU). The GDU creates the opportunity to improve the identification of select gilts with superior reproductive and genetic potential, and lifetime productivity.

Direct boar contact is critical for the detection and stimulation of the onset of puberty. With direct boar contact, commencing at 160 days of age, 95% of gilts were detected in oestrous by 260 days of age and the average age at puberty was 193 days of age (Calderon Diaz et al., 2015). The removal of gilts for reproductive reasons was significantly higher in gilts with delayed onset of first oestrus (Patterson et al., 2010; Roongsitthicaie et al., 2013).

Apparently, a small population of gilts are considered late responders and have a more delayed puberty event (Vallet, 2015). These late responders represent 'problem' animals and likely contribute to poor lifetime productivity. Thus, it is imperative to commence daily direct contact with boars when gilts are 160–170 days of age and to detect the onset of oestrus.

Current recommendations are to mate gilts on their second or third oestrous cycle (heat). This recommendation is to improve the likelihood of the gilts reaching sufficient maturity and having adequate body reserves to sustain reproductive performance over subsequent parities (Cottney et al., 2012). Some argue that this delay increases NPD and feed costs. These negative factors are offset by increased fertility and/or increased lifetime productivity. In contrast, another report indicated that there was no advantage, in terms of farrowing rate and number of piglets born alive, in performing the first mating of gilts if the growth rate is greater than 770 g/d and back fat is greater than 17 mm (Filha et al., 2010).

For producers wishing to breed gilts on the first observed oestrus, age, body weight and back fat thickness of gilts at the first observed oestrus significantly influenced the total number of piglets born per litter (TB) and the number of piglets born alive per litter (BA) in the first three parities (Tummark et al., 2007). The mean litter size (TB and BA) in the first three parities were highest in gilts that had a first observed oestrus between 181 and 200 days with 110.1–120.0 kg body weight (BW) and 13.1–15.0 mm BF (Tummark et al., 2007). Alternatively, others recommend that gilts should be bred at 135–150 kg BW (Williams et al., 2005; Fila et al., 2010). It should also be noted that gilts with a high growth rate tended to show signs of oestrus earlier than gilts with a low growth rate (Tummark et al., 2009). Based on these latter studies, gilts can be mated on their first oestrus, provided that they have the appropriate growth rate, body weight and back fat thickness. Since these parameters often are not achieved on a routine basis, the delayed mating at second or third oestrus provides the gilts with an opportunity to gain sufficient weight and condition.

Currently, considerable attention is focused on the litter of origin and lifetime productivity, and the initial rearing of replacement gilts. The initial birth weight of a gilt appears to influence subsequent reproductive performance. In other words, a gilt with a low birth weight (LBW) phenotype likely will not be suitable for subsequent reproductive performance (Patterson and Foxcroft, 2016; Flowers, 2015). In fact, it was suggested that a minimum birth weight is likely required for a gilt to attain its reproductive potential later in life. Replacement gilts raised in small litters were heavier at weaning and had a significant weight advantage throughout their life compared to gilts raised in large litters (Flowers, 2015). These recent findings illustrate the need to further evaluate the influence of litter size of origin, birth weights and early rearing methods (i.e. in the first two months of life) on the reproductive performance and longevity of gilts. These parameters apply to nucleus and multiplier herds and not necessarily commercial sow farms.

Feed allowance prior to and after mating also deserves consideration. Evidently, pre-mating dietary treatments had no effect on any reproductive measure (Condous et al., 2014). In contrast, embryonic survival was greater in gilts with high feed intake post-mating compared with gilts with low feed intake post-mating (88.4 ± 2.5 versus 77.8 ± 4.0%), resulting in more conceptuses (14.0 ± 0.6 vs 11.7 ± 0.7; Condous et al., 2014). These authors indicated that a reduction in post-mating feed intakes to maintenance levels impaired embryo survival (Condous et al., 2014).

Genetic improvement programmes should emphasize live-born pigs, weight of live-born pigs and weaning weights. Although it is highly desirable to have large litters of pigs with few stillborn or mummified piglets, the previous discussion raises some concerns regarding selection based solely on litter size. Furthermore, birth weight decreased as litter size increased (Johnson et al., 1999), and thus, a small female pig at birth may not be the ideal candidate for subsequent selection as a breeding female.

2.2 Litter size

Selection for hyperprolific sows, as a means of increasing litter size and profit, resulted in an increased number of LBW piglets. These LBW piglets might suffer from increased morbidity and mortality during the early neonatal period. Another, more practical issue, commonly encountered in commercial farms, is that the hyperprolific sows, particularly in their third or greater parity, rarely have sufficient, functional mammary glands to meet the colostrum and milk demands of large litters. Producers are forced to cross-foster piglets, use nurse sows, adjust feeding strategies for sows, wean lightweight pigs or accept higher pre-weaning mortality. Alternatively, producers, such as those in France, also select sows for the number of mammary glands to meet the demands of the large litters (Martineau and Badouard, 2009).

The LBW piglets also have reduced growth performance, meat and carcass quality, which lead to economic losses for the farmer in the postnatal period. Feeding sows additional lipids during gestation reduces the incidence of LBW piglets and improves their survival rates (De Vos et al., 2014). To increase the survival and growth rates of LBW piglets, a producer must focus on ensuring adequate colostrum and milk intake. Interventions include supplementing piglets, split nursing, split weaning and cross-fostering (De Vos et al., 2014). The extra care and management appear to take place more often in smaller herds than in the much larger sow farms. An abundance of skilled labour with stellar stockmanship skills is required to rear the LBW pigs in large litters or to manage sow feed intake during gestation and lactation. Unfortunately, numeric increases in piglet numbers at birth or weaning often ignore piglet quality and vigour.

Currently, special attention is also given to variation in birth weight, as increased within-litter birth weight variation is observed in modern sows, especially with large litters and in older parity sows. This variation is associated with pre-weaning mortality, variable weights at weaning and deteriorated growth performance. Therefore, understanding the factors that can influence the events that occur during gestation and have an impact on foetal growth and development are important to achieve better efficiency. This understanding also is necessary to develop strategies that can be used to achieve increased within-litter uniformity of piglet birth weight. At a given placental size, foetal growth may vary because of differences in placental vascularization and efficiency (Campos et al., 2012). Extra feed or energy given to sows during late gestation marginally improves birth weight, and positive effects are not consistent between different studies (Dwyer et al., 1994; Lawlor et al., 2007; Nissen et al., 2003; Rehfekdt and Kuhn, 2006). Therefore, more research concerning sow nutrition during gestation associated with the provision of balanced diets to meet requirements of sows and foetuses are still required (Laws et al., 2009; Campos et al., 2012)

The non-infectious factors associated with stillborn piglets can be roughly categorized as genetic, maternal, piglet and environmental. Interactions exist between most of these factors. Producers must implement a more effective farrowing management to obtain a high reproductive efficiency. Litter size, parity, sow's body condition and farrowing supervision/birth assistance apparently are the most relevant risk factors associated with stillborn piglets (Vanderhaeghe et al., 2013).

Type of breed and high temperatures in the farrowing unit (≥22°C) were associated with significantly more stillbirths. In contrast, showering sows with water before parturition resulted in a significant reduction in stillbirths (Vanderhaeghe et al., 2010). Supervision of the farrowing process must be frequent to prevent stillbirths. Occasional supervision was no better than a complete lack of supervision (Vanderhaeghe et al., 2010).

Although stillborn piglets contribute to the decreased efficiency, veterinarians and producers must consider the cost and effort to reduce the frequency of stillbirths. The expense associated with decreasing the stillbirth rate from 7% to 5% may be more than the loss of piglets. This is definitely a decision that requires careful consideration before implementing major management changes. The induction of farrowing with prostaglandins with or without oxytocin was used to modify the parturition date and allow the producer to provide supervision of the farrowing process (Dial et al., 1987; Holyoake et al., 1995; De Rensis et al., 2012). It is evident that direct supervision of the farrowing process, with or without pharmacological intervention, reduces the stillbirth rate and improves the liveability of the newborn piglets.

3 Weaning to oestrus interval

The weaning to oestrus interval (WEI) and weaning to first service interval (WSI) contribute to non-productive sow days. Weaning serves as a natural synchronization of oestrus and most sows return to oestrus within three to seven days following weaning (Hurtgen et al., 1980; Belstra et al., 2004). The physiological processes involved in the return to oestrus after weaning were previously described (Almond, 1992). Prolonged WEI represents a classic example of seasonal infertility (described elsewhere) and the problem is more common during the summer, and primiparous sows are more likely affected than multiparous sows (Britt et al., 1983).

Several other factors also influence WEI. These factors include boar exposure (Walton, 1986; Langendijk et al., 2000), lactation length (Walker and Watt, 1979), nutrition during and after lactation, environmental temperature, facility design (Almond, 1992; Kemp and Soede, 2012) and photoperiod. Thus, prolonged WEI truly represents a multi-factorial problem, which interferes with a producer's ability to keep sufficient animals in designated breeding groups. In addition to changes in the management of sows, considerable effort was placed on the use of exogenous hormones (see Section 4.1) to induce a fertile oestrus in weaned sows.

An inverse relationship exists between WEI and lactation length. The days from weaning to oestrus increase as lactation length is decreased. During lactation, the hypothalamic-pituitary-ovarian axis is suppressed by suckling. Weaning brings an abrupt cessation to the suppression and the sow returns to oestrus. With changing production methods, lactation length is 28 days or longer. Consequently, it is not unusual for lactation oestrus in some sows (Kemp and Soede, 2012). The intensity of suckling is diminished in these sows in late lactation, particularly with 'get away' systems and group lactation systems (Pederson et al., 1998; Hulten et al., 2006). The duration and intensity of suckling may also influence rate of return to oestrus following weaning. The removal of piglets for a period of time each day or decreasing litter sizes prior to weaning caused a rapid return to oestrus following weaning (Stevenson and Britt, 1981).

3.1 Facility

The three basic housing systems for weaned sows include extensive (pasture) systems, group housing and individual housing in stalls. As mentioned in Section 5, assessments of facility design on sow performance is confounded by variations in housing systems (including space allowance), stockmanship and suitability of a genetic line or breed of sow for a particular housing system (Almond, 1992). Quite simply, the optimal system for the weaned sow is not clearly defined. The individual stall housing system eliminates fighting between sows, permits feed allowance on an individual basis, and provides easy access to and identification of animals. The most commonly cited problems of individual stalls are the sow's inability to turn around and lack of social interactions with other sows. Group housing and extensive systems allow social interaction; however, excessive fighting is common between sows in pens or on pasture. Although some recommend space allowances of 1.8 to 2.4 m^2/sow, the ideal space allowance for weaned sows remains somewhat debatable despite the clear need to reduce aggression and stress at mixing (Hemsworth et al., 2013).

A previous study revealed that WEIs were 4.25 and 4.5 days in sows penned in groups (4–5/group) and in sows placed into individual stalls, respectively (Schmidt et al., 1985). The onset of oestrus and oestrus expression were equivocal for group housing and individual housing (Kemp et al., 2005). Subsequently, it was demonstrated that the most vulnerable period for group-housed sows was between weeks two and three of pregnancy and not in WEI (Spoolder et al., 2009). In contrast, it was recently shown that sows housed in groups at weaning and regrouped after insemination experienced higher stress than sows housed in individual stalls at weaning and mixed in groups after insemination (Rault et al., 2014). These differences contributed to lower mating success within five days of weaning (Rault et al., 2014).

Owing to implications of elevated ambient temperatures in seasonal infertility, producers use evaporative cooling devices in farrowing rooms and breeding/gestation barns. Evaporative cooling improved feed consumption and consequently decreases lactational weight losses and WEI. In summary, additional research is required to identify the most cost-effective and productive housing system for weaned sows.

3.2 Nutrition

Feed intake during lactation is essential for milk production and piglet growth; however, excessive weight loss during lactation was associated with post-weaning infertility (Varley and Cole, 1976). Another report indicated that a negative deviation in predicted lactation feed intake in late lactation decreased subsequent reproductive performance (Yoder et al., 2013). Both the amount and pattern of feed intake influenced subsequent reproductive performance. A drop in feed intake in either the first or second week of lactation contributed to increased problems with post-weaning anoestrus (Koketsu et al., 1996). In brief, there are four primary feed intake patterns for lactating sows: 1) a steady increase in feed intake, 2) marginal feed intake throughout lactation, 3) marginal feed intake for the first 7 or more days and then a gradual increase and 4) a steady increase followed by diminished feed intake at approximately 11–14 days after farrowing (Fig. 1). It should be no surprise that post-weaning reproductive performance would be superior in the first group of sows. Primiparous sows often fall into the fourth group of sows. It may be related to gut fill: these animals go off feed because they simply cannot consume the

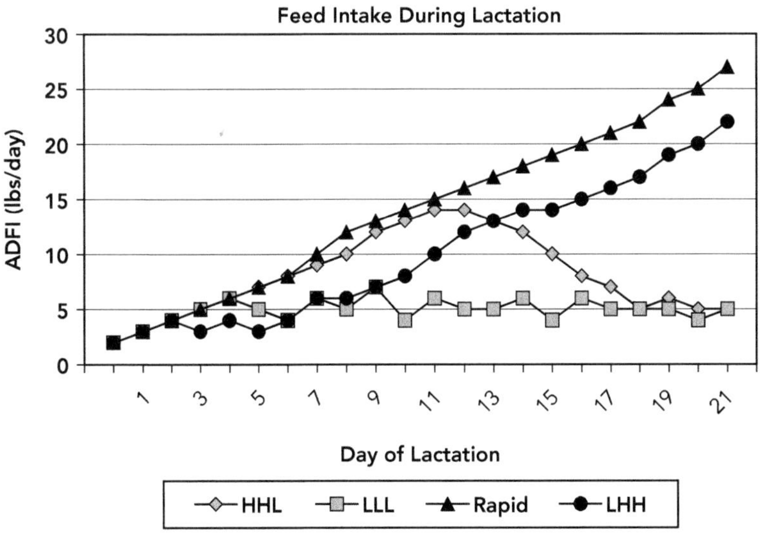

Figure 1 Feed intake patterns during lactation. The four patterns included a rapid increase (Rapid), a low feed intake during lactation (LLL), an initial low intake followed by higher feed intake in the last 10 or more days of lactation (LHH) and high intake for 11 days followed by a marked decrease (HHL). Adapted from Koketsu et al. (1996).

quantity of feed expected by the farrowing house personnel. In warm climates, part of the seasonal infertility problem is attributed to diminished feed intake (not allowance) during the summer months. Despite innovative methods to increase feed intake (balanced with the economics of feed cost), producers continue to observe seasonal infertility.

Sows that lactate at a high rate and consume insufficient quantities of energy lose weight during lactation and frequently fail to gain weight quickly following weaning. In general, it is assumed that the provision of supplemental dietary fat during lactation reduces the problem of diminished reproductive performance after weaning, particularly during the summer months (Shurson et al., 1986). Recently, it was shown that the addition of an animal–vegetable blend or choice white grease to lactation diets improved subsequent litter size, rebreeding interval and reproductive performance during the summer months (Rosero et al., 2012).

Protein intake during lactation also exerts a major influence on WEI in first-litter sows (King and Dunkin, 1986). The effect of protein intake was independent of energy intake; increasing daily protein intakes to at least 700 g crude protein, either alone or in combination with an increase in energy intake, increased the proportion of sows exhibiting oestrus within eight days of weaning (King and Dunkin, 1986). In addition, estimated daily nitrogen loss during lactation was associated with WEI (Noblet and Etienne, 1987). Apparently up to 55 grams crude lysine/day are required for sows during lactation (Dourmad et al., 1998).

Protein intake is required for growth as well as milk production in primiparous sows. Thus, primiparous sows mobilize muscle mass during lactation, whereas the weight loss in older sows is caused by loss of adipose tissue (King and Dunkin, 1986). These observations may explain, at least in part, the differences in the incidence of post-weaning anoestrus in different parity sows. To improve the dietary protein utilization, the dietary amino acid content and ratios can be adjusted by expected feed intakes or parities of sows (Kim et al., 2009). The negative influence of diminished feed intake during the summer months was reduced by including 0.25% to 0.5% spray-dried plasma in the lactation diets of primiparous sows (Crenshaw et al., 2007).

Therefore, there are techniques available to counter the negative influence of diminished feed and protein intakes by sows during the summer months.

4 Managing reproductive functions and fertilization

4.1 Manipulation of reproductive function

It has been decades since the publication of reports on the use of pregnant mare's serum gonadotrophin (PMSG, now called equine chorionic gonadotrophin or eCG) for superovulation in pigs (Hunter, 1964; Guthrie et al., 1974). The initial use of eCG focused on embryo quality and transfer; however, the combination of eCG and human chorionic gonadotrophin (hCG) (PG600®) has proven to be a useful tool to induce oestrus and ovulation in anovulatory females (Britt et al., 1989; Hidalgo et al., 2014).

Weaning of piglets provides a natural synchronization for the onset of oestrus. However, several factors, as described in Section 3, contribute to delayed and variable returns to oestrus after weaning. Despite awareness of these factors, it is common to see considerable variation in WEI and WSI. As shown in Fig. 2, most sows do return to oestrus in one week, but the remainder exhibit oestrus at various times after weaning. These latter sows' group represent the 'problem animals' and interfere with the goal of having 95% of sows bred within seven days after weaning.

To circumvent problem sows and to improve timing of matings after weaning, there are a number of methods to manipulate the onset of oestrus after weaning. One method is the use of PG600® on the day of weaning (Estienne and Hartsock, 1998; Knox et al., 2001). This method appears to be useful for primiparous sows, particularly in the summer months, but the benefits of treating multiparous sows with PG600® at weaning are not as apparent. Another method of oestrus control is the oral administration of a progestogen (altrenogest, Matrix®) for seven days (or other time periods) after weaning (Boyer and Almond, 2014). The breeding barn personnel indicated that the synchrony of oestrus following the cessation of Matrix® administration was greater than the natural return to oestrus. The use of gonadotrophin-releasing hormone (GnRH) analogues by either intra vaginal (triptorelin; Ovugel®) or intramuscular (buserelin; Receptal® and perforelin; Maprelin®) routes of administration has gained considerable attention. Both products appear useful for timed insemination of weaned sows (de Jong et al., 2013; Knox et al., 2011; Martinat-Botte et al., 2010; Driancourt et al., 2013). Finally, the use of porcine luteinizing hormone (LH) at oestrus onset also has proven useful for fixed-time insemination (Fontana et al., 2013). All of these approaches are beneficial for manipulating oestrus and/or ovulation after weaning, and most appear to have few, if any, negative effects on reproductive performance; however, they do not improve poor technique and lazy personnel. In other words, an excellent producer can use most new techniques in a successful fashion, but the methods will have marginal benefit if the breeding barn personnel are 'suboptimal'.

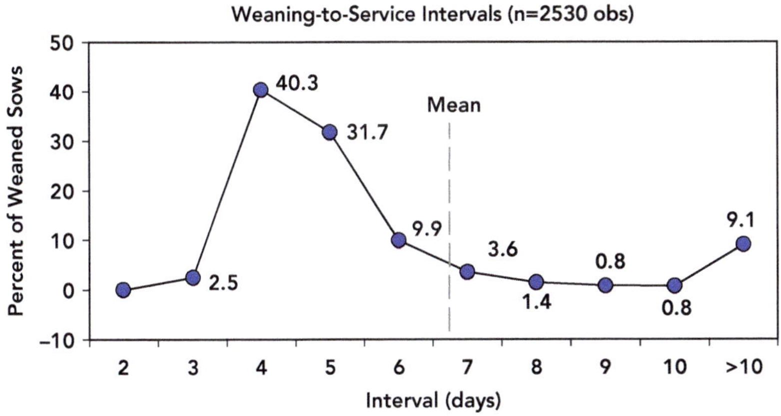

Figure 2 Per cent of weaned sows exhibiting oestrus after weaning. Data were obtained from a 4000-sow farm in North Carolina, USA.

The use of exogenous hormones or other pharmacologic interventions have questionable benefits when they are used during pregnancy. This is due to inconsistent results (Wessels et al., 2014), risks and off-label use of the products. Most embryonic and early foetal losses occur prior to calcification/mineralization of the foetuses. In other words, the disparity between litter size and the number of ovulations is attributed foetal or embryonic losses early in pregnancy. As stated by Wessels et al. (2014), 'Agents affecting embryonic development, oestrogen production, uterine secretions, embryo attachment, placental development and angiogenesis at the maternal–foetal interface need to be assessed for their effects on litter size in order to make great strides in improving reproductive outcomes'.

The introduction of gilts into the breeding herd represents a critical component of a successful reproductive programme. Considerable effort and funding are devoted to raising the most suitable select gilts, and introducing the gilts into breeding groups as replacement females. Consequently, it is not surprising that producers attempt to have the appropriate number of gilts in oestrus on a regular schedule. To achieve the breeding targets, the induction and control of oestrus in gilts have received considerable attention. A fertile oestrous is induced with the PG600® given to gilts as early as 154 days of age (Britt et al., 1989; Hidalgo et al., 2014). To further synchronize oestrus, a progestogen (altrenogest, Matrix®) can be given orally to cycling gilts (15 mg/day for 14–18 days) followed by an injection of PG600® (Estienne and Crawford, 2015). To further synchronize ovulation following altrenogest and eCG treatment, both porcine LH (Degenstein et al., 2008; Ulguim et al., 2014) and a GnRH agonist (Receptal®; Martinat-Botte et al., 2010) administration were shown to improve synchrony of ovulation. The improved synchrony of the onset of oestrus and ovulation provide producers with the opportunity to mate females once, often as a fixed-time insemination, and thus save time, labour and insemination doses.

4.2 Mating management and insemination techniques

The three methods of breeding of female pigs include pen breeding, hand mating and artificial insemination (AI). Pen breeding with the boar is used in smaller farms and limited attention typically is paid to accurate recording of the onset of oestrus or the quality of mating. In contrast, hand mating (supervised mating) involves oestrus detection daily or twice daily in weaned sows or gilts. The mating process is supervised by breeding herd personnel, who confirm the mating and record sow and boar numbers. Obviously, hand mating requires greater labour and skilled personnel than the pen breeding method. Due to financial advantages and genetic improvement, the use of AI has dramatically increased over the last few decades. Heat detection is performed once or twice per day to establish the onset of oestrus. In general, gilts are inseminated within 8–12 hours after the onset of standing oestrus, and inseminated a second time 12–16 hours later. Sows are inseminated at 24 and 48 hours after the onset of oestrus. The timing of AI or natural service often are modified to suit specific sow farms, depending on facilities, labour and overall success of the mating programme. In other words, mating programmes often are customized to suit the needs of a particular farm. As mentioned, breeding programmes previously utilized a strategy of two to three inseminations after the detection of oestrus; however, this strategy is being altered with the use of exogenous hormones to control ovulation and permit single, timed inseminations (see Section 4.1).

Boar contact or exposure improves the detection of oestrus, and can increase the number of sows that ovulate and show oestrus after weaning (Langendijk et al., 2000). WEI influences the duration oestrus and the onset of oestrus-to-ovulation interval (Belstra et al., 2004). Apparently, there is an inverse relation between WEI and the duration of oestrus and the onset of oestrus-to-ovulation interval (Kemp and Soede, 1996). Invariably, sows tend to ovulate at approximately 70% of their oestrus, regardless of duration (Weitze et al., 1994).

It is beyond the scope of this chapter to provide an in-depth review of the problems associated with infertility in boar. In fact, books are available on the topic of boar reproduction and fundamentals of AI (Almond et al., 1998; Bonet et al., 2013). From a practical perspective for on-farm applications, there is increased use of post-cervical AI. The reproductive efficiency is similar between post-cervical AI or intrauterine AI (IUI) and the traditional method of cervical AI (Hernandez-Caravaca et al., 2012; Gonzalez-Pena et al., 2014); however, the post-cervical method requires a lower number of spermatozoa per dose, which represents an opportunity for savings, and increased rates of genetic improvement (Knox, 2016). As the methods for deep intrauterine AI (DUI) improve, its application in the pork industry likely will increase, particularly, if breeding herd personnel receive the appropriate training on the technique. For the IUI, considerable efforts have been placed on hormonal manipulation of ovulation to permit fixed-time inseminations (Fontana et al., 2013).

5 The impact of dry sow housing systems

Housing systems vary widely for weaned and pregnant sows. The systems range from individual stalls to large group systems of over 100 sows per pen. The precise impact of housing system on reproductive performance was reviewed recently (Einarsson et al., 2014; Kemp and Soede, 2012); however, it is apparent that the choice of housing system typically is based on legislation and consumer demands, building costs and a variety of factors, which exclude fertility and reproductive performance (Tuyttens et al., 2011). The overall impact of group-housing systems on reproductive performance appears to be highly variable, and few meaningful conclusions were derived from a large survey of farmers (Spoolder et al., 2009). Other factors, such as stress, floor type, bedding, feeding schedule (Schneider et al., 2007) and day of mixing gestating sows (Knox et al., 2014) likely influence the impact of housing system. Reviews found few remarkable differences among housing systems for gestating sows (McGlone et al., 2004; Einarsson et al., 2014). Conversely, retrofitting facilities from gestation stalls to pens compromised reproductive performance (Johnston and Li, 2013). The transition in housing systems requires modification of animal handling and stockmanship skills.

Although the impact of housing system on pregnancy remains a conundrum, the influence of housing on the weaned sow also represents an issue for the pork industry. The dilemma is whether to house sows in stall or groups after weaning and after insemination. According to Rault and coworkers (2014), regrouping sows after insemination creates higher stress than housing sows in individual stalls at weaning, and then mixing in groups after insemination. One of the concerns was the suppression of oestrus-related behaviour due to housing sows in groups (Kemp et al., 2005; Rault et al., 2014). In contrast, group housing and individual housing after weaning did not affect the oestrus detection rate or duration of oestrus (Langendijk et al., 2000).

It is a challenging task to compare the strengths and weaknesses of the various housing systems. Changes in nutritional requirements, genetic lines, feed delivery systems, insemination techniques and other factors make it difficult to compare housing systems. Due to these changes, producers, scientists and veterinarians face difficult decisions, particularly, on recommendations for a housing system. A functional system for producers in one country may not be appropriate for producers in other countries.

6 Seasonal infertility in sow

Seasonal infertility was characterized as extended WEI, increased pregnancy losses, and decreased conception and farrowing rates (Almond et al., 2006). The loss of pregnancies typically are observed following matings in the summer and early fall. Although seasonal infertility affects both primiparous and multiparous sows, the impact of season on reproduction usually was more pronounced in gilts and primiparous sows (Almond et al., 2006; Almond 1992).

6.1 Contributing factors

Multiple factors were hypothesized as the causes for seasonal infertility. Initially, photoperiod was considered as a major contributor to seasonal infertility. Wild boars historically bred in the winter months, when day length was short (Love et al., 1993; Rozeboom et al., 2000); however, commercial farms need to ensure a consistent level of production, which is not influenced by season, in order to maintain profit margins. Therefore, the number of hours of artificial lighting in barns was extensively investigated. In one study, farrowing rates improved with ten hours of constant lighting during gestation starting in late summer months (Chokoe and Siebrits, 2009). Other studies showed that when the number of hours of light were reduced during periods of high ambient temperatures, then the WEIs were optimal (Rozeboom et al., 2000). Twenty-four hours of light increased the length of standing oestrus, thereby increasing the likelihood of farm employees identifying sows in oestrus (Parera and Hacker, 1984). Due to conflicting results from multiple studies on photoperiod, temperature possibly plays a greater role in seasonal infertility than photoperiod (Rozeboom et al., 2000). These two factors could also work synergistically to cause seasonal infertility (Almond, 1992).

The thermoneutral zone for a sow is approximately 7 to 21°C; temperatures above 27°C cause heat stress in sows (Rozeboom et al., 2000). There are four methods to cool sows: conduction, convection, latent and radiant; however, the two primary methods to cool sows are convection and latent cooling. Convection cooling, commonly used in tunnel ventilation with fans, can be both natural and forced, but heat transfer to the air is forced. Latent cooling involves phase change heat transfer and is used to cool sows by evaporation of water. Foggers and drippers facilitate latent cooling from the sow's surface. However, respiration is the sow's natural form of latent cooling.

The normal respiration rate of sows ranges between 15 and 25 breaths per minute. Sows whose respiration rate is greater than 40 breaths per minute are at risk of suffering the effects of heat stress, and sows with respiration rates of over 60 breaths per minute are heat stressed (Rozeboom et al., 2000). As respiration rates increase to facilitate sow cooling, her body temperature decreases. However, a sow in late gestation with a respiration rate of 186 breaths per minute had a higher body temperature than an open sow with a respiration rate of 64 breaths per minute (Heitman et al., 1951). This indicates that while both sows were heat stressed, gestating sows have a higher metabolic load, which an increased respiration rate alone cannot overcome.

Producers observed that sows spend less time standing when the ambient temperature is high. The reluctance to stand is due to the greater heat production (180 kcal per 100 minutes) while standing compared to a lying position (Noblet et al., 1990). The decreased time for standing reduces the time for eating and drinking. In fact, feed intake decreased by over 13% in sows lying rather than standing (McGlone et al., 1988). Another study indicated that average daily feed intake was 43% lower for sows at 30°C compared to sows at 20°C (Messias de Braganca et al., 1998). When a sow's feed intake is reduced, she does not meet her energy requirements for lactation causing her to use body reserves for milk production, which in turn, leads to weight loss (Noblet et al., 1990).

Follicle development, WEI and ovulation rate are affected in sows with restricted feed intake. In a thermoneutral study, follicles on day two after weaning had a larger mean diameter in sows fed at a high level of energy during lactation compared to sows fed at a low level (van den Brand et al., 2000). Only 15 out of 24 sows (63%) fed at a low level during lactation compared to 23 out of 24 sows (96%) fed at a high level returned to oestrus within ten days after weaning. Another study compared the effects of restricted feeding versus high ambient temperature effects on feed intake. With high temperatures, one-third of sows returned to oestrus in five to six days (similar to sows at 20°C), one-third of sows returned to oestrus at a similar time as the feed restricted sows (at 20°C) and the remaining one-third of sows had a WEI that was delayed even longer (Messias de Braganca et al., 1998). Similarly, a decrease in pulsation of LH was observed because of decreased feed intake from high ambient temperature (Barb et al., 1991). A decrease in LH can increase WEI, as follicles are dependent on LH for maturation to pre-ovulatory size (Williams, 2009).

Farrowing rates also decreased with high ambient temperatures. In one study, farrowing rates ranged from 78% to 82.1% when temperatures were above 35°C compared to a range of 89.1 to 92.1% with temperatures less than 30°C (Almond and Bilke, 2005). These suppressed farrowing rates were attributed to early embryonic loss during the summer months (Tast et al., 2002). In addition, reduced LH secretion during the summer months along with restricted feeding suppresses progesterone secretion leading to inadequate embryonic signals and pregnancy loss (Peltoniemi et al., 2000).

Mycotoxins, especially zearalenone, were associated with adverse effects on swine reproduction. Zearalenone is formed by *Fusarium* spp. moulds in high relative humidity storage conditions. It causes oestrogen-like effects by binding to oestradiol receptors. Pigs are especially susceptible and the effects include anoestrous sows, increased WEI and embryonic loss, and abortion, which can be seen three to seven days after the initial ingestion of mycotoxin. However, these effects are no longer present 14 days after the withdrawal of contaminated feed. Zearalenone concentrations of 5 to 10 ppm can cause these reproductive symptoms, and the higher the concentration, the longer the duration of anoestrus (Kanora and Maes, 2009; Mostrom, 2012).

6.2 Treatment and prevention

As discussed previously, two forms of cooling reduce heat stress in sows: convection and latent cooling. Convection cooling uses fans in tunnel ventilation to transfer heat from the sow's body to the air. Within the warmer pig production areas of the United States, producers often used a standard 500 cubic feet per minute (CFM) per crate for fans in farrowing rooms. In Europe, the recommendations were 500 m^3/500 kg BW (Seedorf et al., 1998). Now, many US producers attempt to utilize 750 CFM per crate during the summer months. For breeding and gestation barns (200 kg BW sows), 300 CFM per sow is considered optimum. In breeding and gestation barns, the original recommendation for tunnel velocity is 250 feet per minute (FPM), but 300 FPM is recommended for warmer climates (Rozeboom et al., 2000). Obviously, these recommendations vary depending on countries and climates. Producers should consult with ventilation experts to determine the appropriate recommendations for their particular farm.

Latent cooling uses foggers and drippers for phase change heat transfer from the sow's skin surface. Fans, drippers and foggers are programmed to turn on between 24°C and 25.5°C in barns. Humidity must be taken into account when programming fans because the higher the humidity is, the hotter the apparent temperature. Cool cells are also used to cool barns and alleviate the negative effects of heat stress (Rozeboom et al., 2000; Aarnink et al., 2006; Justino et al., 2012). Unfortunately, in high humidity these become less effective. Obviously, the aforementioned recommendations are suitable for specific regions in the United States and countries with excessive summer heat.

Other methods to decrease heat stress in sows involve management. One report indicated that sows in individual housing do not show a seasonal decrease in farrowing rate (Peltoniemi et al., 2000). When group housing sows, moving and mixing sows should not take place until after day 30 in gestation to decrease loss of embryos (Rozeboom et al., 2000). Water flow and water consumption are also important management factors. Gestating sows should consume approximately 12–25 L/day, while lactating sows typically drink 10–40 L/day (National Research Council, 2012). A deficiency in water intake obviously will contribute to problems associated with heat stress and seasonal infertility.

7 Stockmanship and managing disease

7.1 Stockmanship

The role of stockmanship is critical to the reproductive performance of sow herds. Early studies showed a negative correlation between fear of humans and reproductive performance, and the variation in fear of humans accounted for approximately 20% of the variation in reproductive performance among farms (Hemsworth et al., 1981; Hemsworth et al., 1989). Farrowing rates were significantly lower in gilts exposed to a negative handling treatment than gilts receiving a positive treatment (Hemsworth et al., 1986). Furthermore, a recent study demonstrated that farmers' attitudes influenced production parameters (Kauppinen et al., 2012). The perceived ease of providing animals with a favourable environment correlated negatively with piglet mortality and positively with the number of weaned piglets in first parity litters (Kauppinen et al., 2012). These observations on pig handling and welfare reinforce the general principles for welfare of animals (Fraser et al., 2013).

Due to the growing attention to welfare, it is apparent that effective training of the stockperson is imperative for improved productivity (Coleman and Hemsworth, 2014). Training is critical as farm size has increased with a greater number of animals, recognition of individual animals is rare and there is greater need to hire more employees with minimal pig handling experience (Seabrook, 1988). Most of the pig production companies have training programmes and standard operating procedures (SOPs) for routine farm activities. However, these programmes often fail to yield consistent results in reproductive efficiency. As shown in Fig. 3, the farrowing rates, subsequent to summer matings, varied among four sow farms within the same production system. The farms were almost identical, with the obvious exception of the farm workers. Thus, the SOPs for the company do not completely account for variation in stockmanship. Re-training or replacement of breeding herd personnel were required to alleviate the deficiency in two of the farms.

7.2 Infectious versus non-infectious pregnancy losses

Historically, veterinarians and producers recognized the onset of an infectious disease outbreak. For example, a dramatic increase in mummies and stillborn pigs would have been attributed to porcine parvovirus (PPV). Another example would be the near-catastrophic outbreak of abortions, stillborns and mummies associated with porcine reproductive and respiratory syndrome virus (PRRSv) infections in naïve herds. Currently, farms have issues with reproductive losses associated with porcine circovirus (PCV-2), swine influenza and consequences of porcine epidemic diarrhoea virus (PEDv) outbreaks. For some agents, vaccines are effective, while for other diseases, producers and veterinarians question vaccine efficacy. Often, producers are simply too cautious to quit vaccinations for fear of an outbreak.

Some clinical signs associated with an infectious agent are subtle or are present only in a particular age group or stage of pregnancy. An interesting case of PCV-2 associated reproductive failure (Pittman, 2008) illustrates the complexity of some of the disease problems. In this particular case, the average number of mummies per litter increased from 0.1 to 0.4. This increase reached a peak of 0.7 mummies per litter in one week. After ten weeks, the frequency of mummies returned to <0.1/litter. The problem was restricted

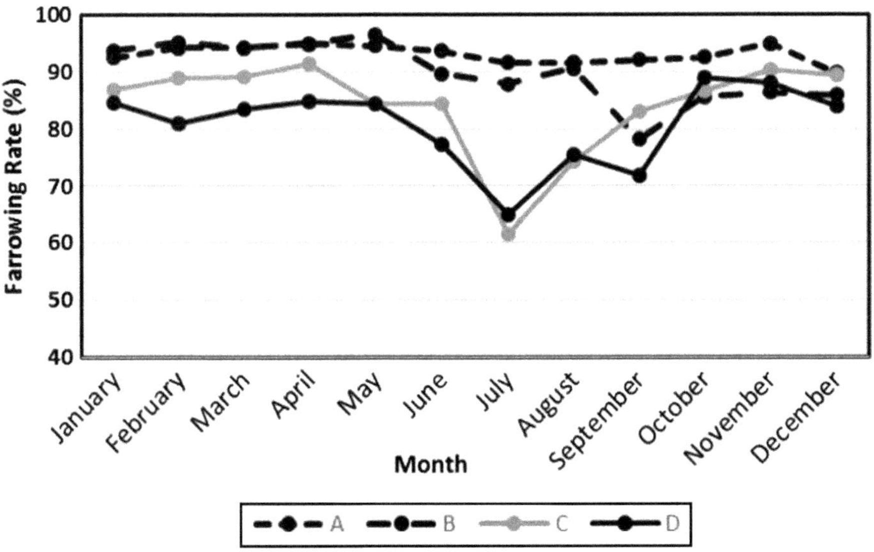

Figure 3 Farrowing rates in four commercial sow farms within the same production system. The months represent the dates of matings, and the rates are subsequent to these matings. All farms had identical facilities, ventilation and cooling systems, feed and genetic lines of pigs. Semen was supplied from the same boar stud and delivered on the same day to each of the four farms.

to gilt litters and only to gilts that entered the herd from an internal source. Concomitantly, the percentage of stillborn and mummified foetuses, and number of abortions were higher in these internally sourced gilts. Gilts entering the herd from an external source were not affected. The diagnostic evaluation confirmed the presence of PCV-2b in the aborted and mummified foetuses. The external source had been vaccinated for PCV-2. Upon the implementation of a vaccination programme of gilts, the clinical problem abated. The particular case emphasizes the need for detailed farm records, including reproduction, vaccination and treatments, and a reliable diagnostic laboratory.

Unfortunately, many problems associated with reproductive failure are attributable to management or mismanagement at one or more levels of personnel. One of the best examples of a management problem, considered an infectious disease, is the sow with a post-mating, vulvar discharge. Affected animals produce a purulent discharge at 13–18 days after mating. Invariably, these animals return to oestrus at a regular interval after mating. The diagnostic evaluation often, but not always, described metritis or endometritis. Often, there were no pathological findings. The underlying cause of the discharging sow was a dilemma and many veterinarians and producers tried various treatments with marginal success. This mystery was elucidated by De Winter and coworkers (1996), who demonstrated that the likelihood of a uterine infection increased dramatically when sows were mated after ovulation, that is late oestrus. Multiple mating schemes with AI and natural matings inadvertently 'forced' producers to use a third or fourth mating in late oestrus. These late matings took place after ovulation, which is a time when the uterus is highly susceptible to infection. Furthermore, discharging sows were more common when the onset of oestrus was missed with poor oestrus detection programmes. In summary, the discharging sow represents an excellent example of a management problem erroneously considered an infectious disease outbreak.

8 Conclusions

In general, the reproductive performance of sow herds improved considerably over the last few decades. Farrowing rates and litter sizes are reaching levels that previously were unattainable; however, reproductive efficiency of sow herd is continually affected by a variety of factors. Economic and societal pressures, such as gilt development, sow longevity, housing systems and lifetime performance, exert greater pressures on sow herd productivity. Concomitantly, producers encounter the age-old problems of seasonal infertility, extended WEIs, nutritional requirements of highly productive sows and nagging issues with infectious diseases. Although pharmacological interventions and technological improvements in AI have created greater opportunities for improved reproductive efficiency, fundamental stockmanship skills remain the foundation for success and profitability of the breeding herd. Without knowledgeable and caring stockpeople, it is unlikely that reproductive efficiency will improve in the pork industry.

9 Where to look for further information

Perhaps the best source of fundamental and scientific information on pig reproduction is the series (I to IX) of books titled *Control of Pig Reproduction*. The books, published as supplements in the journals *Reproduction* and *Society for Reproduction and Fertility*, are reviewed proceedings from the International Conference on Pig Reproduction (ICPR). The ICPR started in 1981 and is held every four years in a variety of international locations.

Applied aspects of reproductive efficiency are included in the *Diseases of Swine* (10th Edition, Wiley-Blackwell, 2012). Previous editions of the textbook are somewhat variable regarding the content on porcine reproduction and efficiency in the breeding herd.

Fundamental information on pig reproduction and efficiency is provided in the *Merck Veterinary Manual* and can be accessed at http://www.merckvetmanual.com/management-and-nutrition/management-of-reproduction-pigs/.

10 References

Aarnink, A. J. A., Schrama, J. W., Heetkamp, M. J. W., Stefanowska, J. and Huynh, T. T. T. (2006), Temperature and body weight affect fouling of pig pens. *J. Anim. Sci.* 84, 2224–31.

Almond, G. W. (1992), Factors affecting the reproductive performance of the weaned sow. *Vet. Clin. N. Am. Food Anim. Pract.* 8(3), 503–15.

Almond, G., Britt, J., Flowers, B., Glossop, C., Levis, D., Morrow, M. and See, T. (1998), *The Swine AI Book*. 2nd edition. Ruth Cronje (ed.), North Carolina State University.

Almond, G. W., Flowers, W. L. Batista, L. and D'Allaire, S. (2006), Diseases of the reproductive system. In: *Diseases of Swine*. Straw, B. E., Zimmerman J. J., D'Allaire, S. and Taylor, D. J. (eds), 9th Edition. Blackwell Publishing Ltd: Oxford, UK, pp. 113–48.

Almond, P. K. and Bilkei, G. (2005), Seasonal infertility in large pig production units in an Eastern-European climate. *Aust. Vet. J.* 83, 344–6.

Barb, C. R., Estienne, M. J., Kraeling, R. R., Marple, D. N., Rampacek, G. B., Rahe, C. H. and Sartin, J. L. (1991), Endocrine changes in sows exposed to elevated ambient temperature during lactation. *Dom. Anim. Endocrinol.* 8, 117–27.

Belstra, B. A., Flowers, W. L. and See, M. T. (2004), Factors affecting temporal relationships between estrus and ovulation in commercial sow farms. *Anim. Reprod. Sci.* 84, 377–94.

Bonet, S., Casas, I., Holt, W. and Yeste, M., Eds. (2013), *Boar Reproduction: Fundamentals and New Biotechnological Trends.* Springer Verlag Berlin Heidelberg: New York. eBook. 632p.

Boyer, P. E. and Almond, G. W. (2014), Use of altrenogest at weaning in primiparous sows. *J. Swine Health Prod.* 22, 134–7.

Britt, J. H., Szarek, V. E. and Levis, G. (1983), Characterization of summer infertility of sows in large confinement units. *Theriogenology* 20, 133–40.

Britt, J. H., Day, B. N., Webel, S. K. and Brauer, M. A. (1989), Induction of fertile estrus in prepuberal gilts by treatment with a combination of pregnant mare's serum gonadotropin and human chorionic gonadotropin. *J. Anim. Sci.* 67, 1148–53.

Calderon Diaz, J. A., Vallet, J. L., Lents, C. A., Nonneman, D. J., Miles, J. R., Wright, E. C., Rempel, L. A., Cushman, R. A., Freking, B. A., Rohrer, G. A., Phillips, C., Dedecker, A., Foxcroft, G. and Stalder, K. J. (2015), Age at puberty, ovulation rate, and uterine length of developing gilts fed two lysine and three metabolizable energy concentrations from 100 to 260 d of age. *J. Anim. Sci.* 93(7), 3521–7.

Campos, P. H. R. F., Silva, B. A. N., Donzele, J. L., Oliveira, R. F. M. and Knol, E. F. (2012), Effects of sow nutrition during gestation on within-litter birth weight variation: a review. *Animal* 6(5), 797–806.

Chokoe, T. C. and Siebrits, F. K. (2009), Effects of Season and regulated photoperiod on the reproductive performance of sows. *S. African. J. Anim. Sci.* 39, 45–54.

Coleman, G. J. and Hemsworth, P. H. (2014), Training to improve stockperson beliefs and behavior towards livestock enhances welfare and productivity. *Rev. Sci Tech. Off. Int. Epiz.* 33(1), 131–7.

Condous, P. C., Kirkwood, R. N. and van Wettere, W. H. E. J. (2014), The effect of pre- and post-mating dietary restriction on embryonic survival in gilts. *Theriogenology* 148, 130–6.

Cottney, P. D., Magowan, E., Ball, M. E. E. and Gordon, A. (2012), Effect of oestrus number of nulliparous sows at first service on first litter and lifetime performance. *Livest. Sci.* 146, 5–12.

Crenshaw, J. D., Moser, R. L., Wilson, M. E., Russell, L. E., Boyed, R. D. and Campbell, J. M. (2007), Lactation feed disappearance and weaning to estrus interval for sows fed spray-dried plasma. *J. Anim. Sci.* 85, 3442–53.

De Jong, E., Kauffold, J., Engl, S., Jourquin, J. and Maes, D. (2013), Effect of a GnRH analogue (Maprelin) on the reproductive performance of gilts and sows. *Theriogenology* 80, 870–7.

De Jong, E., Laanen, M., Dewulf, J., Jourquin, J., de Kruif, A. and Maes, D. (2013), Management factors associated with sow reproductive performance after weaning. *Reprod. Dom. Anim.* 48, 435–40.

De Rensis, F., Saleri, R., Tummaruk, P., Techakumphu, M. and Kirkwood, R. N. (2012), Prostaglandin F2α and control of reproduction in female swine. *Theriogenology* 77, 1–11.

De Vos, M., Che, L., Huygelen, V., Willemen, S., Michiels, J., Van Cruchten, S. and Van Ginneken, C. (2014), Nutritional interventions to prevent and rear low-birthweight piglets. *J. Anim. Phys. and Anim. Nutrition.* 98, 609–19.

De Winter, P. J. J., Verdonck, M., de Kruif, A., Coryn, M., Deluyker, H., Devriese, L. and Haesebrouck, F. (1996), The relationship between the blood progesterone concentration at early metoestrus and uterine infection in the sow. *Anim. Reprod. Sci.* 41, 51–9.

Degenstein, K. L., O'Donoghue, R., Patterson, J. L., Beltranena, E., Ambrose, D. J., Foxcroft, G. R. and Dyck, M. K. (2008), Synchronization of ovulation in cyclic gilts with porcine luteinizing hormone (pLH) and its effects on reproductive function. *Theriogenology* 70, 1075–85.

Dial, G. D., Almond, G. W., Hilley, H. D., Repasky, R. R. and Hagan J. (1987), Oxytocin precipitation of prostaglandin-induced farrowing in swine: determination of the optimal dose of oxytocin and optimal interval between prostaglandin F2 alpha and oxytocin. *Am. J. Vet. Res.* 48, 966–70.

Dial, G. D. (1990), Computerized records: Use in troubleshooting reproductive problems of commercial swine herds. Proceedings of Swine Day, Agricultural Experiment Station and Cooperative Extension Service, Kansas State University, pp. 1–13.

Distl, O. (2007), Mechanisms of regulation of litter size in pigs on the genome level. *Reprod. Dom. Anim.* 42(Suppl 2), 10–16.

Dourmad, J. Y., Etienne, M. and Noblet, J. 1998. Effect of protein and lysine supply on performance, nitrogen balance, and body composition of sows during lactation. *J. Anim. Sci.* 76, 542–50.

Driancourt, M. A., Cox, P., Rubion, S., Harnois-Milon, G., Kemp, B. and Soede, N. M. (1980), Induction of an LH surge and ovulation by buserelin (As Receptal) allows breeding of weaned sows with a single fixed-time insemination. *Theriogenology* 80, 391–9.

Dwyer, C. M., Stickland, N. C. and Hetcher, J. M. (1994), The influence of maternal nutrition on muscle fiber number development in the porcine fetus and on subsequent postnatal growth. *J. Anim. Sci.* 72, 911–17.

Eckhardt, O. H. O., Martins, S. M. M. K., Pinese, M. E., Horta, F. C., Russeto, A. C., Torres, M. A., De Andrade, A. F. C., Muro, B. B. D., Marino, C. T., Rodrigues, P. H. M. and Moretti, A. S. (2014), Gonadotropin-induced puberty does not impair reproductive performance of gilts over three parities. *Reprod. Dom. Anim.* 49, 964–9.

Einarsson, S., Sjunnesson, Y., Hulten, F., Eliasson-Selling, L., Dalin, A. -M., Lundeheim, N. and Magnusson, U. (2014), A 25 years experience of group-housed sows-reproduction in animal welfare-friendly systems. *Acata Vet. Scand.* 56, 37.

Estienne, M. J. and Hartsock T. G. (1998), Effect of exogenous gonadotropins on the weaning-to-estrus interval in sows. *Theriogenology* 49, 823–8.

Estienne, M. J. and Crawford, R. J. (2015), Characteristics of estrous cycles in gilts treated with gonadotropins after estrus or treatment with a progestogen. *Theriogenology* 83, 459–63.

Filha, W. S. A., Bernardi, M. L., Wentz, I. and Bortolozzo, F. P. (2010), Reproductive performance of gilts according to growth rate and backfat thickness at mating. *Anim. Reprod. Sci.* 121, 139–44.

Flowers, B. (2015), Birth weight, neonatal management and sow longevity potential. National Hog Farmer Blueprint No. 60. Pentron Media, Inc: Overland Park, Kansas, pp. 8–10.

Fontana, D. L., Ulguim, R. R., Sbardella, P. E., Bernardi, M. L., Wentz, I. and Bortolozzo, F. P. (2013), Fixed-time post-cervical insemination in sows receiving porcine luteinizing hormone at oestrus onset. *Anim. Reprod. Sci.* 144, 109–14.

Fraser, D., Duncan, I. J. H., Edwards, S. A., Grandin, T., Gregory, N. G., Guyonnet, V., Hemsworth, P. H., Huertas, T. M., Huzzey, J. M., Mellor, D. J., Mench, J. A., Spinka, M. and Whay, H. R. (2013), General principles for the welfare of animals in production systems: The underlying science and its application. *The Vet. J.* 198, 19–27.

Gonzalez-Pena, D., Knox, R. V., Pettigrew, J. and Rodriguez-Zas, S. L. (2014), Impact of pig insemination technique and semen preparation on profitability. *J. Anim. Sci.* 92, 72–84.

Gordon, I. R. (1997), *Controlled Reproduction in Pigs*. Wallingford, Oxon, UK: CAB International.

Guthrie, H. D., Henricks, D. M. and Handlin, D. L. (1974), Plasma hormone levels and fertility in pigs induced to supervoulate with PMSG. *J. Reprod. Fert.* 41, 361–70.

Heitman, H., Hughes, E. H. and Kelly, C. F. (1951), Effects of elevated ambient temperature on pregnant sows. *J. Anim. Sci.* 10, 907–15.

Hemsworth, P. H., Brand, A. and Willems, P. J. (1981), The behavioural response of sows to the presence of human beings and their productivity. *Livest. Prod. Sci.* 8, 67–74.

Hemsworth, P. H., Barnett, J. L. and Hansen, C. (1986), The influence of handling by humans on the behavior, reproduction and corticosteroids of male and female pigs. *Applied Anim. Behav. Sci.* 15, 303–14.

Hemsworth, P. H., Barnett, J. L., Coleman, G. J. and Hansen, C. (1989), A study of the relationships between the attitudinal and behavioural profiles of stockpeople and the level of fear of humans and the reproductive performance of commercial pigs. *Applied Anim. Behav. Sci.* 23, 301–14.

Hemsworth, P. H., Rice, M., Nash, J., Giri, K., Butler, K. L., Tilbrook, A. J. and Morrison, R. S. (2013), Effects of group size and floor space allowance on grouped sows: Aggression, stress, skin injuries, and reproductive performance. *J. Anim. Sci.* 91, 4953–64.

Hernandez-Caravaca, I., Izquierdo-Rico, M. J., Mata, C., Carvajal, J. A., Vieira, L., Abril, D., Soriano-Ubeda, C. and Garcia-Vazquez, F. A. (2012), Reproductive performance and backflow study in cervical and post-cervical artificial insemination isows. *Anim. Reprod. Sci.* 136, 14–22.

Hidalgo, D. M., Friendship, R. M., Greiner, L., Manjarin, R., Amezzcua, M. R., Dominguez, J. C. and Kirkwood, R. N. (2014), Influence of gonadotrophin-induced first oestrus on gilt fertility. *Reprod. Dom. Anim.* 49, 899–902.

Holyoake, P. K., Dial, G. D., Trigg, T. and King, V. L. (1995), Reducing pig mortality through supervision during the perinatal period. *J. Anim. Sci.* 73, 3543–51.

Hulten, F., Wallenbeck. A. and Rydhmer, L. (2006), Ovarian activity and oestrous signs among group-housed, lactating sows: influence of behavior, environment and production. *Reprod. Dom. Anim.* 41:448–54.

Hunter, R. H. F. (1964), Superovulation and fertility in the pig. *Anim. Prod.* 6, 189–94.

Hurtgen, J. P., Leman, A. D. and Crabo, B. (1980), Seasonal influence on oestrous activity in sows and gilts. *J. Am. Vet. Med. Assoc.* 176, 119–23.

Johnson, R. K., Nielsen, M. K. and Casey, D. S. (1999), Responses in ovulation rate, embryonal survival, and litter traits in swine to 14 generations of selection to increase litter size. *J. Anim. Sci.* 77, 541–57.

Johnston, L. J. and Li, Y. Z. (2013), Performance and well-being of sows housed in pens retrofitted from gestation stalls. *J. Anim. Sci.* 91, 5937–45.

Justino, E., Naas, I. A., Neves, D. P. and Salgado, D. D. (2012), Heat loss in lactating sows under different ventilation systems and diets during summer. *Am. Soc. Agri. Bio. Eng.* (ASABE) Ninth International Livestock Environment Symposium. Valencia, Spain, pp. 1–6.

Kaneko, M. and Koketsu, Y. (2012), Gilt development and mating in commercial swine herds with varying reproductive performance. *Theriogenology* 77, 840–6.

Kauppinen, T., Vesala, K. M. and Valros, A. (2012), Farmer attitude toward improvement of animal welfare is correlated piglet production parameters. *Livest. Sci.* 143, 142–50.

Kemp, B. and Soede, N. M. (1996), Relationship of weaning-to-estrus interval to timing of ovulation and fertilization in sows. *J. Anim. Sci.* 74, 944–9.

Kemp, B., Soede, N. M. and Langendijk, P. (2005), Effects of boar contact and housing conditions on estrus expression in sows. *Theriogenology* 63, 643–56.

Kemp, B. and Soede, N. M. (2012), Reproductive issues in welfare-friendly housing systems in pig husbandry: a review. *Reprod. Dom. Anim.* 47(Suppl 5), 51–7.

Kim, S. W., Ji, F., Wu, G. and Hurley, W. L. (2009), Ideal amino acid balance for sows during gestation and lactation. *J. Anim. Sci.* 87(E Suppl), E123–32.

King, R. H. and Dunkin, A. C. (1986), The effect of nutrition on the reproductive performance of first litter sows. 4. The relative effects of energy and protein intakes during lactation on the performance of sows and their piglets. *Anim. Prod.* 43, 319–25.

Knox, R., Salak-Johnson, J., Hopgood, M., Greiner, L. and Connor, J. (2014), Effect of day of mixing gestating sows on measures of reproductive performance and animal welfare. *J. Anim. Sci.* 92, 1698–707.

Knox, R. V. (2016), Artificial insemination in pigs today. *Theriogenology* 85, 83–93.

Knox, R. V., Rodriguez-Zas, S. L., Miller, G. M., Willenburg, K. L. and Robb, J. A. (2001), Administration of PG600 to sows at weaning and the time of ovulation as determined by transrectal ultrasound. *J. Anim. Sci.* 79, 796–802.

Knox, R. V., Willenburg, K. L., Rodriguez-Zas, S. L., Greger, D. L., Hafs, H. D. and Swanson, M. E. (2011), Synchronization of ovulation and fertility in weaned sows treated with intravaginal triptorelin is influenced by timing of administration and follicle size. *Theriogenology* 75, 308–19.

Koketsu, Y., Dial, G. D., Pettigrew, J. E., Marsh, W. E. and King, V. L. (1996), Characterization of feed intake patterns during lactation in commercial swine herds. *J. Anim. Sci.* 74, 1202–10.

Konora A. and Maes, D. (2009), The role of mycotoxins in pig reproduction: a review. *Veterinarni Medicina.* 54, 565–76.

Langendijk, P., van den Brand, H., Soede, N. M. and Kemp, B. (2000), Effect of boar contact on follicular development and on estrus expression after weaning in primiparous sows. *Theriogenology* 54, 1295–303.

Langenkijk, P., Soede, N. M. and Kemp, B. (2000), Effects of boar contact and housing conditions on estrus expression in weaned sows. *J. Anim. Sci.* 78, 871–8.
Lawlor, P. G., Lynch, P. R., O'Connell, M. K., McNamara, L., Reid, P. and Stickland, N. C. (2007), The influence of over feeding sows during gestation on reproductive performance and pig growth to slaughter. *Arch. Anim. Breed.* 50, 82–91.
Laws, J., Amusquivar, E., Laws, A., Herrera, E., Lean, I. J., Dodds, P. F. and Clarke, L. (2009), Supplementation of sow diets with oil during gestation: Sow body condition, milk yield and milk composition. *Livest. Sci.* 123, 88–96.
Love, R. J., Evans, G. and Klupiec, C. (1993), Seasonal effects on fertility in gilts and sows. *J. Reprod. Fertil. Suppl.* 48, 191–206.
Martinat-Botte, F., Venturi, E., Guillouet, P., Driancourt, M. A. and Terquie, M. (2010), Induction and synchronization of ovulations of nulliparous and multiparous sows with an injection of gonadotropin-releasing hormone agonist (Receptal). *Theriogenology* 73, 332–42.
Martineau, G. P. and Badouard B. (2009), Managing highly prolific sows. London Swine Conference, London, Ontario, Canada, pp. 3–34.
McGlone, J. J., Stansbury, W. F., Tribble, L. F. and Morrow, J. L. (1988), Photoperiod and heat stress influence on lactating sow performance and photoperiod effects on nursery pig performance. *J. Anim. Sci.* 66, 1915–19.
McGlone, J. J. Borell, V. E. H., Deen, J., Johnson, A. K., Levis, D. G., Meunier-Salaun, M.- C., Morrow, J. L. Reeves, D., Salak-Johnson, J. L. and Sundberg, P. I. (2004), Compilation of the scientific literatures comparing housing systems for gestating sows and gilts using measures of physiology, behavior, performance and health. *Prof. Anim. Sci.* 20, 105–17.
Messias de Braganca, M., Mounier, A. M. and Prunier, A. (1998), Does feed restriction mimic the effects of increased ambient temperature in lactating sows? *J. Anim. Sci.* 76, 2017–24.
Mostrom, M. S. (2012), *Zearalenone*. In: *Veterinary Toxicology*. Ramesh C. (ed.), Gupta. New York: Academic Press, pp. 1266–71.
National Research Council of the National Academies. (2012). *Water*. In: *Nutrient Requirements of Swine*. Lewis, A. J. and Arietis, R. S. (ed.), Washington: The National Academies Press, pp. 66–73.
Nissen, P. M., Danielsen, V. O., Jorgensen, P. F. and Oksbjerg, N. (2003), Increased maternal nutrition of sows has no beneficial effects on muscle fiber number or postnatal growth and has no impact on the meat quality of the offspring. *J. Anim. Sci.* 81, 3018–27.
Noblet, J. and Etienne, M. (1987), Metabolic utilization of energy and maintenance requirements in lactating sows. *J. Anim. Sci.* 64, 774–81.
Noblet, J., Dourmad, J. Y. and Etienne, M. (1990), Energy utilization in pregnant and lactating sows: Modeling of energy requirements. *J. Anim. Sci.* 68, 562–72.
Parera, A. N. M. and Hacker, R. R. (1984), The effects of different photoperiods on reproduction in the sow. *J. Anim. Sci.* 58, 1418.
Patterson, J. L., Beltrana, E. and Foxcroft, G. R. (2010). The effect of gilt age at first estrus and breeding on third estrus on sow body weight changes and long-term reproductive performance. *J. Anim. Sci.* 88, 2500–13.
Patterson, J. L. and Foxcroft, G. R. (2016), Gilt management for improved performance. In: *Current Issues in Swine Reproduction*. Proc. Ann. Mtg. Am Assoc. Swine Vet. Preconference Seminar, pp. 13–17.
Pedersen, L. J., Studnitz, M., Jensen, K. H. and Giersing, A. M. (1998), Suckling behavior of piglets in relation to accessibility to the sow and the presence of foreign litters. *Appl. Anim. Behav. Sci.* 58, 267–79.
Peltoniemi, O. A. T., Tast, A. and Love, R. J. (2000), Factors effecting reproduction in the pig: seasonal effects and restricted feeding of the pregnant gilt and sow. *Anim. Reprod. Sci.* 60, 173–84.
Pittman, J. S. (2008), Reproductive failure associated with porcine circovirus type 2 in gilts. *J. Swine Health and Prod.* 16, 144–8.
Rault, J.- L., Morrison, R. S., Hansen, C. F., Hansen, L. U. and Hemsworth, P. H. (2014), Effects of group housing after weaning on sow welfare and sexual behavior. *J. Anim. Sci.* 92, 5683–92.
Rehfekdt, C. and Kuhn G. (2006), Consequences of birth weight for postnatal growth performance and carcass quality in pigs as related to myogenesis. *J. Anim. Sci.* 84, 113–23.

Rodriguez-Martinez, H., Vallet, J. L. and Ziecik, A. J. (2009), Control of pig reproduction VIII: Proceedings of the Eighth International Conference on Pig Reproduction. Nottingham, UK: Nottingham University Press. 336 p.
Roongsitthichai, A., Cheuchuchart, P., Chatwijitkul, S., Chantarothai, O. and Tummaruk, P. (2013), Influence of age at first estrus, body weight, and average daily gain of replacement gilts on their subsequent reproductive performance as sows. *Livest. Sci.* 151, 238–45.
Rosero, D. S., van Heugten, E., Odle, J., Arellano, C. and Boyd, R. D. (2012), Response of the modern lactating sow and progeny to source and level of supplemental dietary fat during high ambient temperatures. *J. Anim. Sci.* 90, 2609–19.
Rozeboom, K. J., See, M. T. and Flowers, W. (2000), Management practices to reduce the impact of seasonal infertility on sow herd productivity. Animal Science Facts: Extension swine husbandry. North Carolina State University.
Schmidt, W. W., Stevenson, J. S. and Davis, D. L. (1985), Reproductive traits of sows penned individually or in groups until 35 days after breeding. *J. Anim. Sci.* 60, 755–9.
Schneider, J. K., Tokach, M. K., Dritz, S. S., Nelssen, J. L., Derouchey, J. M. and Goodband, R. D. (2007), Effects of feeding schedule on body condition, aggressiveness, and reproductive failure in group-housed sows. *J. Anim. Sci.* 85, 3462–9.
Seabrook, M. F. (1988), The behavior of the pigstockman and its influence on pig performance and behavior – a review. *Pig News and Info.* 9(4), 403–6.
Seedorf, J., Hartung, J., Schroder, M., Linkert, K. H., Pedersen, S., Takai, H., Johnsen, J. O., Metz, J. H. M., Goot Koerkamp, P. W. G., Uenk, G. H., Phillips, V. R., Holden, M. R., Sneath, R. W., Short, J. L., White, R. P. and Wathes, C. M. (1998), A survey of ventilation rates in livestock buildings in northern Europe. *J. Agic. Eng. Res.* 70, 39–47.
Shurson, G. C., Hogberg, M. G., DeFever, N., Radecki, S. V. and Miller, E. R. (1986), Effects of adding fat to the sow lactation diet on lactation and rebreeding performance. *J. Anim. Sci.* 62, 672–80.
Spoolder, H. A. M., Geudeke, M. J., Van der Peet-Schwering, C. M. C. and Soede, N. M. (2009), Group housing of sows in early pregnancy; a review of success and risk factors. *Livest. Sci.* 125, 1–14.
Stevenson, J. S. and Britt, J. H. (1981), Interval to estrus in sows and performance of pigs after alteration of litter size during late lactation. *J. Anim. Sci.* 53, 177–81.
Sulabo, R. C., Jacela, J. Y., Tokach, M. D., Dritz, S. S., Goodband, R. D., Derouchey, J. M. and Nelssen, J. L. (2010), Effects of lactation feed intake and creep feeding on sow and piglet performance. *J. Anim. Sci.* 88, 3145–53.
Tast, A., Peltoniemi, O. A., Virolainen. J. V. and Love, R. J. (2002), Early disruption of pregnancy as a manifestation of seasonal infertility in pigs. *Anim. Reprod. Sci.* 74, 75–86.
Tummaruk, P., Lundeheim, N., Einarsson, S. and Dalin, A.-M. (2001), Effect of birth litter size, birth parity number, growth rate, backfat thickness and age at first mating of gilts on their reproductive performance as slows. *Anim. Reprod. Sci.* 66, 225–37.
Tummaruk, P., Tantasuparuk, W., Techakumphu, M. and Kunavongkrit, A. (2007), Age, body weight and backfat thickness at first observed oestrus in crossbred Landrace x Yourkshire gilts, seasonal variations and their influence on subsequence reproductive performance. *Anim. Reprod. Sci.* 99, 167–81.
Tummaruk, P., Tantasuparuk, W., Techakumphu, M. and Kunavongkrit, A. (2009), The association between growth rate, body weight, backfat thickness and age at first observed oestrus in crossbred Landrace x Yorkshire gilts. *Anim. Reprod. Sci.* 110, 108–22.
Tuyttens, F. A. M., Van Gansbeke, S. and Ampe, B. (2011), Survey among Belgian pig producers about the introduction of group housing systems for gestating sows. *J. Anim. Sci.* 89, 845–55.
Ulguim, R. R., Fontana, D. L., Rampi, J. Z., Bernardi, M. L. Wentz, I. and Bortolozzo, F. P. (2014), Use of porcine luteinizing hormone at oestrous onset in a protocol for fixed-time artificial insemination in gilts. *Reprod. Dom. Anim.* 49, 756–60.
Vallet, J. L. (2015), *Growth and Nutrition Effects on Gilt Development*. National Hog Farmer Blueprint No. 60. Pentron Media, Inc: Overland Park, KN, pp. 12–14.
Van den Brand, H., Dieleman, S. J., Soede, N. M. and Kemp, B. (2000), Dietary energy source at two feeding levels during lactation of primiparous sows: I. Effects on glucose, insulin, and luteinizing

hormone and on follicle development, weaning-to-estrus interval, and ovulation rate. *J. Anim. Sci.* 78, 396–404.
Vanderhaeghe, C., Dewulf, J., De Vliegher, S., Papadopoulos, G. A., de Kruif, A. and Maes, D. (2010), Longitudinal field study to assess sow level risk factors associated with stillborn piglets, *Anim. Reprod. Sci.* 120, 78–83.
Vanderhaeghe, C., Dewulf, J., de Kruif, A. and Maes, D. (2013), Non-infectious factors associated with stillbirth in pigs: a review. *Anim. Reprod. Sci.* 139, 76–88.
Varley, M. A. and Cole, D. J. A. (1976), Studies in sow reproduction. 4. The effect of level of feeding in lactation and during the interval from weaning to remating on the subsequent reproductive performance of the early weaned sow. *Anim. Prod.* 22, 71–7.
Walker, N. and Watt, D. (1979), The effect of weaning at 10, 25, or 40 days on the reproductive performance of sows from the first to the fifth parity. *J. Agric. Sci. Camb.* 92, 449–56.
Walton, J. S. (1986), Effect of boar presence before and after weaning on estrus and ovulation in sows. *J. Anim. Sci.* 62, 9–15.
Weitze, K. F., Wagner-Rietschel, H., Waberski, D., Richter, L and, Krieter, L. (1994), The onset of heat after weaning, heat duration and ovulation as major factors in AI timing in sows. *Reprod. Dom. Anim.* 91, 349–63.
Wessels, J. M., Khalaj, K., Kridli, R. T., Edwards, A. K., Bidarimath, M. and Tayade, C. (2014), Are pharmacological interventions between conception and birth effective in improving reproductive outcomes in North American swine: *Reprod. Dom. Anim.* 49, 536–42.
Williams, N., Patterson, J. and Foxcroft, G. R. (2005), Non-negotiables in gilt development. Advances in Pork Production 16. Univ. Alberta, Edmonton, Alberta, Canada, pp. 1–16.
Williams, A. (2009), Effects of heat stress on reproduction and productivity of primiparous sows and their piglets' performance. MS Thesis. University of Missouri-Columbia.
Yoder, C. L., Dixon, P. M., Duttlinger, V. M., Baas, T. J., Schwag, C. R., Fix, J. S. and Stalder, K. J. (2013), Estimation of deviations from predicted lactation feed intake and the effect on reproductive performance. *Livest. Sci.* 154, 184–92.

Chapter 4

Factors affecting the reproductive efficiency of boars

M. L. W. J. Broekhuijse, Topigs Norsvin Research Center B.V., The Netherlands

1 Introduction

Undocumented tales mentioned artificial insemination (AI) for the first time around 1400. An Arabic chief inseminated his own mares with the semen of a stallion belonging to a rival tribe. Antonie van Leeuwenhoek (1632–1723) was a Dutch tradesman and is known to have made over 500 microscopes. Fewer than 10 have survived to the present day. His discoveries on the presence and movement of 'animalculi', which were sperm cells, were reported in 1677 (Leeuwenhoek 1678). In the eighteenth century, an Italian priest, Spallanzani (1784) inseminated a bitch which whelped three pups (Siebenga 1937). Human artificial insemination was first attempted around 1800 by Hunter, followed by Sims in 1866. Successful AI in rabbits, dogs and horses was first reported by Heape (1897). The Russian professor Elie Ivanov started working with sperm cells at the end of the nineteenth century (Ivanov 1907, 1922). His work stimulated research on AI around the world (Milovanov 1964; Nishikawa 1962, 1964). Significant advances were made in the 1940s in the United States, leading to established procedures used worldwide (Salisbury et al. 1978). This period also saw worldwide studies on AI in pigs: in the United States (McKenzie 1931), Japan (Niwa 1958) and Western Europe (Polge 1956). Boars were trained on mounting dummies (Polge 1956). Other techniques to collect semen included artificial vaginas to apply pressure to the glans (McKenzie 1931; Polge 1956) or the use of a gloved hand (Hancock and Hovel 1959). This minimized the amount of bacterial contamination in the collected semen (Althouse and Lu 2005).

Storing semen long enough for shipment and use in the field was initiated in the United States. A yolk-phosphate semen extender was developed (Philips and Lardy 1940). Storing semen at a temperature of 15–20°C was first recommended by Ito et al. (1948). With

http://dx.doi.org/10.19103/AS.2017.0013.30

Table 1 Breeding pyramid

Animals in production	Goal
1. Great Grand Parents (GGP)	Improving genetic lines and breeds
2. Grand Parents (GP)	Production of: – breeding sows (mainly F1 sow) – end boars
3. Parent Stock (PS)	Production piglets → finishers
4. Finisher production level	

the expansion of AI, the demand for boar semen also increased. A simple solution was diluting the semen. The first extender to be developed was the Beltsville Thawing Solution (BTS) (Pursel and Johnson 1975). This increased the storage time and made the dilution of semen possible. BTS is still the main extender used. The main reason to use AI was to prevent the spread of contagious diseases. The first pig AI stations in the Netherlands were independent cooperative boar stations. Breeding companies worked closely with AI companies to make the boars with the highest genetic index available for AI and the exclusive use of those boars for the nucleus breeders (Table 1).

2 Monitoring pig fertility

2.1 The importance of monitoring fertility

Pregnancy rates of 60% and litter sizes of six piglets were common in the beginning of the use of Dutch pig AI (Feitsma 2009). Results improved with more research and development. This included more knowledge on dilution and treatment of boar semen, the oestrus cycle of the sow and timing of the insemination. By achieving results similar to natural mating, commercial farms started using AI and the use of pig AI increased rapidly. In the beginning, the results depended on trained technicians (for collecting and inseminating semen), but later the demand for self-service AI increased in order to save labour (Strikwerda 2007). Economic losses due to reproductive inefficiency in male animals can be substantial (Roberts 1986; Lunenfeld and Insler 1993). Common problems in subfertile and infertile human patients and animals include low number of sperm cells and/or low semen quality. It is generally accepted that lower motility ejaculates have lower or limited potential to fertilise oocytes (Flowers 1997; Donadeu 2004).

Minimal differences in fertility in pigs can lead to lower efficiency in producing piglets with consequent economic losses. Although the efficiency of piglet production is dependent on many factors, one of them that can be measured is the quality of semen. The most critical factors are specific, precise and accurate fertility tests and precise and accurate fertility data. It is very difficult to obtain reliable field fertility data. The problems that can occur are (1) the boars or sows are not representative for the population, and are too few in number; (2) an insufficient number of sows are inseminated with sperm cells from each ejaculate; (3) too few ejaculates are assessed per boar, (4) a large number of cells is used for each insemination, which makes it difficult to identify the link between semen quality and fertility; and (5) the fertility outcome is not reported or documented properly (Broekhuijse et al., 2011a). For successful implementation of AI, field fertility results of semen used are critical.

2.2 Use of large data sets

From a commercial perspective, it is important for an AI company to monitor its results in practice. In general, variation in fertility in sows is large (Broekhuijse et al. 2011a; Broekhuijse et al. 2012a,b) while the effect of semen factors is relatively small and therefore impossible to find in smaller data sets (Broekhuijse et al. 2011a). By using large data sets, one can perform reliable statistical corrections on both sow- and boar-related parameters to get normalised and representative comparisons. Remaining sow fertility variation can then be assigned to semen quality parameters. Working in pig reproduction means realising that there are parameters one has to take in for granted (e.g. different boar lines, effect of parity of the sow), and there are other parameters that can be optimised to get an optimal fertility result.

To analyse the variation in fertility (e.g. farrowing rate (FR) or total number born (TNB)), at first the results need to be analysed according to model 1, analysing the effects of farm- and sow-related factors. The boar- and semen-related effects on the residuals are then analysed using model 2. The least square means for the two fertility traits is used, because per ejaculate multiple insemination doses are produced, which resulted in multiple fertility results. The method of least square means was first described by Carl Friedrich Gauss around 1794 and is widely used in agricultural science.

Model 1

Y = μ + parity + line♀ + farm + 1st/remating + purebred/crossbred litter + number of inseminations + int-wean-1 + gest_length + year*season + weekday + age semen + error

Where Y is the value of FR or TNB; μ is the mean value of FR and TNB; parity is the effect of the actual parity of the sow; line♀ is the effect of the genetic line of the sow, farm is the effect the farm; 1st/remating is the effect of a first or a remating of the sow; purebred/crossbred litter is the effect of having a purebred of crossbred litter; number of inseminations is the effect of the number of inseminations per heat; int-wean-1 is the effect of the interval between weaning and the first insemination; gest_length is the effect of the gestation length; year*season is the effect of the combination between year and season; weekday is the effect of the day of the week of insemination; age semen is the effect of the age of the semen (days after production) at insemination; and error is the random residual effect that could not be explained by the variables in the model.

Model 2

Y* = μ* + line♂ + boar(line♂) + age + days-ejac + AI station + vol-conc + dose-conc + semen quality + error*

Where Y* is the residual value of FR or TNB; μ* is the mean value of Y*; line♂ is the effect of the genetic line of the boar; boar(line♂) is the effect of the individual boar within the genetic line of the boar; age is the effect of the age of the boar at ejaculation; days-ejac is the effect of the number of days between current and previous ejaculation; AI station is the effect of the AI station; vol-conc is the effect of the number of cells in an ejaculate, dose-conc is the effect of the number of cells in the insemination dose used; semen quality is the effect of measured semen parameters; and error* is the random residual effect that could not be explained by the variables in the model.

3 Considerations when producing AI doses

3.1 Parameters affecting fertility

For an AI station there are many parameters affecting fertility, with the main parameter being the individual boar itself. The ability to identify and cull boars with reduced fertility is of considerable economic interest for AI centres. The effect of sub-fertile boars on field fertility is very large. It is important to identify them as soon as possible, of course preferably before semen is sold, but also retrospectively based on fertility records. The breeding company Topigs Norsvin (in The Netherlands) performs a weekly ranking, based on the corrected FR and TNB, which is corrected for farm- and sow-related parameters (the so called direct boar effect on FR or TNB). The direct boar effect represents the corrected difference from 0 (average) for the boar population of Topigs Norsvin with known records. For the worldwide population the spread in results was between −5.98% and +4.05% for FR and between −1.01 and +0.97 for TNB (n = 1193 boars, 116,749 inseminations, Figure 1a). In the Netherlands, this effect is even smaller: between −5.31% and +3.60% FR and between −0.68 and +0.83 TNB (n = 616 boars, 43,141 inseminations, Fig. 1b).

Table 2 shows the parameters affecting fertility, with Fig. 2 showing the percentages of explained variation in FR (A) and TNB (B). The genetic line of the boars explained the largest part of the boar- and semen-related variation in FR and TNB (Broekhuijse et al.

Table 2 Boar-related sources of variation (direct boar effect) and their effect on farrowing rate (FR) and total number of piglets born (TNB). *P*-value per variable for variation in FR and variation in TNB. Adapted from Broekhuijse et al. (2012b)

Parameter, *P*-values	FR, %	TNB, #
Individual boar (within line)	<0.0001	<0.0001
Line of the boar	0.0012	<0.0001
Age of the boar at ejaculation	0.0008	0.3746
Number of days between ejaculations	0.6370	0.4086
Number of cells in ejaculate	0.7523	0.5212
Number of cells in insemination dose	0.2703	0.4393
AI station (11 locations)	0.9143	0.4013
Motility	0.5259	0.0041
Progressive motility	0.0134	0.3795

$P \leq 0.05$ means that the variable had a significant effect on FR or TNB.

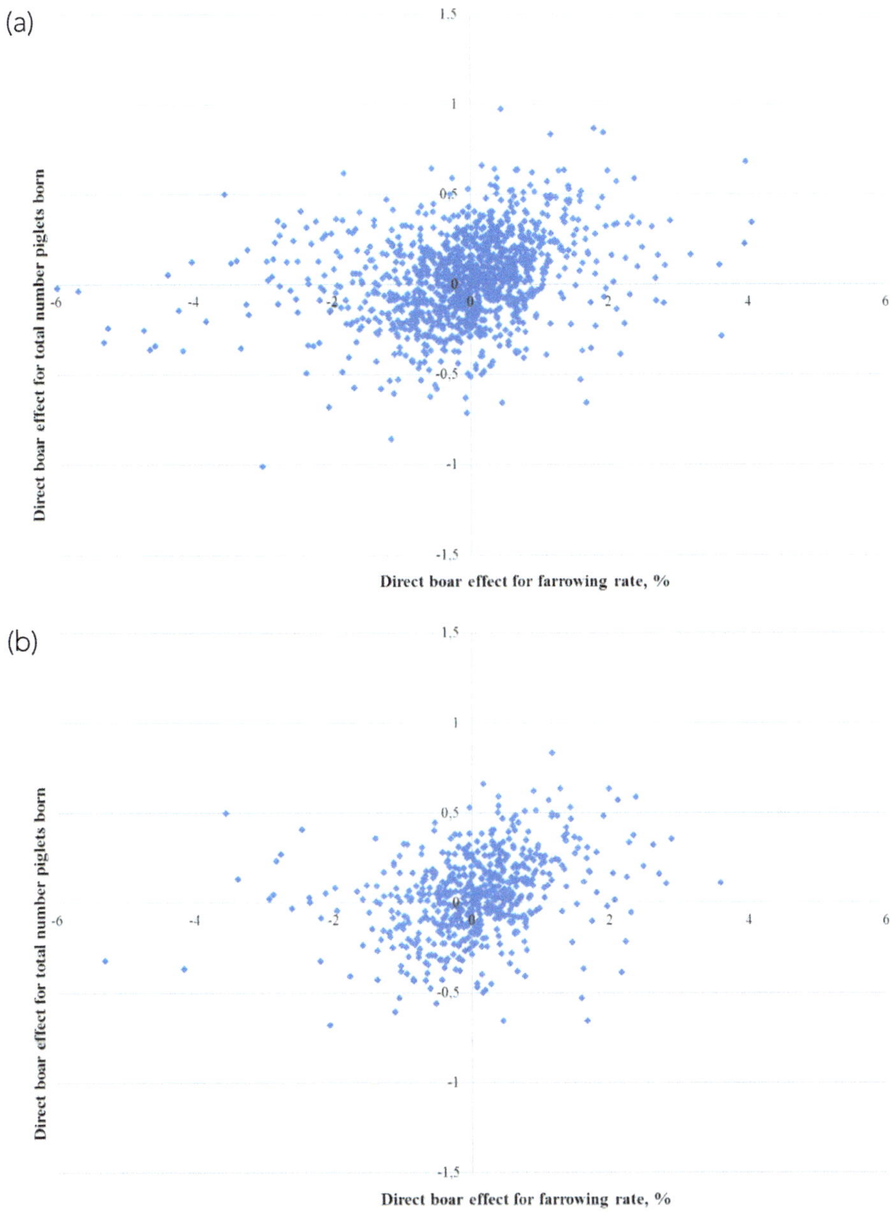

Figure 1 Direct boar effect for farrowing rate (%) vs. direct boar effect for total number of piglets born. (A) worldwide records, n = 1193 boars. (B) the Netherlands records, n = 616 boars. Topigs Norsvin, adapted from Broekhuijse et al. (2015).

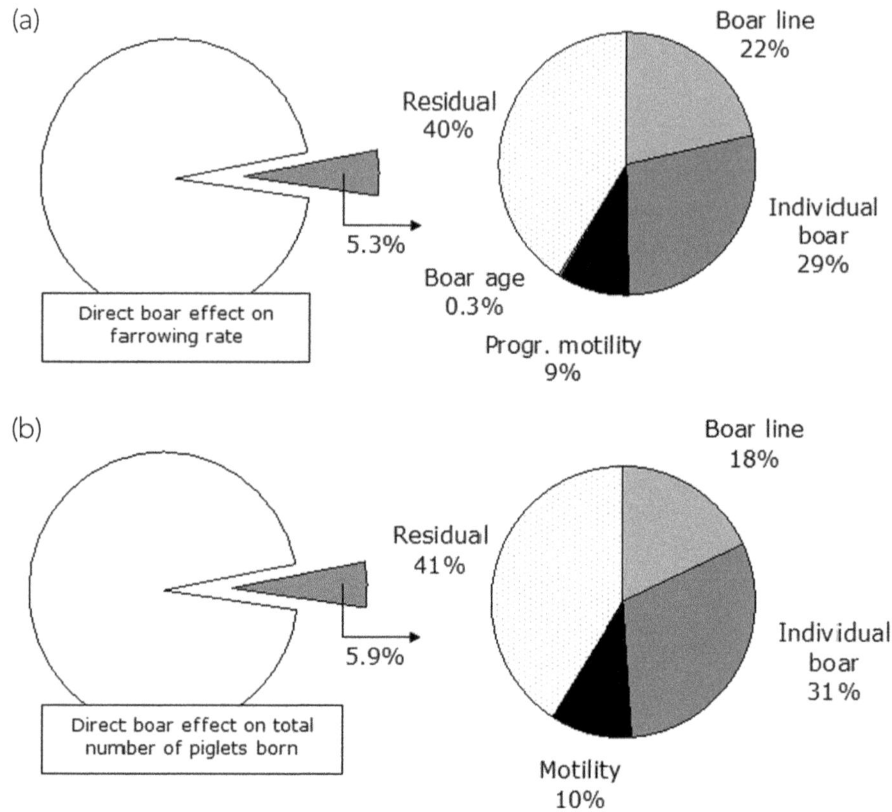

Figure 2 Boar- and semen-related sources of variation (direct boar effect) and their effects on farrowing rate (a) and the total number of piglets born (b). The left-hand chart shows the total variation in fertility. Part of this variation is due to the direct boar effect. This direct boar effect is explained by the parameters in the right-hand chart. Adapted from Broekhuijse et al. (2012b).

2012a,b). This is in accordance with several studies (Kommisrud et al. 2002; Kondracki 2003; Gadea et al. 2004; Janett et al. 2005; Sondermann and Luebbe 2008). Most authors agree that no breed excels in all semen characteristics (Kennedy and Wilkins 1984; Oh et al. 2003). In general, the choice for a genetic line is not based on fertility results but on market perspective of the offspring. AI centres need to include breed differences into their decision-making processes to ensure adequate use of the boar.

3.2 Subfertility identified by reciprocal translocations

Behind many cases of subfertility there is a genomic component. It is known that genetic disorders may not affect semen quality but can ultimately impair fertility (McLachlan and O'Bryan 2010). An example is reciprocal translocations (RT), where two broken-off chromosome pieces of non-homologous chromosomes are exchanged. A monitoring project showed a prevalence of <2% in the population (Feitsma 2009; Rodrıguez et al.

2010). This is difficult to identify in traditional assessments: boars show normal sexual behaviour and semen quality results. However, in terms of fertility, records of decrease of 20% litter size are shown (Mäkinen et al. 1999; Rodríguez et al. 2010). Furthermore, RT is a heritable genetic disorder (Robinson and Buhr 2005), so its negative impact on swine industry would go beyond loss in field fertility.

4 Tools for AI laboratories: semen quality assessment

Introducing boars at AI centres usually means via quarantine and strict health controls. Boars are put in separate collection pens and semen collected using dummy sows and the gloved-hand technique (Hancock and Hovel 1959) or in an automated semen collection system. Commonly, each semen quality assessment starts with macroscopic evaluation: colour, smell, contamination with dirt, blood or urine and viscosity. After volume determination, the concentration and the motility are measured. Concentration of semen can be measured with a wide range of systems.

4.1 Low-dose inseminations

Over the years, the total number of sperm cells in an AI dose decreased in the Netherlands from 4.0 billion in 100-ml disposable bottles (1985) down to 1.5 billion in 80-ml disposable tubes (Feitsma 2009). The relatively high number of sperm cells in a dose used in commercial AI practice likely masks reduced fertility (Foxcroft et al. 2008). Through the years, experiments were performed using low-dose inseminations. These data showed that, in some cases, it was feasible to decrease the number of sperm cells in an AI dose without noticing a significant effect on fertility results. As long as the number of cells in an AI dose is above the threshold, it has been concluded that FR and TNB are independent of the total number of cells. This does not hold true automatically for all AI stations. Many other factors, such as hygiene, extender, transport conditions and health status of boars, might affect the threshold for the number of sperm cells per AI dose (Waberski et al. 1994; Althouse et al. 2000; Thibier and Guerin 2000; Vyt et al. 2004). When a pig breeder is using a dose of boar sperm for AI, he expects the highest quality of semen to ensure maximum fertility.

4.2 Microscopic semen motility assessment

One of the important parts of semen quality evaluation in human and different animal species has always been microscopic semen motility assessment. Traditionally semen motility is assessed by a technician who scores by eye when looking through a phase contrast microscope. Usually there are two parameters for motility: quantity and quality of motile sperm cells. The cut-off values are set arbitrarily. Visual estimation of semen motility is a common laboratory test, but is the subject of discussion regarding the predictive value of fertility, and the subjectivity of the method. In one of our published studies (Broekhuijse et al. 2012a) an eight years' time data set of semen insemination doses from 110 186 ejaculates of 7429 boars were merged to fertility parameters of inseminations of 165 000 sows and these records were used for analysis. Only 6% of the total variation in fertility was due to boar- and semen-related parameters. Although semen motility is considered to be an important parameter to validate the quality of the ejaculate processed, microscopically assessed, it only minimally relates to fertility results under the current Dutch AI practice.

4.3 Computer-assisted semen analysis

To overcome this variability in technician-based semen motility assessment results, systems such as turbidimetry, laser-Doppler spectroscopy and photometric methods have been proposed (Verstegen et al. 2002). In the 1940s, scientists started recognising the need for objective data on the movement of sperm cells (Amann and Katz 2004). Dott and Foster (1979) first proposed computer-assisted semen analysis (CASA) to obtain an overall objective semen analysis. CASA is a computerised system which visualises and digitises the image of sperm cell movement. Using stroboscopy with a frequency of 60 Hz, pictures were taken which record the trajectory of each individual cell in the observed microscopic field. Information on the kinematics of individual cells and ejaculate summary statistics can be calculated from CASA data. The main advantage of CASA is that it allows an objective analysis of sperm motility in samples, and that identical parameters can be used in serial sample analysis. Moreover, the technique provides a continuous value of the exact number of sperm cells that were considered to be motile or immotile. There are several reports that claim CASA to be objective and more standardised when compared to sperm motility analysed microscopically by eye (Farrell et al. 1995; Krause 1995; Farrell et al. 1996; Abaigar et al. 1999; Hoflack et al. 2007; Broekhuijse et al. 2011, 2012a,b). In our study (Broekhuijse et al. 2012b) insemination records and semen parameters from a total of 45 532 ejaculates were collected over a three-year period. The boar- and semen-related variance (direct boar effect) was corrected for the effects of individual boar, genetic line of the boar, age of the boar, days between ejaculations, number of sperm cells in an ejaculate, number of sperm cells in an insemination dose and AI station. Boar- and semen-related parameters explained 5–6% of the variation in field fertility. Motility parameters, measured by CASA, explained 9–10% of the boar- and semen-related variation in field fertility. There were no significant differences between the effects of AI stations on fertility outcome, underscoring the objectivity of the CASA system used. Measuring motility parameters with CASA can be used to assess sperm motility in an objective manner and it enables one to discriminate the fertilising capacity of ejaculates based on the motility pattern depending on genetic line of the boar in AI stations.

4.4 Analysis of other sperm cell properties

There are many other ways for objective assessments of sperm samples, such as flow cytometry. Sperm cells are fluorescently labelled for certain cell characteristics. The principle of this technique is that a sperm cell suspension is introduced in fluid stream through a laser excitation source. Cells can be fluorescently labelled for certain cell characteristics. The cell size morphology is determined by scattered excitation light detection while the fluorescent properties per cell can be determined by fluorescence detectors. This technique allows rapid analysis of multiparametric sperm cell properties (for a review of flow cytometry, see Nunez 2001). Flow cytometry is used not only for semen assessment but also routinely in the diagnosis of any given cell suspension, for instance, in human health for assessing blood cancer. The first flow cytometer was developed in 1953 by Wallace H. Coulter. The first fluorescent flow cytometer was developed in 1968 by Wolfgang Gohde and commercialised in 1969. The strength of the flow cytometry technique compared to microscopic sperm assessments is that measurements on large numbers of sperm cells can be made within a short time frame. Different subsets of cells can be identified and quantified and sperm integrity parameters can be measured. Semen motility only quantifies the movement of the cell, but does not give an indication about any other sperm cell characteristics. Other sperm quality characteristics such as DNA fragmentation and membrane or acrosome integrity could be assessed by flow cytometry and related to field fertility results. Different parameters are assessed for thousands of sperm cells per second. Initially flow cytometry technology was considered to be complicated, not suitable for laboratory routine and more research based. As a result, flow cytometry is not often used as the routine semen quality assessment method for pig AI. The system is relatively expensive and trained technicians are necessary to operate the system. But more and more routine AI laboratories have implemented flow cytometry as a valuable tool to use in the daily routine of AI laboratories, where many ejaculates have to be processed on one production day. Flow cytometry allows assessment of different semen quality characteristics.

To use flow cytometry as part of a daily AI laboratory routine, it is important to see if sperm integrity assessments using flow cytometry correlate with field fertility data. For this purpose 20 boars were followed for a 20-week period. From these ejaculates sperm cell integrity was assessed with respect to DNA and to membrane integrity, acrosome intactness, acrosome responsiveness and mitochondrial potential using established flow cytometric assays. This was done on freshly produced semen and on semen stored for up to 15 days (Broekhuijse et al. 2012c). Remarkably, none of the individual membrane integrity parameters were significantly related to fertility results. In contrast, the amount of DNA damage as assessed after 14 days of semen storage significantly related to field fertility. Thus, the detection of the degree of DNA damage in stored boar semen samples can be used to predict the fertilising capacity of boar ejaculates.

Damage in the DNA is considered to affect the first cleavage divisions and therefore reduce embryo development (Larson-Cook et al. 2003; Fatehi et al. 2006). It has also been shown that sperm cells with compromised chromatin organisation show a reduced capability to bind to oviductal epithelium (Ardon et al. 2008) while spermatozoa that bind to oviduct cells have superior DNA integrity. This suggests that females can select DNA intact sperm to fertilise the oocyte (Ellington et al. 1999). This selection makes sense as it has been demonstrated that oocytes that are fertilised with sperm that carried damaged DNA do not develop (Larson-Cook et al. 2003; Fatehi et al. 2006) and thus affect TNB and possibly even FR. In general, a relatively low level of DFI has been reported in both fresh (Evenson et al. 1994; Rybar et al. 2004) and liquid stored (Waberski et al. 2002; Boe-Hansen et al. 2005; De Ambrogi et al. 2006; Waberski et al. 2011) boar semen. DNA fragmentation of the boars used in the current study was below 10% which is representing high fertility potential. This is demonstrated in other studies (Evenson et al. 1999; Waberski et al. 2011). The literature suggests cut-off levels for boar sperm cells between 2% and 18% for DFI (Rybar et al. 2004; Boe-Hansen et al. 2008; Didion et al. 2009; Waberski et al. 2011).

Fertility can be categorised as compensable or uncompensable (Saacke et al. 2000; DeJarnette 2005). Defective chromatin structure (increased DFI) is an uncompensable trait that affects the function of sperm cells during later stages of fertilisation and in embryonic development (Saacke et al. 2000). The failing of membrane integrity parameters to relate to fertility is therefore categorised as a compensable trait. Defects in compensable traits can be overcome by a large number of sperm cells per insemination. Apparently, the number of motile cells per insemination dose is still compensating for deficiencies in membrane, acrosome or mitochondria.

4.5 Use of genomics

Thousands of sows are inseminated with semen from a given boar. The economic impact of the boar fertility is obvious. The pig must adapt to the local environment, in terms of climate, housing system, health and availability of nutrients (Bouwman et al. 2010; Camerlink et al. 2014; Mathur et al. 2014). A limited number of purebred animals of sire and dam lines in nucleus herds are the basis for cross-breeding for the production level. Information from genetic and genomic selection gives us knowledge on how to decrease the distance between nucleus herds and production level. The available genetic variation is large, but with more efficient AI dose production, this variation will be reduced. Breeding goals have evolved from heritable traits, such as growth and feed efficiency, to sustainability-related traits, such as litter size, mothering ability and piglet vitality. Breeding is becoming more and more technologically intensive. Sequencing the porcine genome and genomic selection is a good example. With these new tools, increasing AI research efforts would boost management of genetic diversity and evolution. Traditional boar breeding programmes have based their selection mainly on production traits, neglecting the potential benefits of selection for improved boar semen traits on fertility. Commercially important traits, such as growth rate, feed efficiency and lean yield, should not be the unique criteria for selection, because this approach can result in reduced semen production due to the negative correlations between semen production and semen traits (Oh et al. 2003). As mentioned before, sperm motility is one of the most widely used parameters to evaluate boar semen quality. However, this trait can only be measured after puberty, which prolongs the generation interval. Therefore, the use of genomic information is as an alternative to evaluate and improve selection for boar fertility traits earlier in life. Genome-wide association studies (GWAS) can overcome these limitations using genotyping technologies to assay the single-nucleotide polymorphisms (SNPs) and relate them to traits of interest (Pearson and Manolio 2008). GWAS have been performed in pigs for many traits. GWAS applied to boar sperm motility have not reported so far, though such studies are very relevant for both pig breeding and reproduction. In our own study (Diniz et al. 2014); we identified six SNPs associated with sperm motility in a large white population. This is the first time that a QTL region for this trait was found between 117.26 and 119.56 Mb on SSC1. Combining the information on the QTL and the candidate gene identified in our study provides useful information that could be used for marker-assisted selection or genomic selection applications in commercial pig populations.

5 Conclusion

The main goal for both breeding and AI companies is efficient dissemination of genes of high-indexed boars. The effect of these high-indexed boars in the cost efficiency of meat production is evident. Results show that the effect of semen quality characteristics on the variation in fertility of breeding boars is relatively small. Despite the fact that these effects seem small, the economic value in a high-producing pig industry is large, with the objective to minimise the variation between farms. Hence the more sows bred per boar, the higher the impact on production costs. Optimisation of the fertilising capacity of boars in this context is critical. At the level of finisher production it allows increased production efficiency, feed efficiency and uniformity. This results in reducing the genetic lag between the top and the bottom of the breeding pyramid.

Since fertility is affected by several factors, achieving high fertility is not simple. Predominantly, AI is used for breeding pigs. The advantage of AI is that you can dilute semen from high fertile breeding boars and in this way inseminate many sows. In the last few years, the pig industry has increased in scale and specialization. Identifying and excluding terminal sires with relatively low fertility becomes more and more critical in the economic context. With knowledge on predicting porcine male fertility, the AI companies can make a step forward in efficient and guaranteed high semen quality production of insemination doses. Collecting field fertility data and merging these with ejaculate records is a very strong tool for both AI and breeding companies and can be used to identify sub-fertile boars. The data analyses performed are based on actual data, representing the relation between semen quality characteristics and field fertility. The AI company can use the results as a tool for efficient production and processing of insemination doses.

Results show a clear relation with fertility, and with standardised objective methods such as CASA even more than with microscopic methods. Other optional tests (flow cytometry) are suggested. DNA fragmentation can be used as a predictive test; membrane integrity tests are also relevant in boar populations where less indirect selection by boar fertility has been achieved. Unfortunately, there is no golden standard concerning the requirements for qualitative good semen. Finding a minimum set of tests with maximal functional coverage is an on-going process, along with the understanding of the quality characteristics essential in sperm cells for optimal fertilisation. The value of accurate and reproducible semen analysis is clear; this knowledge can be extremely helpful in improving semem quality to improve fertility in the field.

For the customers of AI companies, it is important to achieve the highest reproductive efficiency per purchased AI dose. Reproductive performance has a genetic basis; however, it is significantly affected by environmental factors. For optimal genetic expression, the environmental effects should be reduced; one of the variables is the semen quality. Hence, reducing the variation in fertility caused by variation in semen quality will enhance genetic progress.

One important technical aspect to observe when working in pig reproduction is to realise that there are parameters one has to take in for granted (just because there are, e.g. different boar lines, and there is an effect of parity of the sow), and there are other parameters that can be optimised to obtain an optimal fertility result. The genetic line of the boars explained the largest part of the boar- and semen-related variation in FR and TNB (Broekhuijse et al. 2012a,b). In general, the choice for a genetic line is based not on fertility results but on market requirements of the offspring. AI centres need to include breed differences into their decision-making processes to ensure adequate use of the (genetic line of the) boar. Of course, there are many more parameters affecting fertility, with the main parameter being the individual boar itself. The ability to identify and cull boars with reduced fertility is of considerable economic interest for AI centres. The effect of sub-fertile boars on field fertility is very large. Therefore, it is important to identify them as soon as possible, preferably before semen is sold, but also retrospectively based on fertility records.

AI is a key factor for pig production worldwide and genetic development, aiming at maximum boar exploitation and success rates. Fertilisation is a complex process involving a large number of events. Fertility research should be concerned not only with testing new assessment methods *in vitro*, but also with knowing the impact of genetic improvement and the value of predicting field fertility *in vivo*. There will always remain a variation in field fertility results, but the pig industry requires increased efficiency in piglet production to remain competitive. This means opportunities for improvement. Using the best boars means making use of the genetic progress made by genetic companies from their breeding structure. The pig industry will benefit faster and more broadly.

6 References

Althouse, G. C., Kuster, C. E., Clark, S. G. and Weisiger, R. M. 2000. Field investigations of bacterial contaminants and their effects on extended porcine semen. *Theriogenology* 53, 1167–76.

Althouse, G. C. and Lu, K. G. 2005. Bacteriospermia in extended porcine semen. *Theriogenology* 63(2): 573–84.

Ardon, F., Helms, D., Sahin, E., Bollwein, H., Topfer-Petersen, E. and Waberski, D. 2008. Chromatin-unstable boar spermatozoa have little chance of reaching oocytes in vivo. *Reproduction* 135: 461–70.

Boe-Hansen, G. B., Christensen, P. and Ersboll, A. K. 2005. Increasing storage time of extended boar semen reduces sperm DNA integrity. *Theriogenology* 63: 2006–19.

Boe-Hansen, G. B., Christensen, P., Vibjerg, D., Nielsen, M. B. F. and Hedeboe, A. M. 2008. Sperm chromatin structure integrity in liquid stored boar semen and its relationships with field fertility. *Theriogenology* 69: 728–36.

Bouwman, A. C., Bergsma, R., Duijvesteijn, N. and Bijma, P. 2010. Maternal and social genetic effects on average daily gain of piglets from birth until weaning. *J. Anim. Sci.* 88: 2883–92.

Broekhuijse, M. L. W. J., Feitsma, H. and Gadella, B. M. 2011a. Field data analysis of boar semen quality. *Reprod. Domest. Anim.* 46: 59–63.

Broekhuijse, M. L. W. J., Gaustad, A. H., Guillen, B. and Knol, E. F. 2015. Efficient boar semen production and genetic contribution: the impact of low-dose artificial insemination on fertility. *Reprod. Dom. Anim.* 50: 103–9.

Broekhuijse, M. L. W. J., Sostaric, E., Feitsma, H. and Gadella, B. M. 2012a. Application of computer assisted semen analysis to explain variations in pig fertility. *J. Anim. Sci.* 90: 779–89.

Broekhuijse, M. L. W. J., Sostaric, E., Feitsma, H. and Gadella, B. M. 2012b. Relation of flow cytometric sperm integrity assessments with boar fertility performance under optimized field conditions. *J. Anim. Sci.*, Epub ahead of print 24 July 2012.

Camerlink, I., Bolhuis, J. E., Duijvesteijn, N., Van Arendonk, J. A. M. and Bijma, P. 2014. Growth performance and carcass traits in pigs selected for indirect genetic effects on growth rate in two environments. *J. Anim. Sci.* 92: 2612–19.

De Ambrogi, M., Ballester, J., Saravia, F., Caballero, I., Johannison, A., Wallgren, M., Andersson, M. and Rodriguez-Martinez, H. 2006. Effect of storage in short- and long-term commercial semen extenders on the motility, plasma membrane and chromatin integrity of boar spermatozoa. *Int. J. Androl.* 29: 543–52.

DeJarnette, J. M. 2005. The effect of semen quality on reproductive efficiency. *Vet. Clin. N. Am.* 21(2): 409–18.

Didion, B. A., Kasperson, K. M., Wixon, R. L. and Evenson, D. P. 2009. Boar fertility and sperm chromatin structure status: a retrospective report. *J. Androl.* 30: 655–60.

Diniz, D. B., Lopes, M. S., Broekhuijse, M. L. W. J., Lopes, P. S., Harlizius, B., Guimaraesa, S. E. F., Duijvesteijn, N., Knol, E. F. and Silva, F. F. 2014. A genome-wide association study reveals a novel candidate gene for sperm motility in pigs. *Anim. Reprod. Sci.* 151: 201–7.

Donadeu, M. 2004. Advances in male swine artificial insemination (AI) techniques. *Pig. J.* 54: 110–22.

Ellington, J. E., Evenson, D. P., Wright Jr., R. W., Jones, A. E., Schneider, C. S., Hiss, G. A. and Brisbois, R. S. 1999. Higher quality human sperm in a sample selectively attach to oviduct (fallopian tube) epithelial cells in vitro. *Fertil. Steril.* 71: 924–9.

Evenson, D. P., Thompson, L. and Jost, L. 1994. Flow cytometric evaluation of boar semen by the sperm chromatin structure assay as related to cryopreservation and fertility. *Theriogenology* 41: 637–51.

Evenson, D. P. 1999. Loss of livestock breeding efficiency due to uncompensable sperm nuclear defects. *Reprod. Fertil. Dev.* 11: 1–15.

Fatehi, A. N., Bevers, M. M., Schoevers, E., Roelen, B. A., Colenbrander, B. and Gadella, B. M. 2006. DNA damage in bovine sperm does not block fertilization and early embryonic development but induces apoptosis after the first cleavages. *J. Androl.* 27: 176–88.

Feitsma, H. 2009. Artificial insemination in pigs, research and developments in the Netherlands: a review. *Acta Sci. Vet.* 37(1): 61–71.

Flowers, W. L. 1997. Management of boars for efficient semen production. *J. Reprod. Fertil.* 52: 67–78.

Foxcroft, G. R., Dyck, M. K., Ruiz-Sanchez, A., Novak, S. and Dixon, W. T. 2008. Identifying useable semen. *Theriogenology* 70: 1324–36.

Hancock, J. L. and Hovel, G. L. R. 1959. The collection of boar semen. *Vet. Rec.* 71: 664–5.

Heape, W. 1897. The artificial insemination of mammals and subsequent possible fertilization or impregnation of their ova. *Proc. R Soc. Lond. B.* 61: 52–63.

Ito, T., Niwa, T. and Kudo, A. 1948. Studies on artificial insemination in swine. *Zootech Exp. Sta Res. Bul.* 55: 1–74.

Ivanov, E. I. 1907. De la fecondation artificielle chez les mammiferes. *Arch. Sci. Biol.* 12: 377–511.

Ivanov, E. I. 1922. On the use of artificial insemination for zootechnical purposes in Russia. *J. Agric. Sci.* 12: 244–56.

Larson-Cook, K. L., Brannian, J. D., Hansen, K. A., Kasperson, K. M., Aamold, E. T. and Evenson, D. P. 2003. Relationship between the outcomes of assisted reproductive techniques and sperm DNA fragmentation as measured by the sperm chromatin structure assay. *Fertil. Steril.* 80: 895–902.

Leeuwenhoek, A. 1678. De natis e semine genitali animalculis. *R Soc. (Lond) Philos. Trans.* 12: 1040–3.

Lunenfeld, B. and Insler, V. 1993: Infertility: the dimension of the problem. In V. Insler B. Lunenfeld (eds), *Infertility: Male and Female*. Churchill Livingstone, Edinburgh, pp. 3–7.

Mäkinen, A., Andersson, M., Häkkinen, A. and Kuosmanen, S. 1999. A reciprocal translocation between autosomes 8 and 10 in a boar used for artificial insemination service and its effects on litter size. *Anim. Reprod. Sci.* 56: 237–43.

Mathur, P. K., Herrero-Medrano, J. M., Alexandri, P., Knol, E. F., ten Napel, J., Rashidi, H. and Mulder, H. A. 2014. Estimating challenge load due to disease outbreaks and other challenges using reproduction records of sows. *J. Anim. Sci.* 92: 5374–81.

McKenzie, F. F. 1931. A method for collection of boar semen. *J. Am. Vet. Med. Assoc.* 78: 244–6.

McLachlan, R. I. and O'Bryan, M. K. 2010. Clinical Review#: state of the art for genetic testing of infertile men. *J. Clin. Endocrinol. Metab.* 95: 1013–24.

Milovanov, V. K. 1964. Artificial insemination of livestock in the U.S.S.R. (Trans. by A. Birron and Z. S. Cole and S. Monson), Washington, DC: Jerusalem Technical Services, U.S. Department of Commerce.

Nishikawa, Y. 1962. *Fifty Years of Artificial Insemination of Farm Animals in Japan*. English Bull 2, Kyoto, Japan: Kyoto University.

Nishikawa, Y. 1964. History and development of artificial insemination in the world. In: Proceedings of the 5th International Congress on Animal Reproduction and Artificial Insemination, Trento, Italy, Vol. 7, 163–259, Veterinary Institute, Skopje.

Niwa, T. 1958. *Artificial Insemination with Swine in Japan*, Chiba-Shi, Japan: National Institute of Agricultural Science.

Nunez, R. 2001. Flow cytometry: principles and instrumentation. *Curr. Issues Mol. Biol.* 3(2): 39–45.

Oh, S. H., See, M. T., Long, T. E. and Galvin, J. M. 2003. Genetic correlations between boar semen traits. *J. Anim. Sci.* 81: 317.

Pearson, T. A. and Manolio, T. A. 2008. How to interpret a genome-wide association study. *JAMA* 299(11): 1335–44.

Philips, P. H. and Lardy, H. A. 1940. A yolk-buffer pabulum for the preservation of bull semen. *J. Dairy Sci.* 23: 399–404.

Polge, C. 1956. The development of artificial insemination service for pigs. *Anim. Breed. Abstr.* 24: 209–17.

Pursel, V. G. and Johnson, L. A. 1975. Freezing of boar spermatozoa: fertilizing capacity with concentrated semen and a new thawing procedure. *Anim. Sci.* 40(1): 99–102.

Roberts, S. J. 1986. Infertility in male animals (andrology). In S. J. Roberts (ed.), *Veterinary Obstetrics and Genital Diseases: Theriogenology*, 3rd edn. Edwards Brothers, North Pomfre, VT, pp. 752–896.

Robinson, J. A. B. and Buhr, M. M. 2005. Impact of genetic selection on management of boar replacement. *Theriogenology* 63: 668–78.

Rodriguez, A., Sanz, E., De Mercado, E., Gomez, E., Martın, M., Carrascosa, C., Gomez-Fidalgo, E., Villagomez, D. A. and Sanchez-Sanchez, R. 2010. Reproductive consequences of a reciprocal chromosomal translocation in two Duroc boars used to provide semen for artificial insemination. *Theriogenology* 74: 67–74.

Rybar, R., Faldikova, L., Faldyna, M., Machatkova, M. and Rubes, J. 2004. Bull and boar sperm DNA integrity evaluated by sperm chromatin structure assay in the Czech Republic. *Vet. Med.* 49: 1–8.

Saacke, R. G., Dalton, J. C., Nadir, S., Nebel, R. L. and Bame, J. H. 2000. Relationship of seminal traits and insemination time to fertilization rate and embryo quality. *Anim. Reprod. Sci.* 60–61: 663–77.

Siebenga, J. 1937. Kunstmatige inseminatie bij runderen. [Artificial insemination in cattle]. Universiteit Utrecht, Utrecht.

Spallanzani, L. 1784. Dissertations relative to the natural history of animals and vegetables. (Trans. By T. Beddoes) in *Dissertations Relative to the Natural History of Animals and Vegetables*. London: J. Murray, 195–9.

Strikwerda, R. 2007. Revolutie in het dierenrijk. [Revolution in the animal kingdom] Nationaal Veeteelt Museum, Beers.

Thibier, M. and Guerin, B. 2000. Hygienic aspects of storage and use of semen for artificial insemination. *Anim. Reprod. Sci.* 62: 233–51.

Vyt, P., Maes, D., Dejonckheere, E., Castryck, F. and Van Soom, A. 2004. Comparative study on five different commercial extenders for boar semen. *Reprod. Domest. Anim.* 39: 8–12.

Waberski, D., Helms, D., Bayerbach, M., Weitze, K. F., Bollwein, H., Bluemig, P., Wileke, H. and Acevedo, N. 2002. Sperm chromatin structure in boars used in artificial insemination. *Reprod. Domest. Anim.* 37(4): 257 (abstr).

Waberski, D., Meding, S., Dirksen, G., Weitze, K. F., Leiding, C. and Han, R. 1994. Fertility of long-term-stored boar semen: influence of extender (Androhep and Kiev), storage time and plasma droplets in the semen. *Anim. Reprod. Sci.* 36: 145–51.

Waberski, D., Schapmann, E., Henning, H., Riesenbeck, A. and Brandt, H. 2011. Sperm chromatin structural integrity in normospermic boars is not related to semen storage and fertility after routine AI. *Theriogenology* 75: 337–45.

Xu, X., Pommier, S., Arbov, T., Hutchings, B., Sotto, W. and Foxcroft, G. R. 1998. In vitro maturation and fertilization techniques for assessment of semen quality and boar fertility. *J. Anim. Sci.* 76: 3079–89.

Chapter 5

Genetic factors affecting feed efficiency, feeding behaviour and related traits in pigs

Duy Ngoc Do, McGill University, Canada; and Haja N. Kadarmideen, Technical University of Denmark, Denmark

1 Introduction

A global population already in excess of 7 billion people has led to increased demand for food and water, especially the demand for meat. While the need to increase animal production to meet global needs seems obvious, limited availability of resources is an important challenge in improving meat production. Moreover, the cost of feed used in animal production has also increased dramatically, which requires animal breeders to find alternatives to reduce feed consumption while maintaining animal productivity. One available alternative to alleviate both these problems is to emphasize feed efficiency in animal production. Genetic selection of feed-efficient animals via animal breeding approaches will also provide a sustainable way of improving production while reducing feed intake and hence putting less pressure on the limited amount of resources that are

http://dx.doi.org/10.19103/AS.2017.0013.06

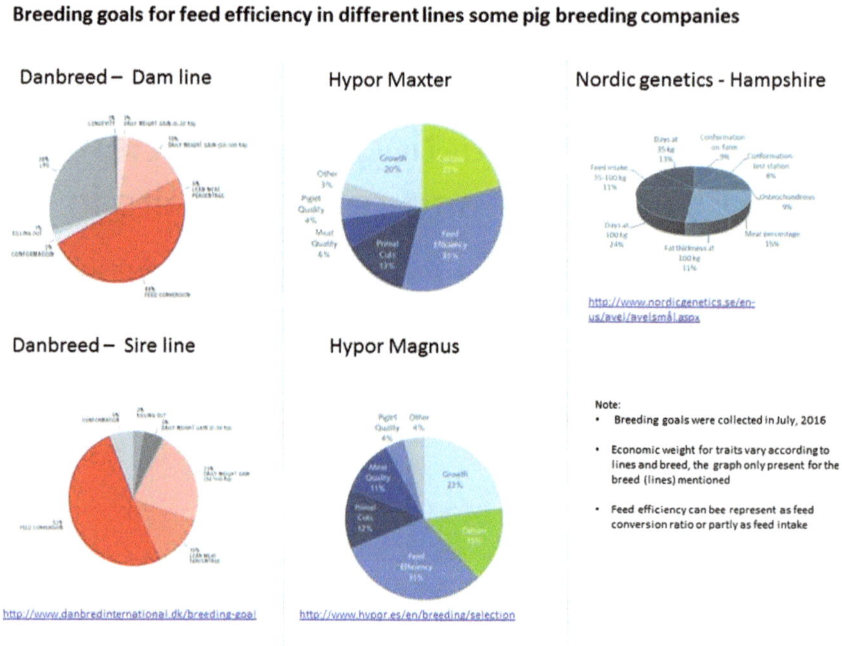

Figure 1 Example of breeding goals with feed efficiency traits in selected breeding companies.

required for animal production. Since around 34% of global meat consumption is pig meat, increasing pig production is very important to meet the demand for meat. The improvement of feed efficiency in pigs, therefore, is an important goal for livestock industries in many countries. Feed costs account for a significant proportion of the total costs incurred in livestock production, accounting for almost half of the total production cost depending on production systems and livestock species. Therefore, feed efficiency is an important objective in breeding programmes for all breeds. Despite the significance weightage given to feed efficiency in breeding programmes, it is a trait that is difficult to measure. While raw feed intake and utilization traits are easily recorded and are also heritable, selection for reduced feed intake (FI) is not optimal because of its strong negative correlation with production traits such as body weight (BW), composition and average daily gain (ADG). Therefore, alternative measures of feed efficiency such as food conversion ratio are required for use in breeding schemes. Breeding for feed conversion (also called food conversion ratio (FCR)) is routinely done in many pig-breeding companies and the economic weight for this trait accounts for a large proportion of total economic weight in their breeding programmes (Fig. 1)

2 Measures of feed efficiency

The ability to measure feed efficiency accurately is important for genetic improvement. Traditionally used FCR does not optimally measure feed efficiency, and hence, there is significant interest in developing alternative measures for feed efficiency. In growing animals, ratio or residual (regression) traits can be used to describe feed efficiency (Berry and Crowley, 2013). Currently, many breeding companies use FCR or feed:gain ratio as an indicator for feed efficiency. FCR is computed as the total amount of FI divided into the total amount of weight gain. However, FCR is not an ideal indicator (Gunsett, 1984) as a ratio trait for feed efficiency because it is not normally distributed and has no real mean and variance (Atchley and Anderson, 1978). Inclusion of ratio traits also causes problems when using linear selection indices (Werf, 2004) which are normally applied in the livestock industry. Since FCR is influenced by the variability in growth and maintenance requirements, it does not accurately measure efficiency of feed conversion (Aggrey and Rekaya, 2013). Moreover, it has been demonstrated that the genetic improvement of FCR observed in pigs is actually a function of genetic improvement of these constituent traits (Knap and Wang, 2012). In light of these deficiencies, Koch et al. (Koch et al., 1963) introduced the concept of residual feed intake (RFI) and defined it as the difference between an animal's actual FI and its predicted FI, the prediction usually being estimated on the basis of energy requirements for production, maintenance and body condition change. Therefore, a low RFI animal is more feed efficient than an animal with high RFI. Selection on RFI has been proposed to improve feed efficiency because of its phenotypic independence of growth and maintenance requirement. Besides RFI, Berry and Crowley (Berry and Crowley, 2012) proposed other indicators for feed efficiency including residual BW gain (the residuals from a multiple regression model regressing ADG on both dry matter intake (DMI) and BW). Some indirect measures of feed efficiency such as relative growth rate (growth relative to instantaneous body size) (Fitzhugh and St CS, 1971) and the Kleiber ratio (ADG per unit metabolic BW) (Kleiber, 1961) have also been proposed. Both these measures can be considered as feed efficiency indicators provided that all animals on test are fed the same restricted diet (Berry and Crowley, 2013). Therefore, these measures are rarely used in breeding programmes and the current breeding programme in livestock production mainly uses either FCR or RFI as an indicator for feed efficiency.

3 Residual feed intake

RFI can also be calculated on the basis of the regression at the phenotypic, genetic and nutritional level (Hoque and Suzuki, 2009). The common equation is used in cattle as

$$\text{RFI} = \text{FI (or DMI)} - \beta_1 \times \text{MBW} - \beta_2 \times \text{ADG},$$

where RFI is the residual feed intake, FI is the feed intake (also called DMI), MBW is the mid-test metabolic BW (MBW), ADG is the average daily gain and β_1 and β_2 are partial regression coefficients of animal feed intake on MBW and ADG. The equation can be simplified without correcting for carcass composition for growing bull because these animals have a low fat content (Jensen et al., 1992). The same equation has also been used in different studies of other species such as chicken (van Eerden et al., 2004), turkey (Willems et al., 2013), rainbow trout (Silverstein et al., 2005) and rabbit (Drouilhet et al., 2013). In growing pigs, different studies have investigated the efficacy of including MBW in the equation for calculating RFI. In one study, Haer (1992) demonstrated little variation in MBW and no significant effect of it on FI (perhaps because the pig was tested in fixed weight range (Hoque and Suzuki, 2009)), and therefore they suggested replacing it with the weight gain and lean in the equation. Using the random regression model, Strathe et al. (2014) have also demonstrated that the genetic variance of MBW is approximately 19% of genetic FI variance. Therefore, several studies also include the MBW in the calculation of RFI in the model in different pig breeds such as in Duroc (Hoque et al., 2009), Yorkshire (Gilbert et al., 2007; Cai et al., 2008), Piétrain (Saintilan et al., 2011) and Landrace (Hoque and Suzuki, 2008).

Although RFI is based on a simple concept, there are several different ways to calculate RFI in pigs as well as in different species. In a review, Hoque and Suzuki (2009) summarized models for calculation of RFI for beef cattle and recommended the equation: RFIphe = DFI – (β1× ADG + β2 × MWT) where RFIphe = phenotypic RFI, DFI = daily feed intake, MWT = MBW at mid-test, ADG = average daily gain and β1 and β2= partial regression coefficients of animal's DFI on MWT and DG, respectively. In cattle, bulls normally have a relatively low fat content; hence, RFI can be calculated without correction for carcass composition for growing bulls (Jensen et al., 1992). On the other hand, in the case of pigs, variation in maintenance requirements as predicted from MBW is not significantly related to variation in feed consumption, because pigs are usually tested over a relatively fixed weight range (Hoque and Suzuki, 2009). In agreement, we also tested a third RFI measure which, in addition to RFI2, also considered MWT, but results were not significantly different than for RFI2 (results not shown). However, it is important to note that the current RFI calculation methods face some limitations. For instance, the calculation methods are based on linear regression which may not account for all environmental factors affecting calculations of RFI. Moreover, RFI was calculated on the basis of summarized ADG, which cannot explain the possible variances of calculated RFI over time periods. It is also interesting to look at RFI as longitudinal data.

In the study of Do et al. (2013b), RFI was computed as the difference between the observed average daily feed intake (DFI) and the predicted DFI. Two models were used to define RFI. In the first model for RFI1, predicted DFI was estimated using linear regression of DFI on initial test weight (BWd) and ADG from 30 kg to 100 kg (ADG2). In the second model, BWd for RFI2, ADG2 and back fat (BF) were used as regressors:

$$DFI1 = \beta o + \beta_1(BWd) + \beta_2(ADG2) + e$$

$$DFI2 = \beta o + \beta_1(BWd) + \beta_2(ADG2) + \beta_3(BF) + e$$

where βo = the intercept, β_1 = the partial regression coefficient of DFI on BWd, β_2 = the partial regression coefficient of DFI on ADG2, β_3 = the partial regression coefficient of DFI on BF and *e* = the residual error term. RFI1 and RFI2 were calculated in each breed separately.

4 Genetics of residual feed intake (RFI) in pigs

4.1 Heritabilities estimates

Heritability estimates for RFI vary between studies depending upon breed, RFI measure used, population size and other environmental factors. Generally, heritability estimates for RFI are moderate and ranges from 0.10 (Johnson et al., 1999) to 0.47 (Hoque and Suzuki, 2008). The summary of heritabilities recorded for RFI in pigs is shown in Table 1.

In Denmark, for instance, approximately 5,000 boars are measured individually for FI every year at the DanAvl Danish test station (Bøgildgård). FI is measured in every pen

Table 1 Estimated heritability for RFI from the different studies

Trait	Heritability	Breeds[2]	Population size	Reference
RFI1	0.17	LW	26,706	Johnson et al. (1999)
RFI2	0.11	LW	26,706	Johnson et al. (1999)
RFI3	0.15	LW	26,706	Johnson et al. (1999)
RFI4	0.10	LW	26,706	Johnson et al. (1999)
RFI1	0.24	LW	2,884	Nguyen et al. (2005)
RFI2	0.22	LW	2,884	Nguyen et al. (2005)
RFI3	0.23	LW	2,884	Nguyen et al. (2005)
RFI	0.18	LW and LL	3,188	Von Felde et al. (1996)
RFI3	0.14	LW	793	Gilbert et al. (2007)
RFI2	0.24	LW	657	Gilbert et al. (2007)
RFI1	0.33	YY, LL and DD	7.562	Kennedy et al. (1993)
RFI2	0.30	YY, LL and DD		Kennedy et al. (1993)
RFI1	0.41	DD	541	Hoque et al. (2007)
RFI1	0.47	LL	341	Hoque and Suzuki (2008)
RFI2	0.29	LL	341	Hoque and Suzuki (2008)
RFI1	0.34	YY, LL and DD	16,872	Do et al. (2013-
RFI2	0.38	YY, LL and DD	16,872	Do et al. (2013)

[1] RFI1, RFI2, RFI3 and RFI4 regressed on the basis of ADG, ADG and BF, ADG, BF and MBW, ADG, BF, MBW and loin eye area, respectively. [2] LW = Large White, LL = landrace, YY = Yorkshire and DD = Duroc.

(around 10–12 pigs) using automatic feeding machines (ACEMO). According to the Danish Pig Research Centre, 2013, the improvement in feed conversion in the breeding system has remained at around 0.03 FUpig/kg growth (1 FU ~ 1g) per year. The improvement will reduce 0.4 Euro feed cost per finisher each year, thereby improving pig production environment.

Thus, the study used phenotypic records on a total of 16,872 boars from three breeds:- Duroc (DD) (7388), Landrace (LL) (4773) and Yorkshire (YY) (4751), and estimated genetic parameters (Do et al., 2013b). Data were recorded at a central test station (Bøgildgård) for a period of four years (2008–11) and were supplied by the Pig Research Centre, Danish Agriculture and Food Council. Pedigrees were created for these pigs. Average DFI was derived as the total amount of recorded FI divided by the number of corresponding days at the feeder. Average daily gain traits (ADG2) were calculated as linear regressions of BW from 30 kg to 100 kg on test days. FCR was calculated as the ratio of DFI relative to ADG2 on test. Backfat (BF) was recorded by using real-time ultrasound at the end of the fattening periods.

Estimated heritability for FCR was 0.30 ± 0.04 in DD and 0.32 ± 0.05 in both LL and YY. Estimated RFI1 and RFI2 were 0.34 ± 0.04 and 0.38 ± 0.04 in DD, 0.34 ± 0.05 and 0.36 ± 0.05 in LL, 0.39 ± 0.04 and 0.40 ± 0.04 in YY, respectively. Heritabilities of ADG2 and BF have been estimated in many studies and it is well known that ADG2 has moderate heritability, while BF has moderate to high heritability. The estimated heritabilities of ADG2 and BF were close to the estimates of 0.36 ± 0.07 for ADG2 and 0.54 ± 0.09 for real-time ultrasonic BF in American Duroc and Landrace pigs (Lo et al., 1992) or estimated values of 0.42 for ADG2 and 0.68 for BF in Yorkshire by Cai et al. (2008). Estimated heritabilities for feed efficiency were moderate in this study and generally in agreement with previous studies. The heritability estimate for FCR (0.30–0.32) was consistent with previous reports (Hoque et al., 2009 and Mrode and Kennedy, 1993), but lower than that of 0.42 as reported by Kadarmideen et al. (2004). The heritabilities of RFI1 (0.34–0.39) and RFI2 (0.36–0.40) were greater than that of FCR (0.30–0.32) in the study, in agreement with results in Duroc by Hoque et al. (2007) or in Yorkshire by Cai et al. (2008). RFI2 had greater heritability than RFI1, which is in agreement with a report by Johnson et al. (1999), but in contrast to the results of Hoque and Suzuki (2008). Heritability of RFI can be highly variable depending on the genetic and phenotypic parameters among its component traits (Kennedy et al., 1993). Heritability estimates for measures of RFI2 in DD were 0.38 and similar to RFI2 in Japanese Duroc (0.22–0.38; Hoque et al., 2009). In LL, the estimated heritability for RFI2 was greater than in other studies (Hoque and Suzuki, 2008; Kennedy et al., 1993; VonFelde et al., 1996). The estimated heritability in YY for RFI2 was also greater than that reported by Nguyen et al. (2005) and Mrode and Kennedy (1993).

4.2 Genetic correlation of RFI with feed intake

To include any new trait in a selection index, estimating genetic and economic parameters associated with it is a prerequisite, for example, trait heritability, genetic correlations with other economically important traits, the economic value of the trait (economic weight) and response to current selection methods. Hence, genetic correlation estimates are needed. Generally, high and positive genetic correlations were found between different indicators for feed efficiency. Hoque et al. (2007) reported that genetic correlation among phenotypic, genetic and nutritional RFI measures and FCR ranged from 0.67 to 0.84. Similarly, high and positive genetic correlations have also been found between RFI and FI in multiple different studies, for example, 0.78 and 0.81 in DD and LL pigs, respectively (Hoque and Suzuki, 2008). Since RFI is phenotypically computed from growth and performance traits, it has no phenotypic correlation with these predictors. However, it is possible to estimate genetic correlations, and the estimated genetic correlation between RFI and ADG ranges from zero to low values depending on population size and statistical methods. The estimates of genetic correlation between RFI with ADG and BF from the literature are shown in Table 2. Johnson et al. (Johnson et al., 1999) reported genetic correlations between ADG and RFI (adjusted for ADG and BF) were 0.17. Kennedy et al. (1993) reported that the correlation was 0.21, while Hoque and Suzuki (2008) reported that the correlation was close to zero (0.02). Similarly, RFI also had zero to low genetic correlation with BF (if RFI was adjusted for BF).

The genetic correlations estimated between RFI and DFI or FCR were favourable. The genetic correlations estimated between RFI and ADG (or BF) were low or close to zero depending on breeds, and these results are in agreement with the results reported from the previous studies (Kennedy et al., 1993; Johnson et al.; 1999; Nguyen et al., 2005). As mentioned previously, heritability and genetic correlation estimates in different studies are obtained using different regression models and resource populations. Hence, the differences in estimated genetic parameters are to be expected. In addition, we also found that both RFI1 and RFI2 had favourable genetic trends (EBV reducing over the year) in current breeding programmes in all Danish pig breeds (Do et al., 2013b), and these results are also

Table 2 Genetic correlation between the RFI measures and ADG and BF from the different studies

Trait[1]	ADG(± SE)	BF(± SE)	Breeds[2]	Reference
RFI1	0.22 ± 0.05	0.77 ± 0.04	DD	Hoque et al. (2009)
RFI1	0.03 ± 0.09	0.46 ± 0.15	LL	Hoque and Suzuki (2008)
RFI2	0.16 ± 0.07	0.11 ± 0.07	DD	Hoque et al. (2009)
RFI2	0.02 ± 0.09	0.06 ± 0.14	LL	Hoque and Suzuki (2008)
RFI1	0.11 ± NA	0.67 ± NA	LW	Johnson et al. (1999)
RFI2	0.17 ± NA	0.22 ± NA	LW	Johnson et al. (1999)
RFI1	0.18 ± 0.10	0.34 ± 0.07	YY, LY and DD	Mrode and Kennedy (1993)
RFI2	0.21 ± 0.11	0.15 ± 0.09	YY, LL and DD	Mrode and Kennedy (1993)
RFI1	0.23 ± 0.12	0.76 ± 0.08	DD	Hoque et al. (2007)
RFI2	−0.35 ± 0.13	−0.10 ± 0.11	YY	Gilbert et al. (2007)
RFI3*	0.00 ± 0.12	0.33 ± 0.16	YY	Gilbert et al. (2007)
RFI3	0.18 ± NA	0.20 ± NA	LW	Johnson et al. (1999)
RFI2	0.41 ± 10	0.11 ± 0.09	LW and LL	Von Felde et al. (1996)
RFI2	−0.05 ± 0.07	0.22 ± 0.04	DD	Jiao et al. (2014a)
RFI1	0.16 ± 0.09	–0.07 ± 0.08	DD	Do et al. (2013b)
RFI2	0.04 ± 0.09	–0.19 ± 0.08	DD	Do et al. (2013b)
RFI1	–0.04 ± 0.03	0.17 ± 0.09	LL	Do et al. (2013b)
RFI2	0.02 ± 0.10	–0.08 ± 0.11	LL	Do et al. (2013b)
RFI1	0.25 ± 0.09	0.43 ± 0.12	YY	Do et al. (2013b)
RFI2	0.20 ± 0.09	–0.01 ± 0.10	YY	Do et al. (2013b)

[1] RFI1, RFI2 and RFI3 regressed on the basis of ADG, ADG and BF and ADG, BF and MBW, respectively. [2]: LW = Large White. * The model also used lean meat percentage to compute RFI.

in agreement with the genetic trends for RFI in Japanese DD pigs (Hoque and Suzuki, 2008). Overall, these results indicate that RFI is heritable and has favourable genetic correlation with other economically important traits in Danish pig breeds, which together support its inclusion in selection indices and in current multi-trait breeding goals in Danish pigs.

4.3 Quantitative trait loci for feed efficiency in pigs

Quantitative trait loci (QTL) mapping is the first step to detect chromosomal regions affecting complex traits. QTL detection experiments in livestock started in the 1990s, when Andersson et al. (1994) detected a QTL for fatness on pig chromosome (SSC) 4. Many QTLs were detected initially by using genetic linkage maps in either cross breeds for highly divergent traits of interest or commercial populations where half-sib families were available. During this period, QTL experiments were based on resource populations with a few hundred animals, but over time resource population size has increased to thousands of animals coupled with an increasingly large number of markers. As a consequence, the

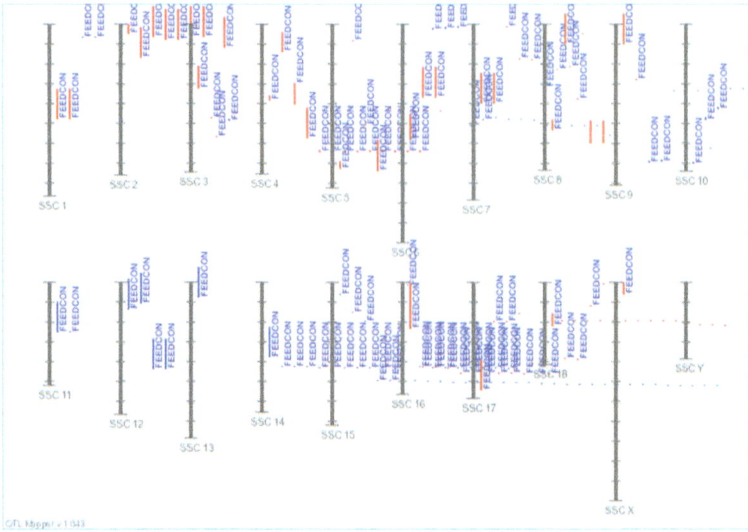

Figure 2 QTL for FCR from Pig QTLdb (April 2016).

number of detected QTLs has also increased rapidly, and around 15,108 pig QTLs are curated into the Pig QTLdb from over 537 publications representing around 660 different traits (http://www.animalgenome.org/cgi-bin/QTLdb/SS/index (updated on 29 April 2016)). However, a majority of these QTLs influence production and meat quality traits that are easy to measure, for example, ADG, BF and loin muscle area. Although feed efficiency is an important trait for pig production, relatively few QTLs have been detected for this phenotype (Houston et al., 2005; Liu et al., 2007; Duthie et al., 2008; Zhang et al., 2009; Gilbert et al., 2010) compared to other production traits. The majority of identified QTLs are for FCR, except for one study in which Gilbert et al. (2010) detected a QTL on SSC5 and 9 for RFI in growing pigs in a Piétrain–Large White backcross. Another candidate gene study by Fan et al. (Fan et al., 2010) indicated that single-nucleotide polymorphisms (SNPs) in the *FTO* and *TCF7L2* gene were significantly associated with RFI. However, most of these QTLs have been detected by using around 100–200 microsatellites (Ernst and Steibel, 2013) and have large confident intervals. Since the ultimate goal of identifying QTLs is to locate the gene responsible (causal mutation) for a particular phenotype, QTLs with confidence intervals that span a region of about 20 centimorgan limit the ability to identify the casual genes. Different approaches like fine mapping and increasing the detection power by increasing the size of the resource population, numbers of markers and meta-analysis have achieved limited success thus far.

5 Genome wide association studies (GWAS) of feed efficiency in pigs

Genome-wide association studies (GWAS) are widely used to identify genetic variants associated with traits of interest via markers spread throughout the genome of many individuals. It has evolved over the last ten to fifteen years to become a powerful method to investigate the genetic architecture of both human diseases and complex traits in different species. Availability of the pig genome sequence (Groenen et al., 2012) and a commercial SNP chip (Ramos et al., 2009) has facilitated better understanding of genetic mechanisms underlying complex porcine traits; provided important insights into evolution, domestication and breed formation; and significantly improved selection methods for production and disease traits in the pig industry. With the availability of PorcineSNP60 BeadChip, several GWAS have been performed to scan SNP associated with both production and disease traits in crossbred and commercial pig populations. Several GWAS have been performed in pigs for feed efficiency so far. Sahana et al. (2013) used a resource population of 3,071 Duroc pigs to detect 79 significant ($P < 0.0001$) SNPs for FCR on the basis of a linear mixed model (LMM) approach and 44 significant SNPs (Bayes factor ≥ 10) on the basis of the Bayesian variable approach. The authors suggested Hypoxia inducible factor 1, alpha subunit inhibitor (*HIF1AN*) and ladybird homeobox 1 (*LBX1*) as putative positional candidate genes affecting FCR on the basis of their location near significantly associated SNPs. Onteru et al. (2013) used 1,400 pigs from divergently selected RFI lines to identify several QTLs on SSC 7 and 14 for RFI in pigs using both Bayesian approaches and frequentist approaches. Jiao et al. (2014b) detected a significant QTL for FCR on SSC 4 that explains 2.65% of the genetic variance for the trait, but did not find any significant QTLs for RFI in the same study. Sanchez et al. (2014) also detected one QTL for RFI (SSC 6) and three QTL for FCR (SSC 3, 11 and 18) in a recent study.

Two independent GWAS were conducted in our populations because knowledge of the genomic regions and mechanisms affecting RFI in pigs is lacking. In Duroc population (Do et al., 2014b), phenotypic and genotypic records (using the Illumina Porcine SNP60 BeadChip) were available on 1272 boars. Deregressed estimated breeding values were used as response variables in association analyses. Fifteen and twelve loci were significantly associated ($p < 1.52 \times 10^{-6}$) with RFI1 and RFI2, respectively. Among them, 10 SNPs were significantly associated with both RFI1 and RFI2, implying the existence of a common mechanism controlling the two RFI measures. Four and three linkage disequilibrium blocks were found on the two most interesting chromosome regions: 30.5–31.5 Mb on pig chromosome (SSC) 1 and 120.5–121.5 Mb on SSC 9 for both RFI. The SNPs within *MAP3K5* and *PEX7* on SSC 1, *ENSSSCG00000022338* on SSC 9 and *DSCAM* on SSC 13 might be interesting markers for both RFI measures. Functional annotation of nearby genes to significant SNPs indicated regulation of protein, and lipid metabolic process, gap junction, inositol phosphate metabolism and insulin signalling pathways are significant biological processes and pathways for RFI, respectively. Meanwhile, in Yorkshire population (Do et al., 2014c), a total of 596 Yorkshire boars with phenotypes for two different measures of RFI (RFI1 and 2) and 60k genotypic data were used for GWAS and 12 and 7 SNPs were found to be significantly associated with RFI1 and RFI2, respectively. While these analyses identified a number of significant SNPs and positional candidate genes, the effect sizes were relatively small (e.g. highest phenotypic variance explained by a single marker were only 0.21% of phenotypic variances of RFI in Duroc pigs (Do et al., 2014b)). The genetic architecture (numbers of top SNPs and their genomic positions) were very similar between two different measures of feed efficiency in both breeds. For instance, approximately 80% of suggestive SNPs (124 SNPs) were also found to be associated with both traits in Duroc pigs (Do et al., 2014b) or two significant SNPs found common in Yorkshire pigs (Do et al., 2014c). The GWA analysis identified QTL regions spanning 30.5–31.5 Mb on SSC 1 and 120.5–121.5 Mb on SSC 9 as the most interesting regions for RFI in Duroc pigs, and post-GWAS analyses indicated several positional candidate genes (*MAP3K5, PEX7, ENSSSCG00000022338* and *DSCAM*) (mitogen-activated protein kinase kinase kinase 5) that could be interesting for further investigation in Duroc pigs. Meanwhile, we identified several QTL regions on SSC 1, 3, 7, 8, 9, 10, 14 and 15 that were associated with either RFI1 or RFI2 in Yorkshire pigs. We used two different significance thresholds to determine genome-wide significance level to account for the fact that Duroc resource population was almost twice in size compared to the Yorkshire resource. Two overlapping QTLs located at approximately 87 Mb on SSC 8 and at approximately 136 Mb on SSC 15 were also identified in these studies. The genomic insights on possible function of positional candidate genes and how they link to RFI have been discussed (Do et al., 2014b, Do et al., 2014c). The overlap between our finding and other RFI studies have also been indicated in these publications, and are therefore, not discussed further.

Besides positional candidate genes, Do et al. (2014b,c) also described several biological pathways that might regulate RFI. These pathways (e.g. insulin regulation-related pathways and cellular communication pathways) were identified via enrichment analyses in both breeds. While identification of common genes and pathways underlying a particular phenotype is indicative of existence of common biological mechanisms, these mechanisms can be regulated by different set of genes in specific breeds. As mentioned in Do et al. (2014c), determining candidate gene individuals in different breeds is particularly important in the context of developing marker-assisted and genomic selection strategies, because different lines/breeds of livestock may require different sets of markers for genetic improvement. Taken together, these studies are indicative of the fact that RFI is highly polygenic and the application of marker- or QTL-assisted selection will be of limited use because marker/QTL are not able to capture enough phenotypic variances of the traits.

Improved annotation of the pig genome allowed us to allocate SNPs to different functional groups. Since some studies have demonstrated that different genomic regions can contribute differently to additive genomic variances, and consequently influence prediction accuracy. In the study (Do et al., 2015), we addressed the question if different (14) genomic regions contribute differently to total genomic variances. The contribution of different annotated classes was linearly associated with the number of SNPs in each class for RFI as well as for its component traits. The majority of the genomic variance (61.4%) was explained by intergenic SNPs located outside of genic regions regardless of phenotypes used.

Given the linear association between the number of SNPs and variance contribution, the average contribution per SNP was very similar among groups. These results indicate that structural variants are less useful in variance partitioning for genomic prediction, at least for RFI and component traits in the current dataset. The predictive ability was also very similar among different genomic regions. Other studies have demonstrated different results in other species. Yang et al. (2011) demonstrated that SNPs in or near genes explain more variance than SNPs between genes for complex traits such as height and body mass index. Kindt et al. (2013) suggested that SNPs that were further away from the transcription start site were less likely to be significantly associated with trait. However, as mentioned in Do et al. (2015), results from these studies depend on quality of genome annotation, design of SNP chips (mostly based on common variants), LD among SNPs, etc.

6 Component traits of residual feed intake and genetic architecture

Understanding the genetics underlying component traits of RFI helps in the prioritizing of candidate SNP/genes for further investigation. It also helps assure that marker-assisted selection based on candidate genes for RFI identified in this population does not influence daily gain of pigs. In this respect, an important finding was the QTL spanning 64–65 Mb region on SSC 1 which contained several important SNPs for DFI (Do et al., 2014b). This region contained gamma-aminobutyric acid receptor subunit rho-2 (GABRR2) gene that encodes an important inhibitory neurotransmitter in the vertebrate central nervous system and plays key roles in feed/food intake. While no significant SNP was found for ADG is this projects, some genes (e.g. CBLN4, CDH2, NCOA5 and SLC35C2) were in the vicinity of SNPs with suggestive associations with it. We also detected an important QTL region for BF on 82–86 MB on SSC 2 which contains many genes that have been shown to have an important role in fat deposition and lipid metabolism such as HMGCR gene on SSC 2 regulating the synthesis of cholesterol in humans and other species.

BF thickness is one of the important traits that determine meat quality, which is highly relevant for both livestock producers and consumers. Therefore, identification of SNPs linked to BF thickness can provide an alternative tool to improve carcass and meat quality

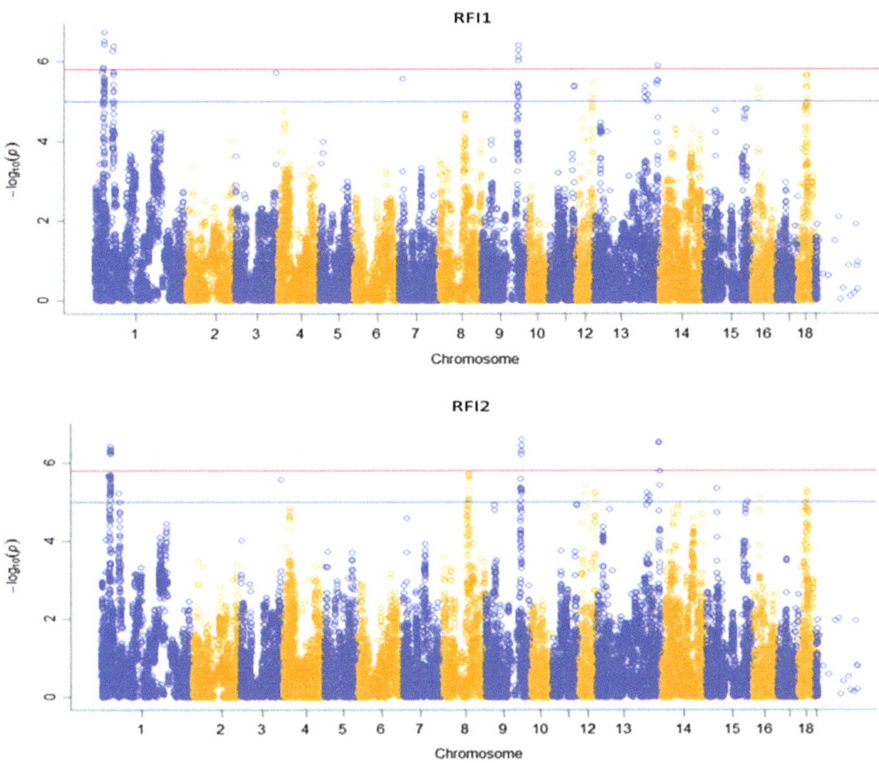

Figure 3 Manhatan plot for genome wide p values for different measures of RFI in Duroc pigs.

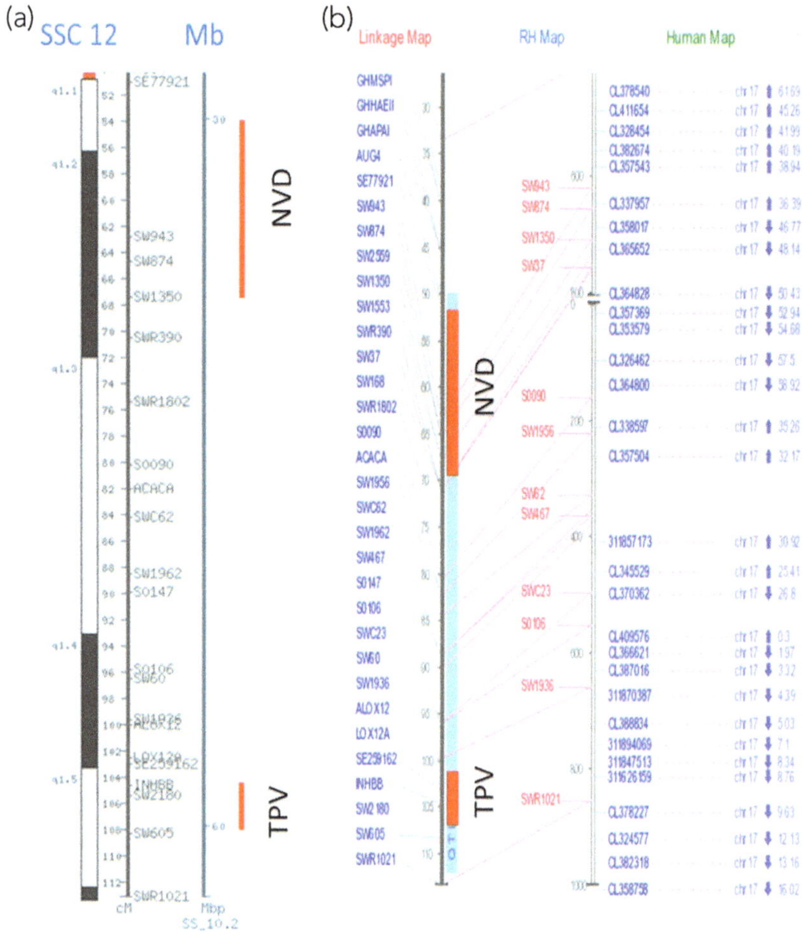

Figure 4 (a) Comparative mapping pig chromosome 12 and (b) human chromosome 17.

traits. We have conducted a GWAS for BF in a DD breed (Do et al., 2014b) and identified six important SNPs (ALGA0014061, ALGA0014028, ASGA0010625, ALGA0014098, DRGA0003063, MARC0074327) on a region of 82–86 Mb on SSC 2 for the trait. Identified positional candidate genes included regulator of G-protein signalling 14 (RGS14), collagen type IV alpha-3-binding protein (COL4A3BP), ankyrin repeat domain 31 (ANKRD31), glucosaminyl (*N*-acetyl) transferase 4, core 2 (GCNT4), family with sequence similarity 169, member A (FAM169A) and 3-hydroxy-3-methylglutaryl-CoA reductase (HMGCR) genes. Some of these genes have been reported important for BF in different species, for instance, HMGCR gene was important for both regulation of synthesis of cholesterol in humans and fat metabolism (http://www.genecards.org/cgi-bin/carddisp.pl?gene=HMGCR). Moreover, we also conducted GWAS to identify genomic regions controlling BF thickness in Yorkshire pigs. After quality control, a total of 37,192 SNPs and 596 pigs remained in the final dataset for GWA analyses. Deregressed EBVs were used as response variables in GWAS which

Table 3 List of SNPs affecting back fat, detected by both linear model and Bayesian methods

SNP	Chr	Position	P value	Bay.Fac	Regions
ASGA0013249	3	9410336	4.3E-05	6.86	intron
H3GA0008606	3	9491312	1.9E-04	5.54	upstream
ASGA0014717	3	58240635	3.6E-05	5.98	downstream
ALGA0019149	3	58257265	3.8E-05	6.42	intergenic
ASGA0014869	3	62638741	5.3E-04	5.10	intron
ASGA0015248	3	90696439	2.9E-05	6.86	intron
ALGA0113932	3	91605127	6.0E-05	7.75	intergenic
ALGA0101441	5	66015525	5.7E-05	5.10	intron
ALGA0111801	5	66156897	7.8E-06	11.86	intergenic
H3GA0016495	5	66327153	2.5E-04	6.86	upstream
DIAS0003118	10	19853519	9.6E-05	9.10	synonymous
ASGA0046875	10	20259974	4.5E-04	5.10	intergenic
CASI0009135	10	21907460	6.0E-04	5.54	upstream

were performed by a LMM using DMU package and a Bayesian variable selection (BVS) model using Bayz package. The results are plotted in the form of Manhattan plots in Fig. 3 for LMM and Fig. 4 for BVS methods. Nine and one SNPs were detected to be genome-wide significantly associated with the traits at the threshold of $p < 5e\text{-}05$ and Bayes factor (BayFac) >10 using LMM and BVS, respectively. Among the top 100 associated SNPs, only 25 SNPs were found in common between two methods (Table 3 shows selected SNPs). The mutation (A/G) at the locus ALGA0111801 located in the intergenic regions on chromosome 5 showed the strongest association with the trait by both methods ($p = 7.77e\text{-}06$ and BayFac = 11.86). Approximately 17kb apart from this locus, the lysophosphatidylcholine acyltransferase 3 (*LPCAT3*) gene involving phospholipid metabolisms can be an interesting candidate gene for fatness traits. Other regions on chromosome 5, 17 and 18 also harbour significant SNPs for the trait and will be further characterized.

7 Selection for feed efficiency

Genomic selection was first applied in Durocs in 2010 by the Danish Pig Research Centre. Subsequently in 2011, it was implemented in Danish Landrace and Large White pigs. Simulations have indicated that genomic prediction can increase progress in the range of 20–25% compared with the traditional breeding methods in LL and YY but around 10–15% for Duroc pigs (Danish Pig Research Centre, 2012). Currently, genomic predictions are performed in nucleus herds; however, the application of genomic prediction on commercial herds is being evaluated. The accuracy of genomic prediction for some traits has also been reported, and prediction accuracy was around 0.226 and 0.352, respectively, for FCR and ADG for using adjusted single step in genotyped Duroc pigs(Christensen et al., 2012). The reliabilities of genomic prediction for two litter size traits (total number of piglets born and litter size at day 5 after birth) were 0.16 and 0.06, respectively, in Landrace pigs using single-step methods (Guo et al., 2014). Genomic prediction has also been applied for meat quality traits such as boar taint compounds and also extends to other slaughter traits. Currently, genomic selection has been routinely used on the basis of single-step GBLUP methods to evaluate breeding values for production traits (ADG and FCR), lean meat percentage, sow longevity, litter size and piglet mortality in three Danish pig breeds.

Genomic predictions are currently used for genetic improvement in all Danish pig breeds and FCR is used in the selection index as an indicator of feed efficiency. The correlation between corrected phenotypes and genomic estimated breeding values using the single-step method is reported to be around 0.16 for Duroc pigs (Christensen et al., 2012). The accuracy (correlation of corrected phenotypes and genomic breeding values divided for square root heritability) of genomic predictions for RFI was relatively higher (0.56) (Do et al., 2014a) than that of FCR. However, the selection index is a linear combination of traits; therefore, using the FCR is not suitable as mentioned above. Logically, DFI can be used as an equivalent to RFI in the selection index by keeping genetic changes of the component traits as zero. However, based on genomic prediction studies in cattle, Pryce et al. (Pryce et al., 2014) suggested to choose RFI in order to account for body tissue mobilization, if genetic correlations of RFI with other traits are known. If genetic correlations are not known, the authors indicate that using DMI in breeding schemes will minimize the risk of an uncertain selection response. In pig breeds, selection for lower RFI in different experimental lines has proved to be very effective. However, selection for increased feed efficiency can also render animals more sensitive to health and other issues due to lower idle resources (energy) available to cope with environmental (outside) challenges (Knap and Wang, 2012). Investigations of the impact of divergent selection for RFI in pigs have been mostly performed by two major groups: the INRA Institute in France and the IOWA State University in the United States. The INRA experiment consisted of lines selected for both high and low RFI (Gilbert et al., 2007), while the IOWA State University experiments selected a line with low RFI alongside a random control line (Cai et al., 2008). Both of these experiments indicated that selection for low RFI reduced FCR and off-set BF while concurrently increasing loin muscle area and dressing percentage (Dekkers and Gilbert, 2010). Moreover, both experiments showed that low RFI pigs ate faster, had low fat carcasses and had no detrimental effects on sow performance (Dekkers and Gilbert, 2010). Other studies have also demonstrated that selection for low RFI decreased BF and increased lean meat content without affecting ADG (Kennedy et al., 1993; Johnson et al., 1999). Barea et al. (2010) demonstrated that pigs with low RFI were energetically more efficient than pigs with high RFI due to lower heat production because of physical activity and basal metabolic rate. Faure et al. (Faure et al., 2013) compared muscle energy metabolism and meat quality between pig lines selected for high and low RFI, and concluded that selection for a low RFI could provide opportunities for reducing environmental impact of pig production while improving lean meat content that will aid in sustainable meat production.

8 Genetic architecture of feeding behaviour traits

8.1 Feeding behaviour traits

Differences in feeding behaviours among pig breeds have been recorded in several studies (Baumung et al., 2006; Fernandez et al., 2011; Labroue et al., 1999). The average daily eating times recorded in our study (81.05 min in DD, 71.02 min in LL and 61.94 min in YY) were considerably greater than those reported by Dehaer and Devries (1993) (51.6 min for Great Yorkshire and 62.1 min for Dutch Landrace boars), or by VonFelde et al. (1996) (50 min for Large White and Landrace boars). The explanation for this could be that the pigs in our study, which is more recent, grow faster and therefore also eat more. The means for number of visits per day (NVD) in our study were in agreement with different reports, although there is some variation in the literature. Apart from sampling errors, differences in breed, sex (boars vs. barrows), stocking rate and feeding system used may all be responsible for differences found among these studies. Consistent with Baumung et al. (2006) and Fernandez et al. (2011), NVD was found to be more frequent in YY than in DD and LL. Landrace and YY spent less TPD than DD. In addition, we also found that FR was greater in LL and YY compared to DD.

We have published genetic analysis of feeding behaviour traits also in the same *Journal of Animal Science* (2013) from the same source (Do et al., 2013b). Data were recorded at a central test station (Bøgildgård) for a period of four years (2008–11) and were supplied by the Pig Research Centre, Danish Agriculture and Food Council. Pedigrees were traced back to 1970. The following feeding behaviour traits were defined and calculated for each boar: DFI: total daily feed intake (kg/d), TPD: total time spent eating per day (minute), NVD: number of visits to the feeder per day, TPV: time spent eating per visit (= TPD/NVD), FPV: mean FI per visit (kg) and FR: mean FI rate (g/minute) (=DFI/TPD).

Using the model described in Do et al. (2013), we estimated genetic parameters for the above feeding behaviour traits. Heritabilities of feeding behaviours in LL and YY were moderate in this study and in general consistent with estimated heritabilities ranging from 0.45 to 0.54 for DFI, FPV and TPV in French Landrace and Large White (Labroue et al., 1997). VonFelde et al. (1996) also reported heritability for feeding behaviours, including FR (0.44), FPV (0.51), NVD (0.43), TPV (0.42) and TPD (0.43) in Germany Landrace and Large White. Estimated heritabilities and other parameters may be biased due to data being considered only for selected pigs (i.e. mainly males with high genetic merit enter the testing station).

8.2 Genetic correlation of RFI with feeding behaviours in pigs

The relationship between feeding behaviours and RFI has been investigated much more in cattle [e.g. (Robinson and Oddy, 2004; Bingham et al., 2009; Kelly et al., 2010; Montanholi et al., 2010)] than in pigs. Generally, genetic correlations between RFI and the other feeding behaviour traits were weak to moderate. Rauw et al. (2006) estimated low and positive genetic correlations between NVD and RFI. Von Felde et al (1996) also found that TPV had a very low genetic correlation with RFI. Young et al. (2011) reported that selections for low RFI line have consequent effect on feeding behaviours in pigs such as low RFI lines have a greater feeding rate and less occupation time per day, visit and hour than control groups.

9 Genomics of feeding behaviour

The use of feeding machines allowed recording of different types of feeding behaviours including total time spent at feeder per day, number of visits to the feeder per day, average duration of each visit, mean FI per visit and mean FI rate. All these traits were moderately to highly heritable (Do et al., 2013b), and therefore, we are expected to identify QTLs influencing them. Our analyses (Do et al., 2013a) indicate that several genomic regions – 64–65 Mb on SSC 1, 124–130 Mb on SSC 8, 63–68 Mb on SSC 11, 32–39 Mb and 59–60 Mb on SSC 12 – are important for feeding behaviour traits. Several studies have demonstrated the potential of using pigs to model human obesity and obesity-related traits (Houpt et al., 1979, Spurlock and Gabler, 2008). However, these studies have mostly focused on fat or fat-related traits. Obesity is a consequence of genetics and environment as well as the interaction between them. Eating behaviour represents a significant factor that contributes to obesity, and this has been investigated and reviewed in many human studies (Sobal and Stunkard, 1989; Wing et al., 2001; Ma et al., 2003; Torres and Nowson, 2007) but not in pigs. Therefore, we conducted pig–human comparative gene mapping in order to identify key genomic regions and/or genes on the human genome that may influence eating behaviour in human beings (e.g. QTL for NVD was homologous with HSA 17q21 regions which contain many obesity candidate genes, including *PPY*, *PON1* and 2, *GAST*, *PNMT*, *STAT3* and *HCRT*). The comparative mapping study between pigs and humans is useful because it is very hard to monitor the eating behaviours in very large populations and the eating behaviour is more complex in humans than in pigs. However, the comparative mapping is also limited by the inability to point out exact genes or mutations controlling phenotypic variance.

10 Towards integrative systems genomics of feed efficiency

Integration, modelling and analyses of several 'omic data types from populations measured for phenotypic traits or diseases are called 'systems genomics' which was first proposed by Kadarmideen et al. (2006) in an animal breeding context and recently in Surajvala et al. (2016), and the concept has been widely being applied for several traits in animal production, including dairy, beef, pig and sheep (Surajvala et al., 2016). Current high-throughput technologies within genomics, epigenomics, transcriptomics, metabolomics and proteomics produce 'big data'. For instance, transcriptomic studies alone could help in validating existing and identifying novel genes and biomarkers for feed efficiency and related traits. Transcriptomic analyses could be performed via RNA sequencing, which is a quantitative approach that directly determines and counts the entire mRNA sequence, thereby estimating RNA expression levels in cells or tissues with higher accuracy than intensity-based microarrays (Wickramasinghe et al., 2014). Besides transcriptomics, metabolic phenotypes can also provide important insights into distinct genetically modulated physiological processes that are difficult to obtain when using conventionally measured production phenotypes representing the cumulative impact of a large number of physiological processes. Using blood plasma metabolomics to obtain biomarker profiles for prediction of feed efficiency in pigs is possible.

The most elegant approach is the integration of genomic and transcriptomic data using an expression QTL (eQTL) approach, detecting regions in the genome that are associated with transcript levels. Those eQTLs can be *cis*- or *trans*-acting. In the case of a *cis*-eQTL, the eQTL is near the location of the gene that encodes the transcript, while in the case of a *trans*-eQTL, the eQTL is farther away or even on another chromosome of the gene that

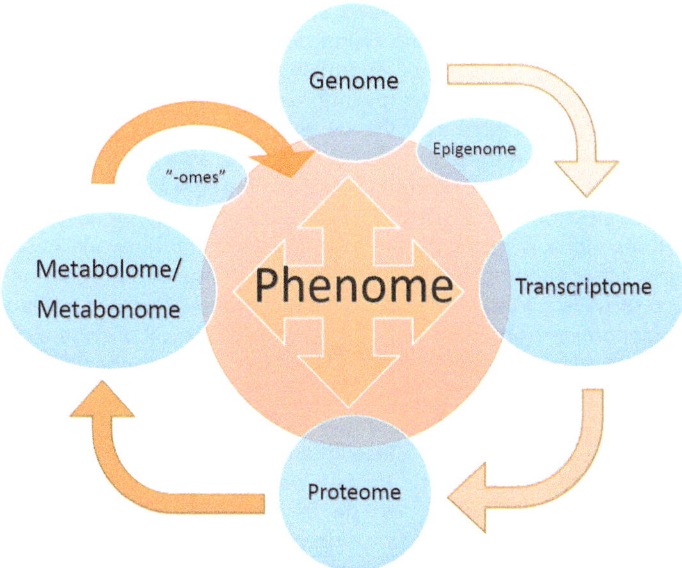

Figure 5 Integrative systems genomics approaches for improving of feed efficiency to include multi-omic profiling of pigs measured for 'phenome' related to all components of feed efficiency.

encodes the transcript. A simple illustration in Fig. 5 shows that the core aspect of such an integrative approach would be based on the 'phenome' (e.g. all observed and metabolic traits that are related to feed efficiency or nutrient utilization) measured on individual pigs. Then, the same pigs would be measured for multi-omic molecular phenotypes using high-throughput biotechnology platforms. Then the data will be modelled keeping the central dogma in biology (DNA-> RNA-> proteins-> metabolites-> phenotype) and many interactions between these levels.

The challenge, though, remains at the level of efficiently removing errors/noise via good quality control methods for each layer of datasets, appropriate data integration as per the defined hypothesis and statistical models, application of advanced statistical–bioinformatic algorithms and meaningful interpretation of results. The clear advantage of these integrative methods is to increase the power of detecting true causal genes, regulatory networks and pathways leading to improved health and/or production. The outcome of such findings on causal genes or variants (QTNs or QTLs), regulator genes, biomarkers and gene networks need to be implemented in genomic breeding programmes, which are becoming more feasible as the newest genomic selection methods are being proposed. In Denmark, there are now new research programmes at the University of Copenhagen (The FeedOMICS project) that apply these approaches to improve feed efficiency in pigs.

11 Conclusion and future trends

In this chapter, the concepts of feed efficiency in pigs were defined and descriptions for different measures of feed efficiency in pigs were given as it exists to date. RFI is an accurate and unbiased estimate of feed efficiency in pigs. However, it is a complex trait that is controlled by many factors. Using both traditional quantitative genetics and modern high-throughput genomic–bioinformatics methods, we were able to conduct our research in a large volume of records in three different pig breeds (Duroc, Landrace and Yorkshire) and gain insights into the genetic architecture and biological pathways underlying feed efficiency in these breeds. We were also able to compute GEBVs for feed efficiency and suggest a potential genomic prediction model for improving feed efficiency using RFI. The results from our work offer many opportunities of follow-up studies. Since genomic analyses alone is not sufficient to understand the network involving complex trait phenotype, other methods to analyse the transcriptome or metabolome or proteome is expected to provide added insights. Ultimately, integrative approaches that combine genomics, epigenomics, transcriptomics and metabolomics will provide a more complete picture for a comprehensive understanding of the biological mechanism underlying feed efficiency (Kadarmideen, 2014). It should also be kept in mind that functional validation of candidate genes and pathways, identified from current studies, using wet-lab approaches is a prerequisite to any application of these findings.

Several methodological improvements may also be considered to improve results in future studies, for example, GWAS and genomic prediction across breeds, using the single-step method to perform GWAS, using random regression to compute RFI and the like. Further, extending the resource population to include phenotypic records from commercial herds and genotyping more animals will always benefit follow-up studies. Including functional information in genomic prediction models might be beneficial in optimizing results. Finally, we recommend an animal trial where either RFI or DFI is included in the breeding programme instead of FCR. Such a study will clarify the breeding objective for feed efficiency and how to use it for genomic improvement of feed efficiency.

12 Where to look for further information

Further information on genetic factors affecting feed efficiency in pigs can be obtained from the authors of this book chapter. We also recommend reading the website: http://www.swinefeedefficiency.com/ for up-to-date information.

13 Acknowledgements

We thank the Department of Breeding and Genetics of the Danish Pig Research Centre for providing all data for the research reported in this study. The work reported in this chapter is mostly a result of a PhD project/thesis of Duy Ngoc Do who was funded by the Faculty of Health and Medical Sciences, University of Copenhagen.

14 References

Aggrey, S. E., and R. Rekaya. 2013. Dissection of Koch's residual feed intake: Implications for selection. *Poultry Science* 92(10):2600–2605.

Atchley, W. R., and D. Anderson. 1978. Ratios and the statistical analysis of biological data. *Systematic Biology* 27(1):71–8.

Barea, R., S. Dubois, H. Gilbert, P. Sellier, J. van Milgen and J. Noblet. 2010. Energy utilization in pigs selected for high and low residual feed intake. *Journal of Animal Science* 88(6):2062–72.

Berry, D. P., and J. J. Crowley. 2012. Residual intake and body weight gain: A new measure of efficiency in growing cattle. *Journal of Animal Science* 90(1):109-15.

Berry, D. P., and J. J. Crowley. 2013. CELL BIOLOGY SYMPOSIUM: Genetics of feed efficiency in dairy and beef cattle. *Journal of Animal Science* 91(4):1594–1613.

Bingham, G., T. Friend, P. Lancaster and G. Carstens. 2009. Relationship between feeding behavior and residual feed intake in growing Brangus heifers. *Journal of Animal Science* 87(8):2685–9.

Cai, W., D. Casey and J. Dekkers. 2008. Selection response and genetic parameters for residual feed intake in Yorkshire swine. *Journal of Animal Science* 86(2):287–98.

Christensen, O. F., P. Madsen, B. Nielsen, T. Ostersen and G. Su. 2012. Single-step methods for genomic evaluation in pigs. *Animal* 6(10):1565–71.

Danish Pig Research Center. 2012. Annual report 2012.

Dekkers, J., and H. Gilbert. 2010. Genetic and biological aspect of residual feed intake in pigs. Proc. 9th World Congr. Genet. Appl. Livest. Prod., Leipzig, Germany 287.

Do, D. N., L. L. Janss, J. Jensen and H. N. Kadarmideen. 2014a. Genomic prediction and variance partitioning for residual feed intake and its component traits in pigs using SNP annotation *Journal of Animal Science* 93 (5):2056–63.

Do, D. N., T. Ostersen, A. B. Strathe, J. Jensen, T. Mark and H. Kadarmideen. 2013a. Genome-wide association study reveals genetic architecture of eating behavior in pigs and its implications for humans obesity by comparative mapping. *PLoS ONE* 8(8).

Do, D. N., T. Ostersen, A. B. Strathe, T. Mark, J. Jensen and H. N. Kadarmideen. 2014b. Genome-wide association and systems genetic analyses of residual feed intake, daily feed consumption, backfat and weight gain in pigs. *BMC Genetics* 15(1):27.

Do, D. N., A. B. Strathe, J. Jensen, T. Mark and H. N. Kadarmideen. 2013b. Genetic parameters for different measures of feed efficiency and related traits in boars of three pig breeds. *Journal of Animal Science* 91(9):4069–79.

Do, D. N., A. B. Strathe, T. Ostersen, S. D. Pant and H. N. Kadarmideen. 2014c. Genome-wide association and pathway analysis of feed efficiency in pigs reveal candidate genes and pathways for residual feed intake. *Frontiers in Genetics* 5.

Drouilhet, L., H. Gilbert, E. Balmisse, J. Ruesche, A. Tircazes, C. Larzul and H. Garreau. 2013. Genetic parameters for two selection criteria for feed efficiency in rabbits. *Journal of Animal Science* 91(7):3121–8.

Duthie, C., G. Simm, A. Doeschl-Wilson, E. Kalm, P. Knap and R. Roehe. 2008. Quantitative trait loci for chemical body composition traits in pigs and their positional associations with body tissues, growth and feed intake. *Animal Genetics* 39(2):130–40.

Ernst, C. W., and J. P. Steibel. 2013. Molecular advances in QTL discovery and application in pig breeding. *Trends in Genetics* 29(4):215–24.

Fan, B., S. Lkhagvadorj, W. Cai, J. Young, R. Smith, J. Dekkers, E. Huff-Lonergan, S. Lonergan and M. Rothschild. 2010. Identification of genetic markers associated with residual feed intake and meat quality traits in the pig. *Meat Science* 84(4):645–50.

Faure, J., L. Lefaucheur, N. Bonhomme, P. Ecolan, K. Meteau, S. M. Coustard, M. Kouba, H. Gilbert and B. Lebret. 2013. Consequences of divergent selection for residual feed intake in pigs on muscle energy metabolism and meat quality. *Meat Science* 93(1):37–45.

Fitzhugh, H., and T. St CS. 1971. Genetic analysis of degree of maturity. *Journal of Animal Science* 33(4):717–25.

Gilbert, H., J.-P. Bidanel, J. Gruand, J.-C. Caritez, Y. Billon, P. Guillouet, H. Lagant, J. Noblet and P. Sellier. 2007. Genetic parameters for residual feed intake in growing pigs, with emphasis on genetic relationships with carcass and meat quality traits. *Journal of Animal Science* 85(12):3182–8.

Gilbert, H., J. Riquet, J. Gruand, Y. Billon, K. Fève, P. Sellier, J. Noblet and J. P. Bidanel. 2010. Detecting QTL for feed intake traits and other performance traits in growing pigs in a Piétrain–Large White backcross. *Animal* 4(08):1308–18.

Groenen, M., A. Archibald, H. Uenishi, C. Tuggle, Y. Takeuchi, M. Rothschild, C. Rogel-Gaillard, C. Park, D. Milan, H. Megens, S. Li, D. Larkin, H. Kim, L. Frantz, M. Caccamo, H. Ahn, B. Aken, A. Anselmo, C. Anthon, L. Auvil, B. Badaoui, C. Beattie, C. Bendixen, D. Berman, F. Blecha, J. Blomberg, L. Bolund, M. Bosse, S. Botti and B. Zhan. 2012. Analyses of pig genomes provide insight into porcine demography and evolution. *Nature* 491:393–8.

Gunsett, F. 1984. Linear index selection to improve traits defined as ratios. *Journal of Animal Science* (USA).

Guo, X., O. F. Christensen, T. Ostersen, D. Sørensen, Y. Wang, M. S. Lund and G. Su. 2014. Improving Genetic Evaluation of Litter Size Using a Single-step Model. in Proc. 10th World Congress on Genetics Applied to Livestock Production (WCGALP).

Haer, L. C. M. d.1992. *Relevance of Eating Pattern for Selection of Growing Pigs*. Landbouwuniversiteit te Wageningen, Wageningen.

Hoque, M., and K. Suzuki. 2009. Genetics of residual feed intake in cattle and pigs: A review. Asian-Aust. *Journal of Animal Science* 22:747–55.

Hoque, M. A., H. Kadowaki, T. Shibata, T. Oikawa and K. Suzuki. 2009. Genetic parameters for measures of residual feed intake and growth traits in seven generations of Duroc pigs. *Livestock Science* 121(1):45–9.

Hoque, M. A., and K. Suzuki. 2008. Genetic parameters for production traits and measures of residual feed intake in Duroc and Landrace pigs. *Animal Science Journal* 79(5):543–9.

Hoque, M. A., K. Suzuki, H. Kadowaki, T. Shibata and T. Oikawa. 2007. Genetic parameters for feed efficiency traits and their relationships with growth and carcass traits in Duroc pigs. *Journal of Animal Breeding and Genetics* 124(3):108–16.

Houpt, K. A., T. R. Houpt and W. G. Pond. 1979. The pig as a model for the study of obesity and of control of food intake: A review. *The Yale Journal of Biology and Medicine* 52(3):307.

Houston, R. D., C. S. Haley, A. L. Archibald and K. A. Rance. 2005. A QTL affecting daily feed intake maps to Chromosome 2 in pigs. *Mammalian Genome* 16(6):464–70.

Jensen, J., I. Mao, B. B. Andersen and P. Madsen. 1992. Phenotypic and genetic relationships between residual energy intake and growth, feed intake, and carcass traits of young bulls. *Journal of Animal Science* 70(2):386–95.

Jiao, S., C. Maltecca, K. A. Gray and J. P. Cassady. 2014a. Feed intake, average daily gain, feed efficiency, and real-time ultrasound traits in Duroc pigs: I. Genetic parameter estimation and accuracy of genomic prediction. *Journal of Animal Science* 92(6):2377–86.

Jiao, S., C. Maltecca, K. A. Gray and J. P. Cassady. 2014b. Feed intake, average daily gain, feed efficiency, and real-time ultrasound traits in Duroc pigs: II. Genomewide association. *Journal of Animal Science* 92(7):2846–60.

Johnson, Z. B., J. J. Chewning and R. A. Nugent. 1999. Genetic parameters for production traits and measures of residual feed intake in large white swine. *Journal of Animal Science* 77(7):1679–85.

Kadarmideen, H. N. 2014. Genomics to systems biology in animal and veterinary sciences: progress, lessons and opportunities. *Livestock Science* 166:232–48.

Kelly, A., M. McGee, D. Crews, A. Fahey, A. Wylie and D. Kenny. 2010. Effect of divergence in residual feed intake on feeding behavior, blood metabolic variables, and body composition traits in growing beef heifers. *Journal of Animal Science* 88(1):109–23.

Kennedy, B., J. Van der Werf and T. Meuwissen. 1993. Genetic and statistical properties of residual feed intake. *Journal of Animal Science* 71(12):3239–50.

Kindt, A. S. D., P. Navarro, C. A. M. Semple and C. Haley. 2013. The genomic signature of trait-associated variants. *BMC Genomics* 14(1):108.

Kleiber, M. 1961. The fire of life. An introduction to animal energetics. The fire of life. An introduction to animal energetics.

Knap, P. W., and L. Wang. 2012. Pig breeding for improved feed efficiency. In *Feed Efficiency in Swine*. J. Patience, ed. Wageningen Academic Publishers, Wageningen, pp. 167–81.

Koch, R. M., L. Swiger, D. Chambers and K. Gregory. 1963. Efficiency of feed use in beef cattle. *Journal of Animal Science* 22(2):486–94.

Liu, G., D. Jennen, E. Tholen, H. Juengst, T. Kleinwächter, M. Hölker, D. Tesfaye, G. Ün, H. J. Schreinemachers and E. Murani. 2007. A genome scan reveals QTL for growth, fatness, leanness and meat quality in a Duroc-Pietrain resource population. *Animal Genetics* 38(3):241–52.
Ma, Y., E. R. Bertone, E. J. Stanek, G. W. Reed, J. R. Hebert, N. L. Cohen, P. A. Merriam and I. S. Ockene. 2003. Association between eating patterns and obesity in a free-living US adult population. *American Journal of Epidemiology* 158(1):85–92.
Montanholi, Y., K. Swanson, R. Palme, F. Schenkel, B. McBride, D. Lu and S. Miller. 2010. Assessing feed efficiency in beef steers through feeding behavior, infrared thermography and glucocorticoids. *Animal* 4(05):692–701.
Mrode, R., and B. Kennedy. 1993. Genetic variation in measures of food efficiency in pigs and their genetic relationships with growth rate and backfat. *Animal Production* 56(02):225–32.
Nguyen, N. H., C. P. McPhee and C. M. Wade. 2005. Responses in residual feed intake in lines of Large White pigs selected for growth rate on restricted feeding (measured on ad libitum individual feeding). *Journal of Animal Breeding and Genetics* 122(4):264–70.
Onteru, S., D. Gorbach, J. Young, D. Garrick, J. Dekkers and M. Rothschild. 2013. Whole genome association studies of residual feed intake and related traits in the pig. *PLoS ONE* 8:e61756.
Pryce, J. E., W. J. Wales, Y. de Haas, R. F. Veerkamp and B. J. Hayes. 2014. Genomic selection for feed efficiency in dairy cattle. *Animal* 8(01):1–10.
Ramos, A. M., R. P. Crooijmans, N. A. Affara, A. J. Amaral, A. L. Archibald, J. E. Beever, C. Bendixen, C. Churcher, R. Clark and P. Dehais. 2009. Design of a high density SNP genotyping assay in the pig using SNPs identified and characterized by next generation sequencing technology. *PLoS ONE* 4(8):e6524.
Rauw, W., J. Soler, J. Tibau, J. Reixach and L. G. Raya. 2006. The relationship between residual feed intake and feed intake behavior in group-housed Duroc barrows. *Journal of Animal Science* 84(4):956–62.
Robinson, D., and V. Oddy. 2004. Genetic parameters for feed efficiency, fatness, muscle area and feeding behaviour of feedlot finished beef cattle. *Livestock Production Science* 90(2):255–70.
Sahana, G., V. Kadlecova, H. Hornshoj, B. Nielsen and O. Christensen. 2013. A genome-wide association scan in pig identifies novel regions associated with feed efficiency trait. *Journal of Animal Science* 91:1041–50.
Saintilan, R., I. Mérour, S. Schwob, P. Sellier, J. Bidanel and H. Gilbert. 2011. Genetic parameters and halothane genotype effect for residual feed intake in Piétrain growing pigs. *Livestock Science* 142(1–3):203–9.
Sanchez, M.-P., T. Tribout, N. Iannuccelli, M. Bouffaud, B. Servin, A. Tenghe, P. Dehais, N. Muller, M. Del Schneider, M.-J. Mercat, C. Rogel-Gaillard, D. Milan, J.-P. Bidanel and H. Gilbert. 2014. A genome-wide association study of production traits in a commercial population of Large White pigs: evidence of haplotypes affecting meat quality. *Genetics Selection Evolution* 46(1):12.
Silverstein, J. T., M. Hostuttler and K. P. Blemings. 2005. Strain differences in feed efficiency measured as residual feed intake in individually reared rainbow trout, Oncorhynchus mykiss (Walbaum). *Aquaculture Research* 36(7):704–11.
Sobal, J., and A. J. Stunkard. 1989. Socioeconomic status and obesity: a review of the literature. *Psychological Bulletin* 105(2):260.
Spurlock, M. E., and N. K. Gabler. 2008. The development of porcine models of obesity and the metabolic syndrome. *The Journal of Nutrition* 138(2):397–402.
Strathe, A., T. Mark, B. Nielsen, D. Do and H. K. J. Jensen. 2014. Deriving genomic breeding values for residual feed intake from covariance functions of random regression models. in Proc. 10th World Congress on Genetics Applied to Livestock Production. Asas.
Torres, S. J., and C. A. Nowson. 2007. Relationship between stress, eating behavior, and obesity. *Nutrition* 23(11):887–94.
van Eerden, E., H. van den Brand, G. De Vries Reilingh, H. K. Parmentier, M. C. M. de Jong and B. Kemp. 2004. Residual feed intake and its effect on Salmonella enteritidis infection in growing layer hens. *Poultry Science* 83(11):1904–10.

Von Felde, A., R. Roehe, H. Looft and E. Kalm. 1996. Genetic association between feed intake and feed intake behaviour at different stages of growth of group-housed boars. *Livestock Production Science* 47(1):11–22.

Werf, J. H. J. v. d.2004. Is it useful to define residual feed intake as a trait in animal breeding programs? *Australian Journal of Experimental Agriculture* 44(5):405–9.

Willems, O. W., S. P. Miller and B. J. Wood. 2013. Assessment of residual body weight gain and residual intake and body weight gain as feed efficiency traits in the turkey (Meleagris gallopavo). *Genetics Selection Evolution* 45(1):26.

Wing, R. R., M. G. Goldstein, K. J. Acton, L. L. Birch, J. M. Jakicic, J. F. Sallis, D. Smith-West, R. W. Jeffery and R. S. Surwit. 2001. Behavioral science research in diabetes lifestyle changes related to obesity, eating behavior, and physical activity. *Diabetes Care* 24(1):117–23.

Yang, J., T. A. Manolio, L. R. Pasquale, E. Boerwinkle, N. Caporaso, J. M. Cunningham, M. de Andrade, B. Feenstra, E. Feingold, M. G. Hayes, W. G. Hill, M. T. Landi, A. Alonso, G. Lettre, P. Lin, H. Ling, W. Lowe, R. A. Mathias, M. Melbye, E. Pugh, M. C. Cornelis, B. S. Weir, M. E. Goddard and P. M. Visscher. 2011. Genome partitioning of genetic variation for complex traits using common SNPs. *Nature Genetics* 43(6):519–25.

Young, J., W. Cai and J. Dekkers. 2011. Effect of selection for residual feed intake on feeding behavior and daily feed intake patterns in Yorkshire swine. *Journal of Animal Science* 89(3):639–47.

Zhang, Z., J. Ren, D. Ren, J. Ma, Y. Guo and L. Huang. 2009. Mapping quantitative trait loci for feed consumption and feeding behaviors in a White Duroc× Chinese Erhualian resource population. *Journal of Animal Science* 87(11):3458–63.

Part 2

Animal nutrition

Chapter 6

Advances in understanding pig nutritional requirements and metabolism

R. J. van Barneveld, R. J. E. Hewitt and D. N. D'Souza, SunPork Group, Australia

1 Introduction

Our knowledge of pig nutritional requirements and metabolism is ever evolving. Linking advances in our understanding to sustainable production of pork requires a more lateral perspective and centres on the primary drivers of sustainability.

Livestock production occupies approximately 75% of agricultural land (Foley et al., 2011), consumes 35% of the world's grain and produces 14.5% of anthropogenic greenhouse gas emissions (Gerber et al., 2013). The demand for meat and dairy products is forecast to increase 60% by 2050 (Alexandratos and Bruinsma, 2012); therefore, it is imperative in meeting this demand that pork production systems do so sustainably. It can be argued that pig production methods have always operated in a sustainable system, whereby the pig, the environment and the consumer have been considered. In the book *Meat: A Benign Extravagance*, the role of the pig has been shown to

http://dx.doi.org/10.19103/AS.2017.0013.05

accumulate resources, otherwise discarded as waste, and to act as a hedge against oscillating availability and price of grain (Fairlie, 2010). However, as the demand for meat increases, we need to continue to advance our sustainable pig production systems, which ensures outcomes that are 'good for the pig, good for the consumer and good for the environment'.

Sustainable agriculture concepts have been developed and continue to be redeveloped to address key issues over time. A three-dimensional model of agricultural sustainability was established by Douglass (1984) and is comprised of the following:

- The first dimension centres on food production, efficiency and profitability – producing sufficient quantities of food for consumers whilst providing sufficient income through the production chain.
- The second dimension examines resources (water, soil, nutrients, etc.) used by agriculture – the quantity and quality of which could impact on production and conversely the impact of production on those resources.
- The third dimension incorporates society – the regulatory environment in which agriculture and all other components of society operate and the expectations of the consumer on what the food should be, how much of it is needed and how it should be produced.

More recently, the United Nations Conference on Sustainable Development, Rio+20, called for an enhancement in food security and nutrition and a more sustainable agriculture, which has led to the formation of five principles for sustainable food and agriculture (FAO, 2014), including:

1. Improving efficiency in the use of resources is crucial to sustainable agriculture.
2. Sustainability requires direct action to conserve, protect and enhance natural resources.
3. Agriculture that fails to protect and improve rural livelihoods, equity and social well-being is unsustainable.
4. Enhanced resilience of people, communities and ecosystems is key to sustainable agriculture.
5. Sustainable food and agriculture requires responsible and effective governance mechanisms.

A key element of these principles for sustainable agriculture is improved efficiency of use of resources. In relation to advances in pig nutritional requirements and metabolism, this can be achieved through:

1. Maintaining consistent sow body condition through gestation and lactation;
2. Reducing variation in pig production systems;
3. Strategic use of metabolic modifiers;
4. Closer matching of nutrient requirements to feed composition;
5. Optimising utilisation of co-products not suitable for human consumption;
6. Optimising gut health and capacity for nutrient utilisation;
7. Enhanced understanding of nutrition and health interactions.

This chapter aims to consider these facets, focusing on advances made in nutritional requirements and metabolism and how these contribute to the sustainable production of pig meat.

2 Maintaining sow body condition through gestation and lactation

In pig production systems, improved efficiency of use of resources can be achieved by limiting nutritional resources directed towards the breeding herd whilst optimising output. Central to this is prevention of catabolism in the sow primarily through optimisation of intake during lactation.

With the above in mind, the apex key performance indicator for efficient use of resources in a pig production system is whole herd feed conversion. This accounts for all nutrients consumed by both the breeding herd and progeny per kilogram of pork produced. Minimisation of sow replacement through optimal gilt development, prevention of lameness and optimisation of reproductive efficiency will contribute significantly to reduced whole herd feed conversion efficiency (Fig. 1) and we advocate that a primary driver in the attainment of these objectives is maintenance of adequate sow body condition and minimisation in variation in this condition across the reproductive cycle.

2.1 Management of gilt weight on herd entry

Improper management of the sow at any stage from her selection into the herd through her subsequent parities has the potential to affect her lifetime productivity and thus her impact on the efficiency of the herd. When gilt management programmes are effective they will result in improved lifetime performance from a smaller gilt pool. Non-negotiable aspects of gilt development have been established (Beltranena et al., 2005; Williams et al., 2005), which included recognising the importance of a dedicated gilt development area, ensuring that you have enough gilts in the pool to meet requirements, breed to

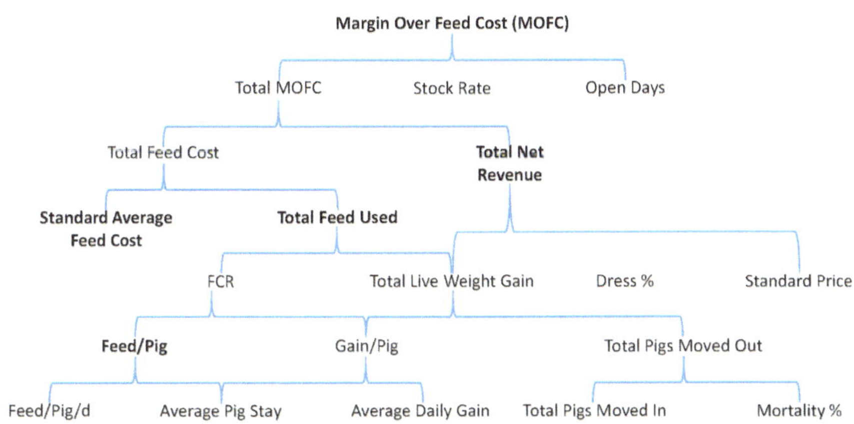

Figure 1 Apex key performance indicator for efficient use of resources in a pork production system.

weight not age and on the second oestrus, and avoid slow-growing gilts. Consequently, feeding programmes during gilt development should be designed with these principles in mind.

2.2 Nutrition to prevent sow lameness

The main nutrients affecting foot health in pigs include the amino acids cysteine and methionine; the minerals calcium, zinc, copper, selenium and manganese (van Barneveld, 2010); and vitamins A, D, E and biotin (Kornegay, 1996). Biotin has been investigated for its effect on the health of pig claws with variable results being reported in response to supplementation (Brooks et al., 1977; Penny et al., 1981; Kopinski and Leibholz, 1989; Greer et al., 1991; Hamilton and Veum, 1984; Lewis et al., 1991). The availability of biotin in feedstuffs for pigs is variable with apparent ileal digestibilities of biotin from diets containing wheat and sorghum being zero (Kopinski et al., 1989).

Skeletal health is also vital to reduce lameness. Osteochondrosis has been suggested as a major cause of leg weakness in growing swine (Nakano et al., 1987) and has been linked to the premature culling of breeder stock (Gresham, 2003). The primary development of osteochondrosis in pigs occurs at a very young age; however, clinical signs of lameness are not observed until much later, up to 18 months of age (Ytrehus et al., 2007), with 80% of pigs marketed estimated to have slight to mild forms of osteochondrotic lesions (Carlson et al., 1988; Ytrehus et al., 2004). Vitamin D plays an integral role in mineral nutrition and is responsible for the intestinal absorption of minerals. Direct dietary supplementation with 25-hydroxycholecalciferol, the activated form of vitamin D that normally occurs in the liver of the pig, has been shown to promote normal endochondral ossification and inhibit the progression of osteochondrosis (Sugiyama et al., 2013).

High-producing sow lines of today are likely to have greater nutrient requirements than those sows used in past studies investigating requirements (Mahan and Newton, 1995), including those of biotin. Change in sow housing from individual stalls to group housing may result in extra attention needing to be paid to nutrient requirements to optimise sow foot health. It should be recognised that the requirement level to maintain optimum foot health may be higher than those required for optimum growth or reproductive performance. For example, Kopinski and Leibholz (1989) suggested that there is no requirement of supplemental biotin for growth of pigs; however, supplementation of 0.050–0.100 mg/kg diet is required for prevention of claw lesions. Future studies investigating the requirements of nutrients for optimum foot health should be standardised with respect to duration of the trials, floor and housing system used, and basal diet. Capacity to develop nutritional strategies to improve sow foot health will require a better understanding of the contributions of individual nutrients, alone or in combination, to foot health.

2.3 Minimising variation in nutrient intake during gestation

Nutrition during gestation is focused on meeting the requirements of the sow for maintenance and the development of maternal tissues and the conceptus (National Research Council, 2012). A higher level of nutrition in early gestation has no impact on embryo survival, but, did however, result in increase in weight and back fat (Hoving et al.,

2012), whilst during late gestation, a higher feeding level also resulted in weight gain, without influencing the birthweight of piglets (Knauer, 2016). The supply of nutrients above requirements during gestation benefits only the sow and not the developing piglets, and further results in a reduction in voluntary feed intake during lactation (Weldon et al., 1994).

2.4 Maintaining body weight during lactation

A reduction in feed intake during lactation is likely to result in the mobilisation of body reserves (Bergsma et al., 2009), as the sow, in contrast to gestation, devotes all her efforts to ensuring the growth of the piglets. Body condition on exit from gestation can have an impact on this. Lean sows tend to have higher protein reserves and a higher voluntary feed intake compared with fat sows. This can be explained by high levels of exogenous protein that may actually stimulate milk production and therefore voluntary feed intake (Revell et al., 1998a). Conversely, fat sows have lower protein reserves, and thus less ability to mobilise them for milk output when supply is limited, especially in early lactation (Revell et al., 1998b). One strategy that has been employed to prevent impaired feed intake in lactation was to restrict feed intake in the immediate period both prior and post parturition (Koketsu et al., 1996); however, more recent research would suggest that *ad libitum* feeding in this period can lower mobilisation of body reserves and increase piglet performance (Cools et al., 2014).

Whilst body tissue catabolism is obviously an immediate issue for the sow and her suckling piglets, the impact on subsequent reproductive performance is considerable. Clowes et al. (2003) investigated the impact of increasing losses of protein on reproductive performance of first parity sows. Ovarian function was suppressed in sows with the greatest protein loss; they had fewer medium-sized follicles, and those follicles contained less follicular fluid, with lower estradiol and IGF-1 concentrations, translating to a lower number of piglets in the subsequent litter (De Bettio et al., 2016).

In summary, the key to a long reproductive life is to optimise the entry of the gilt into the breeding herd through meeting key weight targets, maintaining an adequate level of nutrition throughout gestation to support maintenance of body tissues and the growth of maternal and foetal tissues, and ensuring a high level of feed intake during lactation to reduce losses of body reserves and maintain good ovarian function.

3 Reducing variation in pig production systems

Variation is an often overlooked contributor to inefficient use of resources, and the nutritional interventions that can be applied to reduce variation are poorly understood or limited.

Variation is inherent in any biological system and is a challenge to manage in modern pork production businesses. In the case of the growing herd, inherent variation within a population of pigs represents a significant cost, as a result of the need to select on farm to meet market specifications, poor matching of diet specifications to nutrient requirements, grading losses, higher pre-weaning mortality and challenges associated with health management. As a consequence, any management practice that can be applied to reduce variation at the point of sale has the potential to improve the profitability and overall efficiency of a pig enterprise.

There are many factors that can influence birthweight and variability in birthweight; hence, manipulation by nutritional means alone is unlikely to resolve this. It is, however, relevant that most nutrient specifications used in diet formulations today have been derived from far less prolific sows; hence, it is possible that both the sow and foetuses are compromised, given both have an increase in requirements as gestation progresses (Ball et al., 2008). Lawlor et al. (2007) reported that birthweight, weaning weight and within-litter variation were unaffected when five different dietary digestible energy (DE) levels were provided during different gestation phases. In contrast, Wu et al. (2010) demonstrated that supplementation of gestating sow diets with specific amino acids, free arginine (to bring the lysine to arginine ratio to 2.64) and glutamine, significantly influenced litter birthweight variation. These amino acids were targeted as arginine is extensively catabolised by arginase in the small intestine, and only 60% of dietary arginine enters the portal circulation of gestating gilts (Wu et al., 2007), whilst the uptake of glutamine in the uterus of gestating gilts is the greatest among all amino acids (Wu et al., 1999) and it is abundant in placental and foetal fluids. When Wu et al. (2010) added 0.6% glutamine and 0.4% arginine to corn–soybean meal-based diets, glutamine concentrations in gilt plasma did not decline, but a marked reduction in ammonia and urea in maternal plasma and a decline of variation in birthweights among all piglets born and pigs born alive were observed. Foxcroft (2007) suggested that the smallest pigs within the litter respond the greatest to maternal nutrition interventions, resulting in a reduction in variation in birthweight as evidenced by the significant impact of exogenous somatotropin during early gestation on small pigs (Rehfeldt et al., 2001).

In addition to nutritional interventions that influence variation in birthweight *per se*, it is possible that a considerable amount of the variation in growth performance after birth may be largely determined, and essentially pre-programmed, during foetal development in the uterus (Foxcroft and Town, 2004). Foxcroft (2007) reported that the early period of myogenesis, involving the differentiation of primary muscle fibres, is generally considered resistant to nutritional manipulation, whereas the nutritional effects on differentiation and hyperplasia of secondary fibres have been demonstrated between Day 25 and Day 90 of gestation.

As discussed previously, while the greatest effects of lactation feeding will be on postnatal piglet performance and variation, nutrition of the sow during lactation can also influence variation in litter birthweights. Increased catabolism during the last week of lactation in primiparous sows is known to reduce embryonic survival and development up to Day 30 of gestation in the subsequent litter (Foxcroft, 1997). Quesnel et al. (2008) found significant relationships between sow's body condition and the coefficient of variation (CV) of birthweight. Litter heterogeneity increased with bodyweight at the beginning and end of gestation. The CV for birthweight was not linked with bodyweight gain during gestation, but with back fat gain, which was likely a reflection of changes in feed allowance during gestation as a consequence of body reserve mobilisation during the previous lactation (Quesnel et al., 2008).

Regardless of the variation in birthweight of any litter, all pig production systems should focus on optimising sow milk production and piglet colostrum and milk intake to maximise weaning weights for age. Targets of 75–85 kg of litter weight at 21 days of lactation should be a key focus for production staff, with this target representing the best insight into the effectiveness of any lactation feeding programme.

Douglas et al. (2014) demonstrated that when low-birthweight piglets are grouped and offered supplementary milk, they drink significantly ($P < 0.001$) more than those of mixed group piglets. Further, Quiniou et al. (2002) reported that small birthweights translated into small weaning weights, but this was not connected with reduced growth potential before weaning. Donovan and Dritz (2000) removed larger pigs from the dam for a short period within 24 h after birth to allow smaller pigs in the litter uninhibited access to the dam as a means of reducing variation in weight at weaning. While no differences in average daily gain or weaning weight were detected, there were significant linear improvements in s.d. and CV of the average daily gain, but only for litters with nine or more piglets born alive. Given that smaller piglets respond well while suckling, provision of supplementary milk during the lactation period should increase overall piglet viability and weaning weight and reduce s.d. and CV. Wolter et al. (2002) offered supplemental milk replacer from Day 3 of lactation to weaning at 21 days of age. Supplemental milk significantly ($P < 0.001$) enhanced weaning weights (6.60 kg vs 5.69 kg) and also reduced the CV of weaning weights (6.62% vs 7.04%), although this effect was not significant ($P > 0.05$). Recently, Douglas et al. (2014) reported the converse. These authors were unable to demonstrate any benefit of supplementary milk on the performance of lightweight piglets before or after weaning, but it did affect their drinking behaviour and, as a consequence, the variation in bodyweight of lightweight piglets in mixed litters to slaughter was reduced.

Even when those factors in gestation and pre-weaning that we have already shown to influence variation are addressed, there is still likely to be a level of variation in a population of piglets at weaning, and variation from batch to batch. With the s.d. of the population increasing with age (Patience et al., 2004), we must also focus on efforts to reduce this immediately post-weaning.

A recent Australian study (Hewitt et al., 2015) investigated the ability to influence the performance of lightweight weaners through nutritional intervention in this period of efficient growth. It compared lightweight weaned pigs (3.8 kg at 19 days of age) with heavier pigs (5.2 kg at same weaning age) and a control group that was reflective of the normal population of weaners (4.4 kg). The lightweight pigs remained on the starter diet (15.0 MJ DE/kg, 0.8 g standardised ileal digestible (SID)/MJ DE) until they reached 30 kg liveweight, with the aim of lifting their performance to that of the heavier-weight group. Despite this significant intervention, lightweight pigs ate less and grew slower, taking an extra 13.5 days to reach market weight. Feed conversion did not differ among treatments, suggesting that boosting feed intake should lead to improvements in growth rates. Treating light- or heavier-weight pigs as a distinct group also resulted in an increase in the CV over the growth period, when compared with the control population. This has been reflected in other studies (O'Connell et al., 2005) where uniform groups at weaning became more varied over life time, while the control population became less varied, possibly due to the natural variation in growth response, differing responses to weaning and increased levels of aggression even in groups where there is no obvious hierarchy.

4 Strategic use of metabolic modifiers

The impact of technologies such as metabolic modifiers on pig production is well documented (Dunshea et al., 2005). These technologies that increase feed efficiency and lean tissue deposition while decreasing fat deposition have been developed in an effort to improve the efficiency of animal production. Some of these technologies: somatotropin, immunisation against gonadotropin-releasing factor and orally active dietary additives such as ractopamine, chromium and betaine are discussed in this section.

Perhaps the most researched metabolic modifier for improved efficiency of growth and lactation is somatotropin. The impacts of porcine somatotropin (pST) have been known for more than 70 years (Giles, 1942). Injecting growing pigs with pituitary tissue extracts containing pST increases protein deposition and decreases fat accretion (Turman and Andrews, 1955). Advances in biotechnology allowed production of pST on an industrial scale, and commercial use of bovine somatotropin and pST began in 1994 and 1996 in the United States and Australia, respectively (Bauman and Dunshea, 2010). Exogenous pST treatment has been shown to consistently improve average daily gain, feed conversion efficiency and protein deposition, and reduce fat deposition in pigs (Chung et al., 1985; Etherton et al., 1987; Campbell et al., 1988, 1989; King et al., 2000). As a technology, the growth performance of pST-treated pigs is significant, providing us in many ways a glimpse of what we might expect of pigs 10–15 generations into the future.

As discussed in the previous section, inherent variation in the growing and finishing herd represents a significant cost. Interestingly, one of the few strategies to successfully reduce the CV in finisher pig was the exogenous administration of somatotropin. Dunshea (2005) showed that pigs treated with pST had reduced CV around average daily gain compared with untreated pigs (22.1% vs 19.9% for the control and pST regimes, respectively, $P < 0.001$). The CV around P2 backfat thickness was also reduced by pST administration (16.2% vs 14.2%, $P < 0.001$).

Ractopamine, a β-agonist, is approved for use in many countries as an in-feed ingredient to increase lean tissue growth and improve production efficiency in pigs (Dunshea et al., 2005). Treatment of finishing pigs with ractopamine is well studied and has generally yielded dose-dependent improvements in daily gain, feed efficiency and carcass lean content, along with advantages in increased carcass weight, loin eye area, dressing percentage, carcass cut yields and decreased fat (Dunshea, 1993; Dunshea et al., 2005; Woods et al., 2011). These effects can be seen in all sexes, and unlike for pST, feed intake is typically unchanged or decreased only slightly during β-agonist treatment (Dunshea et al., 2016). A meta-analyses of the effect of ractopamine on carcass weight of pigs found an increase of 1.8 (n = 8 studies) and 2.4 kg (n = 19 studies) with 5 and 10 ppm ractopamine (Apple et al., 2007), which captures the range of doses used commercially. Further modelling suggests that the use of ractopamine in pig diets could have beneficial environmental impacts due to improved efficiency and a reduction in animal numbers of between 5.3 and 6.3% (for 5 or 10 ppm inclusion rates) without an impact on total meat production (Woods et al., 2011). Consequently, it is estimated that between 2.8 and 3.4 billion fewer kilograms (equivalent to 0.29 or 0.35 million ha) of corn and 0.16 or 0.34 billion fewer kilograms (equivalent to 0.059 or 0.127 million ha) of soybeans are required for US pork production each year, respectively (Dunshea et al., 2016).

Castration of entire population of male pigs, a widely practised procedure to eliminate male pheromones, commonly referred to as boar taint, and to reduce the incidence of aggressive behaviours often observed in entire male pigs, results in significant reductions in feed efficiency and excess deposition of fat. An alternative method of inhibiting sexual development and aggressive behaviours in the late finisher phase is immunisation against gonadotropin-releasing factor, referred to as immunocastration (Bonneau and Enright, 1995; Dunshea et al., 2001). This results in a reduction in plasma gonadotropins and testosterone and boar taint compounds in swine (Dunshea et al., 2001). Cronin et al. (2003) reported a reduction in both aggressive and sexual activities in immunised boars who exhibited similar activities as barrows. As a consequence, the immunised pigs spent more time eating, and increased their feed intake. A meta-analysis of 19 studies showed that, compared with non-immunised boars, immunisation against gonadotropin-releasing factor increases feed intake and average daily gain with only small increases in feed conversion rate over the finisher phase. Increases in final carcass weight and back fat depth were also shown when immunised boars were compared with entire males (Dunshea et al., 2013). Sufficient data comparing the growth performance and carcass characteristics of immunised boars with those of barrows is now available to conduct comparative meta-analyses. These analyses of up to 21 studies show that during the finisher phase, immunisation against gonadotropin-releasing factor increases feed intake and average daily gain and decreases feed conversion rate compared with contemporary barrows (Dunshea et al., 2013). From an environmental point of view, the effect of immunocastration on greenhouse gas emissions was reduced by 3.7% compared with a physical castration production system (De Moraes et al., 2013).

Further improvements in growth performance of immunocastrated pigs were also seen when used with other technologies. The growth-promoting effects of pST were enhanced when used in conjunction with immunocastration (McCauley et al., 2003; Oliver et al., 2003). Similarly, there are additive effects of ractopamine and immunocastration in pigs (Moore et al., 2009; Rikard-Bell et al., 2009).

Chromium is an essential mineral element (Mertz, 1969) with trivalent chromium (Cr^{3+}) being the most stable form occurring in nature. However, Cr^{3+} is normally poorly absorbed and utilised with the digestibility of inorganic and organic forms of chromium being 0.5–2% and 10–25%, respectively (Mertz, 1969). It has been suggested that dietary chromium may increase insulin sensitivity in pigs (Steele et al., 1977). A meta-analysis of 31 studies in pigs showed that dietary chromium increased daily gain, feed efficiency, and carcass lean and loin muscle area while decreasing back fat thickness (Sales and Jancik, 2011). The absorption of chromium is poor, with only 0.5–3% of inorganic chromium being absorbed from the gastrointestinal tract (Dowling et al., 1989; Ducros, 1992), resulting in concentration of chromium in the effluent. Although organic forms of chromium are considered more bioavailable than the inorganic forms, only 10–25% of organic chromium is absorbed (Underwood, 1977). An alternative is nano-chromium, which has been shown to increase performance of pigs (Hung et al., 2015), but the use of nano-chromium has become a contentious issue relating to nanoparticles in the food industry (Dunshea et al., 2016).

The use of betaine in pig diets has been shown to improve growth performance by reducing the maintenance energy requirement of the animal, perhaps by reducing the need for ion pumping involved with maintaining intracellular osmolarity (Schrama et al., 2003). In addition, dietary betaine has been reported to increase protein deposition and carcass leanness and to decrease back fat (Suster et al., 2004). There is evidence that betaine has a more pronounced effect when dietary energy is limiting (Suster et al., 2004), and so it offers a means of improving growth performance through ensuring the provision of additional energy. A comprehensive meta-analyses of 15 studies on the effects of dietary betaine in pigs indicates that overall dietary betaine increased feed efficiency and carcass dressing percentage and decreased back fat thickness with no effects on daily gain (Sales, 2011).

The technologies listed in this section are just as, or perhaps more, relevant from a herd efficiency and resource use point of view today as they were when they were first utilised in pig production; however, many have reached their end from a usability point of view. Take pST for example; as a technology to improve resource use efficiency in pigs, it remains the gold standard. Yet pST use today is non-existent as a consequence of consumer and retailer pushback, as well as the need for a better delivery system that utilises less labour and is not reliant on an injection. The search for an orally administered pST delivery system remains an elusive goal.

The consumer perception of 'factory farming' has seen many of the other technologies lumped together as technologies that have been deemed to 'mess with my food' by consumer groups or retailers, with little or no supporting evidence. The resultant effect on pig production has been both the loss of key technologies that reduce environmental impact and an increase in the cost of production (Dunshea et al., 2016).

5 Matching nutrient requirements to diet specifications

Capacity to routinely, accurately and cost-effectively measure variation in the nutritional quality of feed ingredients prior to diet formulation represents a fundamental pillar of sustainable pig production worldwide, as it is the most effective means of matching nutrient requirements to diet specifications. Central to this is the capacity to rapidly measure nutritional quality of feed ingredients and up-to-date definition of nutrient requirements.

5.1 Rapid measurement of nutritional quality

Accepting the premise that measurement of 'available' nutrients will provide the greatest capacity to define the comparative nutritional value of ingredients, there is a need to prioritise which available nutrients are measured in which ingredients. This is due to the considerable cost and effort associated with establishing these analyses and the fact that it is unlikely a single analysis will be suitable for all ingredients. Apart from proximate analysis of all ingredients using either chemical or rapid techniques such as near-infrared spectrophotometry (NIRS), available energy in cereals and available lysine in oilseed meals are the highest priority for analysis of available nutrients, based on their relative proportions in pig diets and their associated variation (van Barneveld et al., 2017).

Significant debate has ensued over the suitability of DE and metabolisable energy (ME) versus net energy (NE) systems as the most appropriate measures of available energy in feed ingredients, particularly cereal grains. The latter has been touted as a better alternative on the basis that DE and ME are poor predictors of animal performance (de Lange and Birkett, 2005). This claim is unsupported by recent research (Cadogan, 2009). The digestibility of a feed is affected little by animal factors, except for young piglets and sows (Noblet and Shi, 1993; Nitrayova et al., 2006). However, NE values for a feed depend on characteristics of the feed, nutrients absorbed and the metabolic status of the animal, with different NE values derived for the same feed when consumed by animals of different genotypes, stages of growth, physiological states and climatic conditions (Kil et al., 2013). The debate about superiority of NE is irrelevant because it is impossible to measure NE at the point of feed manufacture. Furthermore, NE calculations depend on information on the DE content of a feed. Consider a NE equation reported by Noblet (2006):

$$\text{NE (MJ/kg, DM)} = 0.703 \times \text{DE (MJ/kg, DM)} - 0.0041 \times \text{CP} + 0.0066 \times \text{EE} - 0.0041 \times \text{CF} + 0.0020 \times \text{ST}$$

where CP is crude protein, EE is ether extract, CF is crude fibre and ST is starch (all g/kg DM). The equation shows that more than 70% of the NE contribution from an ingredient is derived from DE. Rapid analysis of DE combined with additional chemical components does allow use of NE as an estimate of available energy for an animal, but without the DE component the NE calculation would be impossible.

Further justification for the development of rapid measures of DE in cereals for pigs was provided by van Barneveld (1999). Differences of up to 3.7 MJ/kg dry matter (DM) in DE content were observed following a review of data for more than 70 cultivars of wheat. Similarly, analysis of data for more than 125 cultivars of barley revealed a range in DE estimates from 11.7 to 16.0 MJ/kg DM. The cultivar itself was shown to have a minimal effect on the availability of energy in cereals, while the cultivation conditions and agronomic practices (e.g. fertiliser rate) were reported to have a greater influence than the growing region or the growing year. Many physical and chemical factors were shown to influence DE in cereals, including protein source and type, starch characteristics, fat source and type, non-starch polysaccharide components and anti-nutritional factors. Van Barneveld (1999) concluded that the availability of energy to the animal will depend on the particular combination of these components in a grain and how they behave in the presence of nutrients from other feed ingredients. Hence, a rapid measurement of DE is best based on data derived from *in vivo* measurements.

In feed ingredients such as soybean meal and canola meal that have undergone processing, heat damage can result in an overestimation of nutritional quality due to reactions between the ε-amino group of lysine with other compounds such as carbohydrates (van Barneveld et al., 1999a). It appears that most processors are aware of the potential for heat damage during oil extraction, but quantification of these effects occurs rarely. Rutherfurd et al. (1997) developed the digestible reactive lysine assay as a means of assessing heat damage in feed ingredients. Van Barneveld et al. (1999b) applied this technique to assess the degree of heat damage in cold-pressed, expeller - extracted and solvent-extracted canola meal. Despite similar total lysine concentrations in the canola meals, true ileal reactive lysine digestibility was significantly higher in the cold-pressed meal. Reactive lysine content was shown to be a good indicator of true ileal digestible reactive lysine content (and thus the degree of heat damage). Given this relationship, application of rapid assays for total lysine and reactive lysine would allow more accurate assessment of the nutritional quality of ingredients such as soybean meal and canola meal, and account for this at the time of diet formulation and manufacture.

The need for rapid analysis of reactive lysine in oilseed meals as a routine assessment was demonstrated by Kim and Mullan (2012). A total of 209 samples collected from the major soybean meal producing countries revealed a 27% variation in reactive lysine content (21.9 g–30.1 g/kg as is basis) across all samples and 13% variation within shipment samples. This finding suggests that there is wide variation in reactive lysine content in soybean meal samples, and the use of spot sampling to analyse reactive lysine content can lead to large inaccuracies. Data analysis showed the proportion of reactive lysine relative to total lysine was not different ($P > 0.05$) due to country of origin, while reactive lysine content was significantly different between countries of origin ($P < 0.001$). Kim and Mullan (2012) concluded that the variation in reactive lysine content caused by heat processing is similar between countries but that the total amount of lysine and reducing sugars present in the raw soybeans prior to heat processing are a major source of variation. Nevertheless, the linear relationship between total and reactive lysine was weak (Slope: 6.00, intercept: 0.7152, RSD: 0.913, R2 = 0.52) and cannot be used for robust prediction of reactive lysine from analysed total lysine content.

van Barneveld et al. (1998) developed the first NIRS calibration for the measurement of DE in cereals for growing pigs, using more than 150 samples from across the world that had been subjected to *in vivo* DE analysis (Table 1). Innovative calibration development techniques were employed so that samples from various laboratories could be meaningfully included in a single calibration. This research demonstrated that DE in whole grain cereals can be measured by NIRS to an accuracy of at least ±0.38 MJ/kg, which at the time was comparable to the accuracy obtained when using *in vivo* reference methods. No significant differences in accuracy were observed between milled and whole grain calibrations or between local and global calibrations.

Further to the development of DE calibrations for cereals in pigs, van Barneveld and Ru (2003) also demonstrated that NIRS could be successfully employed to measure the total and reactive lysine content of canola meal, with obvious potential to extend this to soybean meal.

Significant investment has ensued in Australia in the past 18 years to build upon these base calibrations and we now have a practical means of rapid, online assessment of DE content in cereals used in pig diets, together with the option for routine assessment of batches of heat-treated oilseed meals at the point of receival (Black et al. 2003; Table 2).

Recent advances in the development of NIRS calibrations for measurement of digestible nutrients in ingredients for pigs together with more robust NIRS equipment for online applications in feed mills have led to the collection of data that should concern all commercial nutritionists and pig producers. Online measurement of the DE and crude protein contents of wheat in a large Australian feed mill demonstrates why real-time measurement and the capacity to account for variation should exceed all other priorities

Table 1 Summary of calibration statistics for digestible energy (MJ/kg) in whole and milled cereal grain samples for pigs (from van Barneveld et al., 1998)

Sample set	SECV (MJ/kg)	RER	SECV/STD
Whole grain	0.518	8.8	0.49
Whole grain (IV)	0.377	12.1	0.36
Milled grain	0.596	7.7	0.53
Milled grain (IV)	0.428	10.7	0.38

SECV, standard error of cross validation; RER, range error ratio; STD, standard deviation; IV, indicator variables.

Table 2 Summary of calibration statistics for AusScan NIRS calibrations for ileal and faecal digestible energy (MJ/kg, as received in cereals, excluding oats and maize) for pigs, and total and reactive lysine in canola meal (g/kg, as received, Black et al., 2013)

Measurement	*N*	Mean	SD	SECV	1-VR	RPD
Ileal DE (MJ/kg)	236	11.5	1.03	0.48	0.80	2.17
Faecal DE (MJ/kg)	288	13.7	0.69	0.26	0.86	2.65
Total lysine (g/kg)	113	20.19	1.37	0.65	0.78	2.11
Reactive lysine (g/kg)	113	17.97	1.75	0.92	0.73	1.91

DE, digestible energy; *N*, number of samples used in the calibration; SD, standard deviation of values in the population; SECV, standard error of cross validation; VR, variance ratio; RPD, ratio of prediction to deviation (SD/SECV).

in the application of feed evaluation systems (Fig. 2). Over the course of one hour of production, the crude protein content of the wheat varied by 4% units and the DE content by 1.2 MJ/kg as received. Further, the mean DE content was approximately 0.6 MJ/kg lower than the reported book values for wheat (which would be commonly used in the diet formulation). In this particular mill, production rates are approximately 20 tonnes per hour, and it is common for large batches of single diets to be produced over this period. The online data suggests that the first 3 tonnes of feed utilised wheat with a DE content between 12.9 and 13.4 MJ/kg, the next 12 tonnes was fairly consistent at 13.4 MJ/kg, but the last 5–6 tonnes varied between 13.2 and 12.2 MJ/kg as fed. If diets were formulated as a starter feed, some piglets would receive a diet equating to 14.0 MJ DE/kg while others would receive a diet equivalent to 14.7 MJ DE/kg, and the nutritionist would be expecting delivery of feed equating to 15.0 MJ DE/kg to these pigs. This difference in energy intake would suppress production efficiency and represent a major source of variation in pig growth which would increase as the pigs grow.

5.2 Updated information on nutritional requirements

The 11th revised edition of *Nutrient Requirements of Swine* from the National Research Council (NRC) was released in 2012. As in the 10th edition the main emphasis of this publication was to present estimates for nutrient requirements of pigs, and on available nutrient contents in feed ingredients. In addition, information was included on the pigs' need for water, co-products from the corn and soybean industries, estimation of nutrient digestibility, non-nutritive feed additives, feed contaminants, feed processing, means to

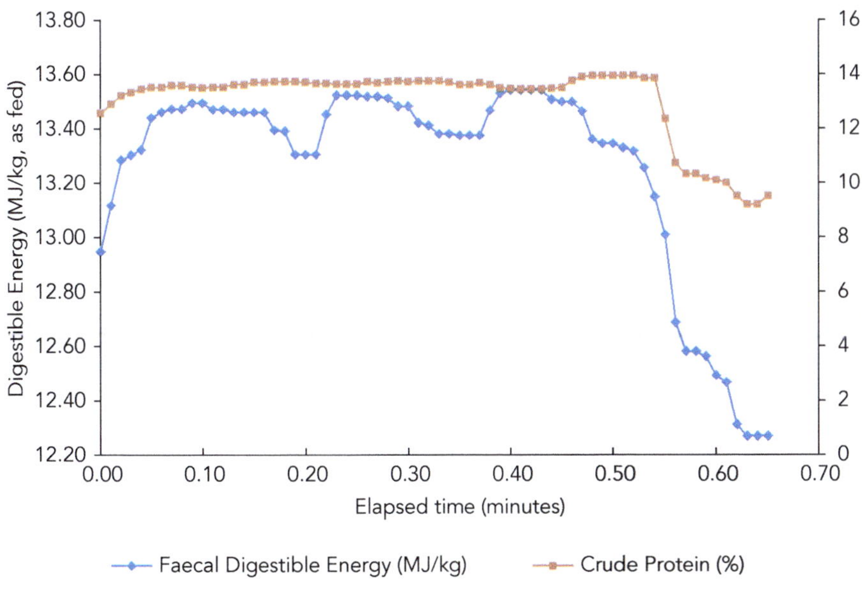

Figure 2 Real-time NIRS analysis of faecal digestible energy (MJ/kg, as fed) and crude protein (%) contents of wheat included in pig diets (D. Pearson, Feedworks Pty Ltd, pers. comm.).

minimise nutrient excretion and research needs. New chapters in this latest edition also included information regarding the nutritional importance of lipids and carbohydrates.

It has been quite some time since the nutritional requirements of the pig were sourced from actual experimental studies to elucidate the nutritional requirement of the growing pig, lactating sow and gestating sows. Instead, mathematical models to estimate nutrient requirements are commonly used (National Research Council, 2012). As summarised by de Lange (2013), models are now used to dynamically represent the partitioning of energy intake and to then estimate amino acid, phosphorus and calcium requirements of swine in a relatively disease- and stress-free environment. Similarly, for the interpretation of nutrient requirements studies, dietary nutrient levels were recalculated and, when needed, estimates of nutrient requirements were established based on a reinterpretation of the data sets. Empirical estimates of requirements for vitamins, fatty acids and minerals other than calcium and phosphorus have been updated from the National Research Council (1998; de Lange, 2013).

6 Optimising utilisation of co-products

In modern intensive production systems focused on providing people with proteins, the animal can be used to 'value-add' raw materials that would otherwise be difficult to include in a human diet. The pig excels in this context and can make use of co-products and grains that have been downgraded from human grain markets. Characterising the nutritional quality of these raw materials often demonstrates their nutritional value exceeds the physical characteristics that may have compromised their suitability as a human food product, making them comparatively valuable components of pig diets.

The FAO estimates that each year, approximately one-third of all food produced for human consumption in the world is lost or wasted (FAO, 2013). This food wastage represents a missed opportunity not only to improve global food security, but also to mitigate environmental impacts and resources use from food chains. Information in terms of this food waste being reused by other industries such as the animal industry is scarce but one can hazard a guess that this is more likely to be low.

Food waste has historically been recycled as livestock feed, particularly for pigs, where cooked food waste (swill) was fed to pigs. As a monogastric species, the pigs' digestive system is well adapted for the conversion of food waste into animal protein (Westendorf, 2000). Swill can be a high-quality animal feed that requires no additional land to be brought into production, and hence has minimal or even positive environmental impact (food waste otherwise disposed). However, the use of swill is prohibited in many countries due to biosecurity and food safety concerns/regulation. In addition, collection and processing costs, nutrient value and even implications on product quality are issues that further hinder the use of food waste as an animal feed ingredient. Hence, for the most part food waste remains a largely untapped resource for the pig industry.

The use of co-products from manufacturing processes (dairy and oilseeds) and surplus human food (bakery products, dairy, confectionary, horticulture, etc.) are, however, well utilised by the pig industry and have been so for quite some time.

The use of distillers dried grains with solubles (DDGS) has been researched for over 50 years. It is now well established for use in just about all pig diets beginning from 2–3 weeks' post weaning in concentrations up to 30% DDGS, through to the lactating and gestating sow up to 30 and 50%, respectively, without any negative effect on growth performance (Stein and Shurson, 2009). The impact on carcass quality, however, is also well documented with an increase in the iodine value of fat (softer fat) from pigs fed DDGS.

There is considerable interest in developing micro- and macroalgae as a sustainable resource that can be used to supply a substantial part of the diet to all phases of pig production. Much of the earlier work has focused on the production of algae as a method of reducing the carbon output of many heavy industries that exhaust significant carbon to the environment. The algal product that is produced as a co-product is available to the animal industries as a feed ingredient. Despite the potential use of algae as a valuable protein source for animals, there has been very little research on the evaluation of algae as a feed ingredient for animals over the past 20–30 years. Lee et al. (1980) fed pigs primarily *Micractinium* sp. and *Scenedesmus* sp. containing ~59% protein, and reported that pig diets containing 8% algae provided comparable performance to corn–soybean meal-fed control groups. Similarly, Taiganides (1992) reported, albeit on a very small study, that steam-boiled algae could replace half of the protein in a pig diet (normally provided by soybean meal) with no significant reduction in growth rate. More recently, Henman (2012) showed that a 10% inclusion of algae meal (as direct replacement for canola meal) in the weaner diet was found to have no negative growth performance effects. As we look to reduce our environmental footprint of pork production, waste management is increasingly being viewed as ways to reduce production costs and generate additional revenue. The ultimate aim here is the exploitation of microalgae for both on-farm wastewater treatment and as a source of feed for pigs. Significant research in Australia is being directed to this very aim. Whilst the challenges of growing algae on on-farm wastewater are significant, the challenges to generate a suitable pig feed ingredient in significant quantities remain a major hurdle. In addition, the food safety and biosecurity risks, similar to those posed by the use of swill in pig production, further complicate the use of such ingredients in pig feed.

The role of a variety of insects that are part of a human diet in many countries is quickly becoming a mainstream but 'micro-niche' protein for human consumption. Insect meal, as it is sometimes referred to by media, may also provide an opportunity to utilise certain food wastes to generate novel pig feed ingredients.

7 Optimising gut health and nutrient utilisation capacity

At a time where antimicrobial resistance (AMR) poses a very real threat to human health, the animal industries, whilst unfairly blamed by media as the sole cause of these issues, do contribute to some of the risk around AMR in pigs, farm workers and the general community. The use of antibiotics as growth promoters was extensive in most pig production systems; however, their continued use irrespective of efficacy is unsustainable. Over the last ten years we are seeing concerted effort to minimise and eventually cease their use globally. Some countries have even gone as far as banning their use in pig production, although the corresponding and significant increase in therapeutic use of antibiotics suggests that perhaps such bans may be less than effective and may take longer for industries to adhere to. Needless to say, the issue of antibiotic use and AMR has caused an increase in both the use and development of alternative nutritional strategies to antibiotic feed additives to improve pig growth performance.

'Gut health' is a term that we define as describing a generalised condition of homeostasis in the gastrointestinal tract (GIT) of the pig (Pluske et al., 2007). As can be expected, the factors and conditions involved in 'gut health' are multifactorial, complex and in many instances poorly described and sometimes incorrectly interpreted, although it is evident that perturbations of the GIT that cause disease, even sub-clinical disease, are a disturbance of this homeostasis. In addition to enteric disease, other factors including the period immediately after weaning, changes in diet and disruptions to meal patterns that disrupt the flow of nutrients through the GIT also affect gut health (Pluske et al., 2007). Gut health can be viewed as an outcome (positive, negative, the status quo) of the complex interactions occurring in the GIT between nutrition (e.g. feed type, feed composition, presence or absence of antibiotic feed additives), the mucosa of the GIT (e.g. type of receptors, mucin, inflammatory activity, cytokine production and barrier function) and the microbiota (e.g. density, location, pathogenicity and extent of colonisation; Pluske et al., 2007). Whilst much of our research focus has been directed to the newly weaned pig, it is important to consider that growing-finishing pigs and sows also suffer from GIT disruptions causing decreased growth performance, morbidity and sometimes death.

In many ways, it is impossible to talk about gut health and growth performance without discussing the role of in-feed antibiotics. In-feed antibiotic additives have been used extensively and successfully in pig diets over many years to improve health and performance. The effect of antibiotic feed additive use in pig production has been variable to say the least, with the relative improvements in growth rate resulting from adding antibiotics to diets for pigs being inversely related to the growth rate of control animals (Braude et al., 1953; cited by Page, 2006). Risks associated predominately with antibiotic resistance of microbes in humans have caused the banning or restriction of antibiotics use as feed additives for pig production in some parts of the world.

8 Understanding nutrition and health interactions

The feed additives/nutritional strategies claiming to alter the microbiota of the GIT to influence 'health' of GIT resulting in improved growth performance have increased since this focus on AMR, and a range of these additives are discussed below.

Zinc oxide (ZnO) is one such product that appears to influence and benefit 'gut health' and growth performance. Numerous studies have shown the production and anti-diarrhoeal benefits of including ZnO at supraphysiological levels (i.e. 2000–3000 mg/kg or ppm) in the diet after weaning, but what is less clear is/are the mechanism(s) whereby ZnO exerts its beneficial effects (Pluske et al., 2007). Reported effects include the increased gene expression of antimicrobial peptides in the small intestine (Wang et al., 2004); positive effects on the stability and diversity of the microflora, particularly with respect to coliforms (Katouli et al., 1999) increased IGF-I and IGF-IR expression in the small intestinal mucosa (Li et al., 2006); bactericidal functions (Jensen-Waern et al., 1998) and reductions in electrolyte secretion *in vitro* from enterocytes (Carlson et al., 2006). Hedemann et al. (2006) found changes in some pancreatic enzymes and mucin staining but concluded that there were no definite answers as to how the growth-promoting and diarrhoea-reducing effects of excess dietary Zn were exerted. Nevertheless, some studies have reported no benefits of feeding ZnO (e.g. Jensen-Waern et al., 1998; Broom et al., 2006), perhaps again a reflection of the 'growth rate of control animals'.

The restrictions on or the banning of in-feed antibiotics has seen an increase in the use of probiotics in the period after weaning, to improve gut health and maintain or improve growth performance. Hillman (2001) showed that intestinal lactobacilli having anti-pathogenic ('probiotic') activity against pathogenic *Escherichia coli* that possessed K88 fimbriae were not evenly distributed across 19 Scottish pig farms. Whilst the need for such probiotics may be warranted, the fact that we are relying on the intake (feed, water) at a time when feed intake is both low and variable (Pluske et al., 1997) certainly has contributed to the success (or not) of such products on improving gut health and growth performance of the young pig. With regard to efficacy of probiotics after weaning, the published data defining the benefits of probiotics for nursery pigs are equivocal, which is not a real surprise given the different species and strains that are used and the wide array of weaning and feeding conditions that products work under (Pluske, 2006).

Work with fermented liquid feed (Demeckova et al., 2002) and *Bacillus cereus* var. *toyoi* (e.g. Taras et al., 2005) suggested that an altered microbiota in the faeces of the dam caused by changing the microbiota of the sow exerts a beneficial influence on both pre- and post-weaning development of the young pig. In the study by Taras et al. (2005), one group of sows were fed for 17 weeks, from Day 24 after mating to Day 28 after farrowing, and the piglets from these sows were fed for six weeks, from Day 15 of lactation to eight weeks of age. The control group of sows/piglets did not receive the probiotic strain. The *Bacillus cereus* var. *toyoi* strain was recovered from the faeces of sows and piglets throughout the trial, including the period 0–14 days of age before introduction of the starter diet occurred, and there was an improvement in FCR of pigs in the post-weaning period derived from sows fed the probiotic during pregnancy and lactation.

Studies by Pluske et al. (2002) and Hampson et al. (2006) showed that feeding a diet low in both soluble non-starch polysaccharides (NSP) and resistant starch (RS) generally afforded protection against *B. hyodysenteriae*. However, the way in which these grains have been processed also appears to be important, especially with cereals inherently low in NSP (<1 g/100 g soluble NSP). The data suggested that a reduction in RS levels (e.g. via extrusion and steam flaking) would only prove effective against SD if the grain in question has an inherently low NSP level to begin with (Durmic et al., 2002). Durmic et al. (2002) reported that colonisation by spirochaetes was highly related to the dietary concentrations of soluble NSP, while development of SD was similarly influenced by the RS content of the diet.

The addition of plasma products to diets has revolutionised the feeding of newly weaned pigs, and especially of pigs weaned before 18 days of age, in the last 25 years. Products such as spray-dried porcine plasma (SDPP), where used, generally enhance performance through increased feed intake and feed efficiency in the immediate post-weaning period. Mechanisms for such effects have been studied (see review by Lallès et al., 2004), including immunoglobulin-independent, glycoprotein-enhanced protection against *E. coli* and the specific protection brought about by plasma immunoglobulins (van Dijk et al., 2001). Spray-dried plasma protein was shown to lower the levels of gene expression of inflammatory cytokines after an Enterotoxic Escherichia coli (ETEC) challenge (Bosi et al., 2004).

Numerous reviews have been written regarding the use of organic and inorganic acids in pig diets (e.g. Partanen, 2001; Decuypere and Dierick, 2003; Mroz et al., 2006). The positive effects of feeding acids to pigs on gut health and development, and indirectly on pig health and productivity, may be attributed to various factors. These include (1) antimicrobial activity of non-dissociated organic acids; (2) lowering digesta pH, in particular in the stomach, aiding protein digestion; (3) lowering stomach emptying rate;

(4) stimulating (pancreatic) enzyme production and activity in the small intestine; and (5) providing nutrients that are preferred by intestinal tissue thereby enhancing mucosal integrity and function. Because of these beneficial and synergistic effects, different combinations of organic and inorganic acids are used widely in diets for newly weaned pigs, and used increasingly in diets for growing-finishing pigs and sows. The effectiveness of feeding acids to pigs will vary with the types and combinations of acid, the animal's state and feed characteristics, in particular the diet's buffering capacity (Blank et al., 1999; Mroz et al., 2006). A relatively recent development is the encapsulation of acids for targeted delivery to different gut segments. Studies such as those conducted by Piva et al. (2007) have shown that relatively simple encapsulation is effective in delaying absorption of dietary acids and allowing more effective delivery of acids to the distal ileum, caecum and colon of piglets.

Halas et al. (2009) tested the hypothesis that feeding diets supplemented with inulin (8% of the diet) and benzoic acid (0.5% of the diet) will reduce faecal ETEC shedding and the incidence of Post-weaning diarrhea (PWD), following oral dosing of pigs after weaning with 3 ml of broth containing 3×10^7, 2×10^9, 1×10^{10} and 5×10^8 colony-forming units (CFU) of a freshly grown strain of ETEC serotype O149:K91:K88 (toxins LT, STa, and STb). Dietary supplementation with inulin, either alone or in combination with benzoic acid, reduced the number of days of diarrhoea in pigs weaned at 21 days of age without significantly affecting average ETEC shedding.

Phosphorylated mannans derived from yeast (*Saccharomyces cerevisiae*) cell wall-stimulated phagocytosis of intestinal macrophages and increased the proportion of CD8+ T lymphocytes in the jejunal lamina propria, in addition to improving growth performance (Davis et al., 2004). The percentages of neutrophils and lymphocytes in plasma were lower and higher, respectively, in the supplemented pigs, suggesting a lower level of stress with mannan in these pigs. In another study, however, plasma haptoglobin was higher in the mannan-supplemented pigs (Burkey et al., 2004). Diet supplementation with live yeast also increased the numbers of macrophages in the ileum (Bontempo et al., 2006).

Carver and Walker (1995) extensively reviewed the role of nucleotides in human nutrition. The potential beneficial effects to the immune system, small intestine growth and development, lipid metabolism and hepatic function highlighted the potential application to animal nutrition. The inclusion of supplemental nucleotides in livestock diets for similar benefits is far from novel; however, using yeast extract as the delivery vehicle has only come about in the last 10–15 years.

Amino acids have come into consideration for specific usage as our ability to measure their influence has increased. Burrin and Stoll (2003) reviewed a number of studies with reference to the amino acid arginine and suggested that given the considerable advances made in the understanding of intestinal nutrient utilisation and metabolism, an alternate or even complimentary goal in nutrition might be to formulate young pig diets with the specific task of optimising the growth, function and health of the gut.

Glutamine is an abundant amino acid found in plant and animal proteins and due to its abundance, it has not traditionally been considered a nutrient needed in livestock diets. However, advances in analytical methods have shown the abundance of glutamine within the fluids and tissues of the body, suggesting that it may be more important than we have traditionally thought, especially for the young piglet. A recent review by Wu et al. (2011) highlighted the important roles of glutamine. This information has led to various studies being conducted in the weaned pig. Johnson et al. (2006) fed 21-day-old weaned pigs diets containing an additional 4.38% glutamine for 14 days. Whilst no effect was seen on feed intake or final weight, they saw significant modifications, potentially beneficial, of immune cells in mesenteric lymph nodes. As a result of these studies and work of their own, Wu et al. (2011) have suggested that specific diets should be supplemented with 1% glutamine. A gestation diet fed from Day 90 onwards would ameliorate foetal growth retardation and pre-weaning mortality. A lactation diet containing 1% glutamine would enhance milk production, and in creep diets it would help maintain gut health, prevent intestinal dysfunction, and increase growth and survival.

The growth performance benefits of products containing yeast extract and peptides to replace plasma in the young pigs have been previously summarised (Tibbetts, 2000). Mahan (1999) demonstrated that peptide protein products in combination with yeast extract could completely replace plasma protein if diets were correctly formulated. In diets not containing blood plasma, these vegetable proteins have been shown to improve feed intake against controls and other animal protein sources (Maribo, 2001). Carlson et al. (2001) conducted a study to evaluate the effect of plasma and yeast extract protein supplementation in the young pigs on lifetime performance. Overall average daily gain during the finisher period and slaughter weight for pigs from the yeast extract protein nursery group was significantly higher compared to gains of pigs from either of the control and plasma treatment groups. Since then, numerous studies have been conducted to further demonstrate the growth performance benefits of yeast extract protein.

9 Future trends and conclusion

In looking back, our understanding of the nutritional requirements and metabolism of the pig has been significantly advanced over the last 10–20 years. From the way we formulate diets, our understanding of the feed ingredients and the nutritional technologies have enabled us to improve the performance of the pig. The magnitude of progress is evident when we compare the performance of pigs today compared to late 1980s; whole of life average daily gain has increased from 450 g/day to >720 g/day, whilst herd feed conversion has decreased from >5.5 to 3.7 kg feed/kg carcass. During this time, significant technologies such as pST and ractopamine that have been key contributors to this improved performance are no longer viable technologies for use by the pork industry. Some of these technologies whilst still effective are no longer viable due to shifting consumer perceptions.

The risk associated with AMR in humans and pigs has seen the use of in-feed antibiotics becoming increasingly unsustainable. This has given rise to a range of new technologies that have positively contributed to both growth performance and animal health. The interaction between nutrition and health is especially apparent in the young pig where technologies to alleviate the growth check is seen as a consequence of weaning.

As we look forward, pig production systems are more than likely to undergo significant change, if the last 10–20 years are anything to go by. Changes in genetics, housing systems, animal health care, as well as systems shaped by changing environmental regulation and consumer attitudes are then likely to affect the nutrient requirements of the modern pig. We will need to pay particular attention to the nutritional requirements of the pig. Our reliance on theoretical data rather than actual nutrient requirement data means we are particularly at risk of 'assuming' we are feeding pigs to grow as closely possible to their genetic potential. The same applies for our estimation of the nutrient specification of feed ingredients. Technologies such as NIR and other 'precision farming' technologies must be better utilised to ensure that the diets we are feeding our pigs actually meet their requirement. Whilst feeder design and innovation to enable us to tailor diets to meet individual pig's requirements, still a distant possibility, does present us with opportunities to advance our understanding of the pigs' nutritional requirements and metabolism.

10 Where to look for further information

This chapter has highlighted the areas within pig nutritional requirements and metabolism that have the potential to impact the sustainability of pig production. When considering these areas, it is imperative to concentrate on the primary drivers of sustainability as outlined by groups like the FAO in documents such as *Building a Common Vision for Sustainable Food and Agriculture – Principles and Approaches* (FAO, 2014).

Our knowledge of nutritional requirements and metabolism continues to evolve, aided by advances in technology and methodology; therefore the most up-to-date information will continue to be found in the scientific journals. A range of references were reviewed in the writing of this chapter and these papers would be an ideal starting point to further understand this topic. In particular, *Metabolic modifiers as performance-enhancing technologies for livestock production* (Dunshea et al., 2016) is a good review of the impact of metabolic modifiers on production, their role in sustainable production and the issues affecting their wider use. *Advances in Pork Production* is the proceedings of the Banff Pork Seminar and has provided a number of references used in this chapter. Covering a range of topics the technology transfer seminars invited papers and abstracts, generally, of high value and have added considerably to our understanding of pig production.

11 References

Alexandratos, N. and Bruinsma, J. (2012). World agriculture towards 2030/2050: the 2012 revision. ESA Working paper No. 12-03. Rome, FAO.

Apple, J. K., Rincker, P. J., McKeith, F. K., Carr, S. N., Armstrong, T. A. and Matzat, P. D. (2007). Meta-analysis of the ractopamine response in finishing swine. *Prof. Anim. Sci.* 23, 179–96.

Ball, R. O., Samuel, R. S. and Moehn, S. (2008). Nutrient requirements of prolific sows. *Adv. Pork Prod.* 19, 223–36.

Bauman, D. E. and Dunshea, F. R. 2010. Somatropin. In Pond, W. G. and Bell, A. W. (Eds), *Encyclopedia of Animal Science*. Second Edition, pp. 995–8. Taylor and Francis: London, UK.

Beltranena, E., Patterson, J., Foxcroft, G. and Pettitt, M. (2005). 12 Non-negotiable aspects of gilt development Western Hog Journal Summer 2004, pp. 45–50.

Bergsma, R., Kanis, E., Verstegen, M. W. A., van der Peet-Schwering, C. M. C. and Knol, E. F. (2009). Lactation efficiency as a result of body composition dynamics and feed intake in sows. *Livest. Sci.* 125, 208–22.

Black, J. L., Flinn, P. C., Diffey, S. and Tredrea, A. M. (2013). An update on near infrared reflectance analysis of cereal grains to estimate digestible energy content for pigs. In Pluske, J. R. and Pluske J. M. (Eds), *Manipulating Pig Production XIV*, p. 54. Australasian Pig Science Association: Werribee, VIC, Australia.

Blank, R., Mosenthin, R., Sauer, W. C. and Huang, S. (1999). Effect of fumaric acid and dietary buffering capacity on ileal and fecal amino acid digestibilities in early-weaned pigs. *J. Anim. Sci.* 77, 2974–84.

Bonneau, M. and Enright, W. J. (1995) Immunocastration in cattle and pigs. *Livest. Prod. Sci.* 42, 193–200 doi: (10.1016/0301-6226(95)00020-I).

Bontempo, V., Di Giancamillo, A., Savoini, G., Dell'Orto, V. and Domeneghini, C. (2006) Live yeast dietary supplementation acts upon intestinal morpho-functional aspects and growth in weanling piglets. *Anim. Feed Sci. Technol.* 129, 224–36.

Bosi, P., Casini, L., Finamore, A., Cremokolini, C., Merialdi, G., Trevisi, P., Nobili, F. and Mengheri, E. (2004). Spray-dried plasma improves growth performance and reduces inflammatory status of weaned pigs challenged with enterotoxigenic Escherichia coli K88. *J. Anim. Sci.* 82, 1764–72.

Braude, R., Wallace, H. D. and Cunha, T. J. (1953). The value of antibiotics in the nutrition of swine. A review. *Antibiot. Chemother.* 3, 271–91.

Brooks, P. H., Smith, D. A. and Irwin, V. C. R. (1977). Biotin supplementation of diets; the incidence of foot lesions, and the reproductive performance of sows. *Vet. Rec.* 101, 46–50.

Broom, L. J., Miller, H. M., Kerr, K. G. and Knapp, J. S. (2006). Effects of zinc oxide and Enterococcus faecium SF68 dietary supplementation on the performance, intestinal microbiota and immune status of weaned piglets. *Res. Vet. Sci.* 80, 45–54.

Burkey, T. E., Dritz, S. S., Nietfeld, J. C., Johnson, B. J. and Minton, J. E. (2004). Effect of dietary mannanoligosaccharide and sodium chlorate on the growth performance, acute-phase response, and bacterial shedding of weaned pigs challenged with Salmonella enterica serotype Typhimurium. *J. Anim. Sci.* 82, 397–404.

Burrin, D. and Stoll, B. (2003). Intestinal nutrient requirements in weanling pigs. In Pluske, J. R., Verstegen, M. W. A. and Le Dividich, H. (Eds), *The Weaner Pig: Concepts and Consequences*, pp. 301–35. Wageningen Academic Publishers Wageningen,: The Netherlands.

Cadogan, D. (2009). Net energy defines lean and fat deposition better than digestible energy. Final Report, Project 2G-104. The Co-operative Research Centre for High Integrity Australian Pork, Roseworthy, SA, Australia.

Campbell, R. G., Steele, N. C., Caperna, T. J., McMurtry, J. P., Solomon, M. B. and Mitchell, A. D. (1988). Interrelationships between energy intake and exogenous porcine growth hormone administration on the performance, body composition and protein and energy metabolism of growing pigs weighing 25 to 55 kilograms body weight. *J. Anim. Sci.* 66, 1643–55.

Campbell, R. G., Steele, N. C., Caperna, T. J., McMurtry, J. P., Solomon, M. B. and Mitchell, A. D. (1989). Effects of exogenous porcine growth hormone administration between 30 and 60 kilograms on the subsequent and overall performance of pigs grown to 90 kilograms. *J. Anim. Sci.* 67, 1265–71.

Carlson, C. S., Hilley, H. D., Meuten, D. J., Hagan, J. M. and Moser, R. L. (1988). Effect of reduced growth rate on the prevalence and severity of osteochondrosis in gilts. *Am. J. Vet. Res.* 49, 396–402.

Carlson, D., Sehested, J. and Poulsen, H. D. (2006). Zinc reduces the electrophysiological responses in vitro to basolateral receptor mediated secretagogues in piglet small intestinal epithelium. *Comp. Biochem. Physiol. A Mol. Integr. Physiol.* 144, 514–19.

Carlson, M. S., Veum, T. L., Turk, J. R., Bollinger, D. W. and Tibbetts, G. W. (2001). A comparison between feeding either peptide or plasma proteins with or without a feed grade antibiotic on pig growth performance and intestinal health. Unpublished. University of Missouri, Columbia, MO, USA.

Carver, J. D. and Walker, W. A. (1995). The role of nucleotides in human nutrition. *J. Nutr. Biochem.* 6, 58–72.

Chung, C. S., Etherton, T. D. and Wiggins, J. P. (1985). Stimulation of swine growth by porcine growth hormone. *J. Anim. Sci.* 60, 118–30.

Clowes, E. J., Aherne, F. X., Foxcroft, G. R. and Baracos, V. E. (2003). Selective protein loss in lactating sows is associated with reduced litter growth and ovarian function. *J. Anim. Sci.* 81, 753–64.

Cools, A., Maes, D., Decaulwe, R., Buyse, J., van Kempen, T. A. T. G., Liesegang, A. and Janssens, G. P. J. (2014). Ad libitum feeding during the peripartal period affects body condition, reproduction results and metabolism of sows. *Anim. Reprod. Sci.* 145, 130–40.

Cronin, G. M., Dunshea, F. R., Butler, K. L., McCauley, I., Barnett, J. L. and Hemsworth, P. H. (2003). The effects of immuno- and surgical-castration on the behaviour and consequently growth of group-housed, male finisher pigs. *Appl. Anim. Behav. Sci.* 81, 111–26.

Davis, M. E., Maxwell, C. V., Erf, G. F., Brown, D. C. and Wistuba, T. J. (2004). Dietary supplementation with phosphorylated mannans improves growth response and modulates immune function of weanling pigs. *J. Anim. Sci.* 82, 1882–91.

De Bettio, S., Majorka, A., Barrilli, L. N. E., Bergsma, R. and Silva, B. A. N. (2016). Impact of feed restriction on the performance of highly prolific lactating sows and its effect on the subsequent lactation. *Animal* 10, 396–402.

de Lange, C. F. M. and Birkett, S. H. (2005). Characterization of useful energy contents in swine and poultry feed ingredients. *Can. J. Anim. Sci.* 85, 269–80.

de Lange, C. F. M. (2013). New NRC (2012) nutrient requirements of swine. *Adv. Pork Prod.* 24, 17–28.

De Moraes, P. J. U., Allison, J., Robinson, J. A., Baldo, G. L., Boeri, F. and Borla, P. (2013). Life cycle assessment (LCA) and environmental product declaration (EPD) of an immunological product for boar taint control in male pigs. *J. Environ. Assess. Policy Manag.* 15, 1350001–1350026.

Decuypere, J. A. and Dierick, N. A. (2003). The combined use of triacylglycerols containing medium-chain fatty acids and exogenous lipolytic enzymes as an alternative to in-feed antibiotics in piglets: concept, possibilities and limitation- An overview. *Nutr. Res. Rev.* 16, 193–209.

Demeckova, V., Kelly, D., Coutts, A. G. P., Brooks, P. H. and Campbell, A. (2002). The effect of fermented liquid feeding on the faecal microbiology and colostrum quality of farrowing sows. *Int. J. Food Microbiol.* 79, 85–97.

Donovan, T. S. and Dritz, S. S. (2000). Effect of split nursing on variation in pig growth from birth to weaning. *J. Am. Vet. Med. Assoc.* 217, 79–81.

Douglas, S. L., Edwards, S. A. and Kyriazakis, I. (2014). Management strategies to improve the performance of low birth weight pigs to weaning and their long-term consequences. *J. Anim. Sci.* 92, 2280–8.

Douglass, G. K. (1984). The meanings of agricultural sustainability. In Douglass, G. K. (Ed.), *Agricultural Sustainability in a Changing World Order*, pp. 3–29. Westview Press: Boulder, CO, USA.

Dowling, H. J., Offenbacher, E. G. and Pi-Sunyer, F. X. (1989). Absorption of inorganic, trivalent chromium from the vascularly perfused rat small intestine. *J. Nutr.* 119, 1138–45.

Ducros, V. (1992). Chromium metabolism. A literature review. *Biol. Trace Elem. Res.* 32, 65–77.

Dunshea, F. R. (1993). Effect of metabolism modifiers on lipid metabolism in the pig. *J. Anim. Sci.* 71, 1966–77.

Dunshea, F. R. (2005). Sex and porcine somatotropin impact on variation in growth performance and back fat thickness. *Aust. J. Exp. Agr.* 45, 677–82.

Dunshea, F. R., Allison, J. R. D., Bertram, M., Boler, D. D., Brossard, L., Campbell, R., Crane, J. P., Hennessy, D. P., Huber, L., de Lange, C., Ferguson, N., Matzat, P., McKeity, F., Moraes, P. J. U., Mullan, B. P., Noblet, J., Quiniou, N. and Tokach, M. (2013). The effect of immunization against GnRF on nutrient requirements of male pigs: a review. *Animal* 7, 1769–78.

Dunshea, F. R., Colantoni, C., Howard, K., McCauley, I., Jackson, P., Long, K. A., Lopaticki, S., Nugent, E. A., Simons, J. A., Walker, J. and Hennessy, D. P. (2001). Vaccination of boars with a GnRH vaccine (Improvac) eliminates boar taint and increases growth performance. *J. Anim. Sci.* 79, 2524–35.

Dunshea, F. R., D'Souza, D. N. and Channon, H. A. (2016). Metabolic modifiers as performance-enhancing technologies for livestock production. *Anim. Front.* 6, 6–14.

Dunshea, F. R., D'Souza, D. N., Pethick, D. W., Harper, G. S. and Warner, R. D. (2005). Effects of dietary factors and other metabolic modifiers on quality and nutritional value of meat. *Meat Sci.* 71, 8–38.

Durmic, Z., Pethick, D. W., Mullan, B. P., Accioly, J. M., Schulze, H. and Hampson, D. J. (2002). Evaluation of large-intestinal parameters associated with dietary treatments designed to reduce the occurrence of swine dysentery. *Br. J. Nutr.* 88, 159–69.

Etherton, T. D., Wiggins, J. P., Evock, C. M., Chung, C. S., Rebhun, J. F., Walton, P. E. and Steele, N. C. (1987). Stimulation of pig growth performance by porcine growth hormone: determination of the dose-response relationship. *J. Anim. Sci.* 64, 433–43.

Fairlie, S. (2010). *Meat: A Benign Extravagance*. Permanent Publications: East Meon UK,.

FAO (2013). *Food Wastage Footprint – Impacts on Natural Resources*. Food and Agriculture Organization of the United Nations: Rome, Italy.

FAO (2014). *Building a Common Vision for Sustainable Food and Agriculture – Principles and Approaches*. Food and Agriculture Organization of the United Nations: Rome, Italy.

Foley, J. A., Ramankutty, N., Brauman, K. A., Cassidy, E. S., Gerner, J. S., Johnston, M., Mueller, N. D., O'Connell, C., Ray, D. K., West, P. C., Balzer, C., Bennett, E. M., Carpenter, S. R., Hill, J., Monfreda, C., Polasky, S., Rockstrom, J., Sheehan, J., Siebert, S., Tilman, D. and Zaks, D. P. M. (2011). Solutions for a cultivated planet. *Nature* 478, 337–42.

Foxcroft, G. R. (1997). Mechanisms mediating nutritional effects on embryonic survival in pigs. *J. Reprod. Fertil.* 52 (Supp.), 47–61.

Foxcroft, G. R. (2007). Pre-natal programming of variation in post-natal performance: how and when? *Adv. Pork Prod.* 18, 167–89.

Foxcroft, G. R. and Town, S. (2004). Prenatal programming of postnatal performance: the unseen cause of variance. *Adv. Pork Prod.* 15, 269–79.

Gerber, P. J., Steinfeld, H., Henderson, B., Mottet, A., Opio, C., Dijkman, J., Falcucci, A. and Tempio, G. (2013). *Tackling Climate Change Through Livestock – A Global Assessment of Emissions and Mitigation Opportunities*. Food and Agriculture Organization of the United Nations (FAO): Rome.

Giles, D. D. (1942). An experiment to determine the effect of growth hormone of the anterior lobe of the pituitary gland on swine. *Am. J. Vet. Res.* 3, 77–86.

Greer, E. B., Leibholz, J. M., Pickering, D. I., Macoun, R. E. and Bryden, W. L. (1991). Effect of supplementary biotin on the reproductive performance, body condition and foot health of sows on three farms. *Aust. J. Agr. Res.* 42, 1013–21.

Gresham, A. (2003). Infectious reproductive disease in pigs. *In Practice* 25, 466–73.

Halas, D., Hansen, C. F., Hampson, D. J., Mullan, B. P., Wilson, R. H. and Pluske, J. R. (2009). Effect of dietary supplementation with inulin and/or benzoic acid on the incidence and severity of post-weaning diarrhoea in weaner pigs after experimental challenge with enterotoxigenic Escherichia coli. *Arch. Anim. Nutr.* 63, 267–80.

Hamilton, C. R. and Veum, T. L. (1984). Response of sows and litters to added biotin in environmentally regulated facilities. *J. Anim. Sci.* 59, 151–7.

Hampson, D. J., Fellstrom, C. and Thomson, J. R. (2006). Swine dysentery. In Straw, B. E., Zimmerman, J. J., D'Allaire, S. and Taylor, D. J. (Eds), *Diseases of Swine*, pp. 785–805. Blackwell Publishing: Iowa, USA.

Hedemann, M. S., Jensen, B. B. and Poulsen, H. D. (2006). Influence of dietary zinc and copper on digestive enzyme activity and intestinal morphology in weaned pigs. *J. Anim. Sci.* 84: 3310–20.

Henman, D. J. (2012). Evaluation of Algal meal as an energy and protein source in pig diets. Final Report, Project 4A-102. The Co-operative Research Centre for High Integrity Australian Pork, Roseworthy, SA, Australia.

Hewitt, R. J. E., Corso, A. and van Barneveld, R. J. (2015). Reducing variation in finisher growth performance through early post-weaning dietary intervention. *Anim. Prod. Sci.* 55, 1573.

Hillman, K. (2001). Bacteriological aspects of the use of antibiotics and their alternatives in the feed of non-ruminant animals. In Garnsworthy, P. C. and Wiseman, J. (Eds), *Recent Advances in Animal Nutrition*, pp. 107–34. Nottingham University Press: Loughborough, UK.

Hoving, L. L., Soede, N. M., Feitsma, H. and Kemp, B. (2012). Embryo survival, progesterone profiles and metabolic responses to an increased feeding level during second gestation in sows. *Theriogenology* 77, 1557–69.

Hung, A. T., Leury, B. J., Sabin, M. A., Lien, T. F. and Dunshea, F. R. (2015). Dietary chromium picolinate of varying particle size improves carcass characteristics and insulin sensitivity in finishing pigs fed low- and high-fat diets. *Anim. Prod. Sci.* 55, 454–60.

Jensen-Waern, M., Melin, L., Lindberg, R., Johannisson, A., Petersson, L. and Wallgren, P. (1998). Dietary zinc oxide in weaned pigs-effects on performance, tissue concentrations, morphology, neutrophil functions and faecal microflora. *Res. Vet. Sci.* 64, 225–31.

Johnson, I. R., Ball, R. O., Baracos, V. E. and Field, C. J. (2006). Glutamine supplementation influences immune development in the newly weaned piglet. *Dev. Comp. Immunol.* 30, 1191–202.

Katouli, M., Melin, L., Jensen-Waern, M., Wallgren, P. and Mollby, R. (1999). The effect of zinc oxide supplementation on the stability of the intestinal flora with special reference to composition of coliforms in weaned pigs. *J. Appl. Microbiol.* 87, 564–73.

Kil, D. Y., Kim, B. G. and Stein, H. H. (2013). Feed energy evaluation for growing pigs. Asian-Australasian *J. Anim. Sci.* 26, 1205–17.

Kim, J. C. and Mullan, B. P. (2012). Quantification of the variability in the amino acid and reactive lysine content of soybean meal and development of a NIR calibration for rapid prediction of reactive lysine content. Final Report, Project 4B–106. The Co-operative Research Centre for High Integrity Australian Pork, Roseworthy, SA, Australia.

King, R. H., Campbell, R. G., Smits, R. J., Morley, W. C., Ronnfeldt, K., Butler, K. and Dunshea, F. R. (2000). Interrelationships between dietary lysine, sex, and porcine somatotropin administration on growth performance and protein deposition in pigs between 80 and 120 kg live weight. *J. Anim. Sci.* 78, 2639–51.

Knauer, M. T. (2016). Effect of increasing sow feeding level in late gestation on piglet quality and sow body condition. *J. Anim. Sci.* 94 (Supp 2), 97.

Koketsu, Y., Dial, G. D., Pettigrew, J. E., Marsh, W. E. and King, V. L. (1996). Characterization of feed intake patterns during lactation in commercial swine herds. *J. Anim. Sci.* 74, 1202–10.

Kopinski, J. S. and Leibholz, J. (1989) Biotin studies in pigs. 2. *Br. J. Nutr.* 62, 761–6.

Kopinski, J. S., Leibholz, J., Bryden, W. L. and Fogarty, A. C. (1989). Biotin studies in pigs. 1. *Br. J. Nut.* 62, 751–9.

Kornegay, E. T. (1986). Biotin in swine production: A review. *Livest. Prod. Sci.* 14, 65–89.

Lalles, J. P., Boudry, G., Favier, C., Le Floc'h, N., Lurona, I., Montagne, L., Oswald, I. P., Pie, S., Piel, C. and Seve, B. (2004). Gut function and dysfunction in young pigs: Physiology. *Anim. Res.* 4, 301-16.

Lawlor, P. G., Lynch, P. B., O'Connell, M. K., McNamara, L., Reid, P. and Stickland, N. C. (2007). The influence of over feeding sows during gestation on reproductive performance and pig growth to slaughter. *Arch. Anim. Breed.* 50, 82–91.

Lee, B. Y., Lee, K. W., McGarry, M. G. and Graham, M. (1980). *Treatment and Resource Recovery.* Report of a Workshop on High-Rate Algae Ponds, 27–29 February 1980. Singapore, p. 27. IDRC, Canada.

Lewis, A. J., Cromwell, G. L. and Pettigrew, J. E. (1991). Effects of supplemental biotin during gestation and lactation on reproductive performance of sows: a cooperative study. *J. Anim. Sci.* 69, 207–14.

Li, X. L., Yin, J. D., Li, D. F., Chen, X. J., Zang, J. J. and Zhou, X. (2006). Dietary supplementation with zinc oxide increases IGF-I and IGF-I receptor gene expression in the small intestine of weanling piglets. *J. Nutr.* 136, 1786–91.

Mahan, D. C. (1999). Comparison of plasma protein and Ultimate Protein in the diets of starter pigs. Report to Alltech. The Ohio State University, Columbus,OH, USA.

Mahan, D. C. and Newton, E. A. (1995). Effect of initial breeding weight on macro- and micromineral composition over a three=parity period using a high-producing sow genotype. *J. Anim. Sci.* 73: 151–8.

Maribo, H. (2001). Commercial products for weaners: NuPro 2000 as an alternative protein source for weaners. Report no 256. The National Committee for Pig Production, Danish Bacon and Meat Council, Danske Slagterier, Denmark.

McCauley, I., Watt, M., Suster, D., Kerton, D. J., Oliver, W. T., Harrell, R. J. and Dunshea, F. R. (2003). A GnRF vaccine (Improvac ®) and porcine somatotropin (Reporcin ®) have synergistic effects upon growth performance in both boars and gilts. *Aust. J. Agr. Res.* 54, 11–20.

Mertz, W. (1969). Chromium occurrence and function in biological systems. *Physiol. Rev.* 49, 163–239.

Moore, K. L., Dunshea, F. R., Mullan, B. P., Hennessy, D. P. and D'Souza, D. N. (2009) Ractopamine supplementation increases lean deposition in entire and immunocastrated male pigs. *Anim. Prod. Sci.* 49, 1113–19.

Mroz, Z., Koopmans, S. J., Bannink, A., Partanen, A. K., Krasucki, W., Overland, M. and Radcliffe, S. (2006). Carboxylic acids as bioregulators and gut growth promoters in non-ruminants. In Mosenthin, R., Zentek, J. and Zebrowska, T. (Eds), *Biology of Nutrition in Growing Animals*, pp. 81–133. Elsevier Limited: Amsterdam, The Netherlands.

Nakano, T., Brennan, J. and Aherne, F. (1987). Leg weakness and osteochondritis in swine: a review. *Can. J. Anim. Sci.* 67, 883–901.

National Research Council (1998). *Nutritional Requirements of Swine*. 10th edition. National Academies Press:, Washington, DC USA.

National Research Council (2012). *Nutritional Requirements of Swine*. 11th edition. National Academies Press: Washington, DC, USA.

Nitrayova, S., Heger, J., Patras, P. and Brestensky, M. (2006). Effect of body weight and pig individuality on apparent ileal digestibility of amino acids and total nitrogen. *Slovac J. Anim. Sci.* 39, 65–8.

Noblet, J. and Shi, X. S. (1993). Comparative digestibility of energy and nutrients in growing pigs fed ad libitum and adult sows fed at maintenance. *Livest. Prod. Sci.* 34, 137–52.

Noblet, J. (2006). Recent advances in energy evaluation of feeds for pigs. In Garnsworthy, P. C. and Wiseman, J. (Eds), *Recent Advances in Animal Nutrition*, pp. 1–26. Nottingham University Press: Nottingham, UK.

O'Connell, N. E., Beattie, V. E. and Watt, D. (2005). Influence of regrouping strategy on performance, behavior and carcass parameters in pigs. *Livest. Prod. Sci.* 97, 107–15 doi: (10.1016/j.livprodsci.2005.03.005).

Oliver, W. T., McCauley, I., Harrell, R. J., Suster, D., Kerton, D. J. and Dunshea, F. R. (2003). A gonadotropin-releasing factor vaccine (Improvac) and porcine somatotropin have synergistic and additive effects on growth performance in group-housed boars and gilts. *J. Anim. Sci.* 81, 1959–66.

Page, S. W. (2006). Current use of antimicrobial growth promoters in food animals: The benefits. In Barug, D., de Jong, J., Kies, A. K. and Verstegen, M. (Eds), *Antimicrobial Growth Promoters: Where Do We Go from Here?*, pp. 19–51. Wageningen Academic Publishers: Wageningen, The Netherlands.

Partanen, K. (2001). Organic acids – Their efficacy and modes of action in pigs. In Piva, A., Bach Knudsen, K. E. and Lindberg, J. E. (Eds), *Gut Environment of Pigs*, pp. 201–18. Nottingham University Press: Loughborough, UK.

Patience, J. F., Engele, K., Beaulieu, A. D., Gonyou, H. W. and Zijlstra, R. T. (2004). Variation: costs and consequences. *Adv. Pork Prod.* 15, 257–66.

Penny, R. H. C., Cameron, R. D. A., Johnson, S., Kenyon, P. J., Smith, H. A., Bell, A. W. P., Cole, J. P. L. and Taylor, J. (1981). Influence of biotin supplementation on sow reproductive efficiency. *Vet. Rec.* 109, 80–1.

Piva, A., Pizzamiglio, V., Morlacchini, M., Tedeschi, M. and Piva, G. (2007). Lipid microencapsulation allows slow release of organic acids and natural identical flavors along the swine intestine. *J. Anim. Sci.* 85, 486–93.

Pluske, J. R. (2006). New thoughts on nutrition of newly weaned pigs. In Murphy, J. M. and Kane, T. M. (Eds), Proceedings of the 6th London Swine Conference, pp. 67–77. London Swine Conference: Ontario, Canada.

Pluske, J. R., Hansen, C. F., Payne, H. G., Mullan, B. P., Kim, J. C. and Hampson, D. J. (2007). Gut health in the pig. In Paterson, J. E. and Barker, J. A. (Eds), *Manipulating Pig Production*. Volume XI., pp. 147–58. Australasian Pig Science Association: Werribee,VIC, Australia.

Pluske, J. R., Pethick, D. W., Hopwood, D. E. and Hampson, D. J. (2002). Nutritional influences on some major enteric bacterial diseases of pigs. *Nutr. Res. Rev.* 15, 333–71.

Pluske, J. R., Williams, I. H. and Hampson, D. J. (1997). Factors influencing the structure and function of the small intestine in the weaned pig: a review. *Livest. Prod. Sci.* 51, 215–36.

Quesnel, H., Brossard, L., Valancogne, A. and Quiniou, N. (2008). Influence of some sow characteristics on within-litter variation of piglet birth weight. *Animal* 2, 1842–9.

Quiniou, N., Dagorn, J. and Gaudré, D. (2002). Variation of piglets' birth weight and consequences on subsequent performance. *Livest. Prod. Sci.* 78, 63–70.

Rehfeldt, C., Kuhn, G., Nürnberg, G., Kanitz, E., Schneider, F., Beyer, M., Nürnberg, K. and Ender, K. (2001). Effects if exogenous somatotropin during early gestation on maternal performance, fetal growth, and compositional traits in pigs. *J. Anim. Sci.* 79, 1789–99.

Revell, D. K., Williams, I. H., Mullan, B. P., Ranford, J. L. and Smits, R. J. (1998a). Body composition at farrowing and nutrition during lactation affect the performance of primiparous sows: I. voluntary feed intake, weight loss, and plasma metabolites. *J. Anim. Sci.* 76, 1729–37.

Revell, D. K., Williams, I. H., Mullan, B. P., Ranford, J. L. and Smits, R. J. (1998b). Body composition at farrowing and nutrition during lactation affect the performance of primiparous sows: II. milk composition, milk yield, and pig growth. *J. Anim. Sci.* 76, 1738–43.

Rikard-Bell, C., Curtis, M. A., van Barneveld, R. J., Mullan, B. P., Edwards, A. C., Gannon, N. J., Henman, D. J., Hughes, P. E. and Dunshea, F. R. (2009). Ractopamine hydrochloride improves growth performance and carcass composition in immunocastrated boars, intact boars, and gilts. *J. Anim. Sci.* 87, 3536–43.

Rutherfurd, S. M., Moughan, P. J. and Morel, P. C. H. (1997). Assessment of the true ileal digestibility of reactive lysine as a predictor of lysine uptake from the small intestine of the growing pig. *J. Agr. Food Chem.* 45, 4378–83.

Sales, J. (2011). A meta-analysis of the effects of dietary betaine supplementation on finishing performance and carcass characteristics of pigs. *Anim. Feed Sci. Technol.* 165, 68–78.

Sales, J. and Jancik, F. (2011). Effects of dietary chromium supplementation on performance, carcass characteristics, and meat quality of growing-finisher swine: a meta-analysis. *J. Anim. Sci.* 89, 4054–67.

Schrama, J. W., Heetkamp, M. J., Simmins, P. H. and Gerrits, W. J. (2003). Dietary betaine supplementation affects energy metabolism of pigs. *J. Anim. Sci.* 81, 1202–9.

Steele, N. C., Althen, T. G. and Frobish, L. T. (1977). Biological activity of glucose tolerance factor in swine. *J. Anim. Sci.* 45, 1341–5.

Stein, H. H. and Shurson, G. C. (2009). The use and application of distillers dried grains with solubles in swine diets. *J. Anim. Sci.* 87, 1292–303.

Sugiyama, T., Kusuhara, S., Chung, T. K., Yonekura, H., Azem, E. and Hayakama, T. (2013). Effects of 25-hydroxy-cholecalciferol on the development of osteochondrosis in swine. *Anim. Sci. J.* 84, 341–149.

Suster, D., Leury, B. J., King, R. H., Mottram, M. and Dunshea, F. R. (2004). Interrelationships between porcine somatotropin (pST), betaine, and energy level on body composition and tissue distribution of finisher boars. *Aust. J. Agr. Res.* 55, 983–90.

Taiganides, E. P. (1992). *Pig Waste Management and Recycling: The Singapore Experience.* International Development Research Centre: Ottawa, Canada.

Taras, D., Vahjen, W., Macha, M. and Simon. O. (2005). Response of performance characteristics and fecal consistency to long-lasting dietary supplementation with the probiotic strain Bacillus cereus var. toyoi to sows and piglets. *Arch. Anim. Nutr.* 59, 405–17.

Tibbetts, G. W. (2000). Biopeptides in post-weaning diets for pigs: results to date. In Lyons T. P. and Jacques K.A. (Eds), *Biotechnology in the Feed Industry.* Proceedings of Alltech's 16th Annual Symposium, pp. 347–56. Nottingham University Press: Loughborough, UK.

Turman, E. J. and Andrews, F. N. (1955). Some effects of purified anterior pituitary growth hormone on swine. *J. Anim. Sci.* 14, 7–18.

Underwood, E. J. (1977). Chromium. In Underwood E. J. (Ed.), *Trace Elements in Human and Animal Nutrition.* Fourth edition, pp. 258–70. Academic Press: New York, NY, USA.

van Barneveld, R., Nuttall, J., Flinn, PC. and Osborne, B. (1998). NIR reflectance measurement of the digestible energy content of cereals for growing pigs. *J. Near Infrared Spec.* 7, 1–7 doi: (10.1255/jnirs.228).

van Barneveld, R. J. (1999). Physical and chemical characteristics of grains related to variability in energy and amino acid availability in pigs: a review. *Aust. J. Agr. Res.* 50, 667–87.

van Barneveld, R. J. and Ru, Y. J. (2003). Measurement of total and reactive lysine content of canola meal using near infrared spectroscopy. In Paterson J. E. (Ed.), *Manipulating Pig Production.* Volume IX, p. 157. Australasian Pig Science Association: Werribee, VIC, Australia.

van Barneveld, R. J. (2010). Mineral supplementation in modern pork production systems for optimum health and efficiency – the need for renewed focus. Proceedings of FeetFirst Sow Lameness Symposium II, pp. 66–81. Minneapolis, MN, USA.

van Barneveld, R. J., Graham, H. and Diffey, S. (2017). Predicting the nutritional quality of feed ingredients for pigs using near infra-red spectroscopy (NIRS) and chemical analysis. *Anim. Prod. Sci.* (In press).

van Barneveld, R. J., Nuttall, J. D., Flinn, P. C. and Osborne, B. G. (1999a). Near infrared reflectance measurement of the digestible energy content of cereals for growing pigs. *J. Near Infrared Spec.* 7, 1–7.

van Barneveld, R. J., Ru, Y. J., Szarvas, S. R. and Wyatt, G. F. (1999b). Effect of oil extraction process on the true ileal digestible reactive lysine content of canola meal. In Cranwell P. D. (Ed.), *Manipulating Pig Production*. Volume VII, p. 41. Australasian Pig Science Association: Werribee, VIC, Australia.

van Dijk, A. J., Everts, B., Nabuurs, M. J. A., Margry, R. J. C. F. and Beynen, A. C. (2001). Growth performance of weanling pigs fed spray-dried animal plasma: a review. *Livest. Prod. Sci.* 68, 263–74.

Wang, Y. Z., Xu, Z. R., Lin, W. X., Huang, H. Q. and Wang, Z. Q. (2004). Developmental gene expression of antimicrobial peptide PR-39 and effect of zinc oxide on gene regulation of PR-39 in piglets. *Asian-Australas. J. Anim. Sci.* 17, 1635–40.

Weldon, W. C., Lewis, A. J., Louis, G. F., Kovar, J. L., Giesemann, M. A. and Miller, P. S. (1994). Postpartum hypophagia in primiparous sows: I. Effects of gestation feeding level on feed intake, feeding behaviour, and plasma metabolite concentrations during lactation. *J. Anim. Sci.* 72, 387–194.

Westendorf, M. L. (2000). Food waste as swine feed. In Westendorf, M. L. (Ed.), *Food Waste to Animal Feed*, pp. 69–89. Iowa State University Press: Ames, IA, USA.

Williams, N. H., Patterson, J. and Foxcroft, G. (2005). Non-negotiables of gilt development. *Adv. Pork Prod.* 16, 281–9.

Wolter, B. F., Ellis, M., Corrigan, B. P. and DeDecker, J. M. (2002). The effect of birth weight and feeding of supplemental milk replacer to piglets during lactation on preweaning and postweaning growth performance and carcass characteristics. *J. Anim. Sci.* 80, 301–8.

Woods, A. L., Armstrong, T. A., Anderson, D. B., Elam, T. E. and Sutton, A. L. (2011). Case Study: Environmental benefits of ractopamine use in United States finisher swine. *Prof. Anim. Sci.* 27, 492–9.

Wu, G., Bazer, F. W., Burghardt, R. C., Johnson, G. A., Kim, S. W., Li, X. L., Satterfield, M. C. and Spencer, T. E. (2010). Impacts of amino acid on pregnancy outcome in pigs: mechanisms and implications for swine production. *J. Anim. Sci.* 88, E195–E204.

Wu, G., Bazer, F. W., Davis, T. A., Jaegar, L. A., Johnson, G. A., Kim, S. W., Knabe, D. A., Meininger, C. J., Spencer, T. E. and Yin, Y.-L. (2007). Important roles for the arginine family of amino acids in swine nutrition and production. *Livest. Sci.* 112, 8–22.

Wu, G., Bazer, F. W., Johnson, G. A., Knabe, D. A., Burghardt, R. C., Spencer, T. E., Li, X. L. and Wang, J. J. (2011). Important roles for l-glutamine in swine nutrition and production. *J. Anim. Sci.* 89, 2017–30.

Wu, G., Ott, T. L., Kanbe, D. A. and Bazer, F. W. (1999). Animal acid composition of the fetal pig. *J. Nutr.* 129, 1031–8.

Ytrehus, B., Carlson, C. S. and Ekman, S. (2007). Etiology and pathogenesis of osteochondrosis. *Vet. Pathol.* 44, 429–48.

Ytrehus, B., Grindflek, E., Teige, J., Stubsjøen, E., Grøndalen, T., Carlson, C. S. and Ekman, S. (2004). The effect of parentage on the prevalence, severity and location of lesions of osteochondrosis in swine. *J. Vet. Med. A Physiol. Pathol. Clin. Med.* 51, 188–95.

Chapter 7

Meeting energy requirements in pig nutrition

J. F. Patience, Iowa State University, USA

1 Introduction

Energy is by far the most expensive component of the diet and, as such, influences the cost of feeding pigs more than any nutrient. Apart from its expense, energy is a critical topic because the level of energy in the diet influences the rate and efficiency of gain, the quality of the resulting carcass and even the quality of the pork produced from the carcass. Unfortunately, as important as energy is, it is not a well-understood topic, due to its complexity. Unlike nutrients in the diet, energy is supplied by numerous dietary constituents, namely amino acids; lipids; simple sugars arising from starch, lactose and sucrose; and complex carbohydrates more commonly known as fibre. Each of these is utilized by the pig as an energy source in different manners, and, indeed, the variation in utilization is also influenced by its metabolic fate.

The study of energy in nutrition goes back about three centuries, and during that period, much has been learned about the different sources of energy and how to formulate diets to meet the energy needs of the pig. Evolving from this research have been energy systems, such as digestible energy (DE), metabolizable energy (ME), net energy (NE) and, most recently in Denmark, the potential physiological energy (PPE) system. Each has advantages and disadvantages, and is used to a significant extent in various countries around the world.

http://dx.doi.org/10.19103/AS.2017.0013.07

Defining the energy requirement of the pig, as a unique and distinct quantity, is not possible and rarely attempted. This is due to the fact that the pig responds to dietary energy in a continuous fashion, like it does to nutrients, but the profile of the response differs. In the case of nutrients, a level in the diet can be defined at which pig performance is maximized. Feeding higher levels will not result in any improvement in performance; of course, the requirement may differ depending on the outcome being measured, such as growth rate or feed efficiency. Once the requirement for that outcome is reached, however, feeding higher levels will provide no further improvement. In the case of energy, no such maximum exists. A level of energy can be defined to maximize growth rate, but feeding higher levels of energy will continue to improve feed efficiency in an almost limitless fashion. Like nutrients, the level of energy that maximizes growth rate will be dependent on environmental influences, including health, housing and ambient conditions, as well as the genotype of the pig. Thus, the response to energy is viewed more as a continuum than is the case with nutrients – although a continuum exists therein as well. The level of energy desired in a diet – the so-called requirement – will be established based ultimately on feed cost in balance with revenue.

This chapter will discuss the basics of energy metabolism, energy systems, energy sources, energy requirements and methods to improve the utilization of energy by the pig. It will end with a brief discussion on the future of research on this topic and will give guidance about where interested readers can obtain further information on energy.

2 Pig energy requirements: importance and challenges

Meeting the energy specifications of a diet represents the largest proportion of its total cost. Also, energy provided by the diet affects the level of growth performance, such as rate and efficiency of gain, and will also impact the quality of the carcass. Consequently, selecting the optimal concentration of energy in the diet is critical to success in pork production. If the selected energy concentration in the diet is too low, pig performance will suffer. If it is too high, the cost of feeding the pig will be excessive, and there is also the possibility of a poor quality carcass.

Yet, few tables of nutrient requirements specify a 'requirement' for energy per se (e.g. NRC, 2012). There are many reasons for this, but the most important one is that selection of the concentration of energy in the diet represents a critical balance among factors such as cost, the energy level of available ingredients and the level of growth performance desired. If the diet contains a higher level of energy, feed efficiency will improve and rate of growth may or may not increase (more on this later) but the cost of the diet will almost certainly escalate. Depending on the cost of energy, the improvement in feed efficiency may result in an equal or even lower feed cost per pig; however, in most situations, the improvement in feed efficiency is not sufficient to overcome the increased cost of the diet, so the feed cost per pig increases. It then comes down to the improvement in growth rate. If a heavier pig results from the faster growth rate, then the added feed cost may be offset by the increased value of the market hog.

As important as energy concentration is to growth performance, daily feed intake is even more important, at least as it relates to growth rate. Recent data comparing growth performance between a region feeding primarily corn and soya bean meal and one feeding primarily wheat, barley and canola meal revealed that diets were typically higher in energy concentration in the corn–soya bean meal region (+8%) resulting in better feed efficiency (+8%), but growth rate was actually lower (–6%). The basis for more rapid growth in the region with the lower energy diets was a much higher daily feed intake (+16%). The higher feed intake could be due to health status, more generous floor space in the barn, or different genetics. Interestingly, energy efficiency, expressed as Mcal/kg body weight gain, was only 1% higher in the region feeding the higher energy diet. This example illustrates the extreme complexity involved in selecting the optimal energy level for pig diets.

Energy is a very complex subject for nutritionists for other reasons as well. Unlike amino acids, vitamins and minerals, which are all supplied by a specific diet component, energy is supplied by numerous constituents that differ in chemical structure and metabolic fate. Furthermore, the efficiency with which each is used differs among sources and is dependent on their ultimate metabolic fate (Patience, 2012). While there is general agreement on the best ways to represent vitamin, mineral and amino acid content of the diet, this is not the case with energy. Some regions use NE, some use DE and others use ME. Some countries have even developed their own national energy system.

As a result, meeting the energy requirements of the pig is no small feat, and depends on the nutritionist's ability to understand the relevant nutrition and biochemistry, as well as the economic implications of their decisions. Small errors can result in significant economic losses; multiplied across thousands or even millions of pigs, the impact on a single producer can be substantial.

The unit of measurement for energy should be the joule, according to the international system of units (SI). A joule is the energy required to move a one-kilogram mass in one minute by a force of one Newton. While the joule has found widespread use globally, the calorie remains in common use in some regions. For the purpose of conversion, 1 kJ = 0.239 kcal and 1 kcal = 4.184 kJ.

3 Energy metabolism

Energy is the fuel of life, and as such, supports the various functions in the pig's body including maintenance, deposition of body protein (lean) and deposition of lipid. In adult animals, it is used to sustain lactation or growth of the litter *in utero*. The universal currency of energy in the body is adenosine triphosphate or ATP; it is generated by oxidation and allows energy to be transformed from one form into another for the purpose of driving metabolic processes.

The first law of thermodynamics states that within a closed or isolated system, energy can be neither created nor destroyed, but it can be transformed from one form into another. The animal body is not a closed system, because it is constantly consuming feed and water, inspiring air, and concurrently eliminating urine, faeces and sweat, as well as expiring air. Thus, energy is in a constant state of flux, with energy in one form or another being transformed in the processes associated with what is broadly called metabolism. Animal energetics concerns itself with five different forms of energy: chemical, thermal, electrical, radiant and mechanical (Baldwin and Bywater, 1984; Botham and Mayes, 2006). According to Baldwin and Bywater (1984), pigs, like other mammals, obtain energy from the feed in chemical form, which they in turn use for the purposes of basal metabolism (heat), transfer to offspring as in lactation and pregnancy (chemical), store as tissue in the body (chemical) and use to perform work (heat, mechanical).

The quantity of energy present in a mixed diet, in an ingredient or even in the animal's body, can be quantified according to the heat produced upon its complete combustion. Consequently, bomb calorimetry evolved as a relatively inexpensive, highly accurate and highly repeatable measurement of energy as it relates to nutrition.

While the total quantity contained in the diet can be determined through the use of bomb calorimetry, not all energy present in feed is available to the pig. Some is undigested and lost via the faeces, and compounds that are derived from incomplete metabolism of protein are lost in the urine. In addition, the conversion of dietary energy into ATP is not a perfectly efficient process and losses occur during intermediary metabolism. Some energy is lost through the skin to the surrounding environment. A more complete explanation of the categories of energy supplied in the feed is provided in the following section of this chapter.

Energy is supplied by oxidation of three dietary constituents: proteins and other nitrogenous compounds, carbohydrates and related compounds, and fats (Boisen and Verstegen, 2000). This makes energy unique, because in the case of nutrients such as amino acids, vitamins and minerals, there is a distinct chemical structure for each. Thus, the study of calcium, for example, involves the study of this single mineral element. The same applies to vitamins and to amino acids. In the case of energy, one can use bomb calorimetry to determine the sum of all energy sources in a feed or an ingredient, but to truly understand energy metabolism, one must understand how it is supplied by each individual dietary component. This is critical, because the pig uses energy from fats, proteins and carbohydrates differently, and indeed, the efficiency with which each source is used depends on its metabolic fate – maintenance versus protein deposition versus lipid deposition (Birkett and de Lange, 2001a).

While different substrates can be used by the pig to generate energy in the body, the common denominator in all instances is ATP, the so-called energy currency in the body. In turn, ATP is formed from one of three processes: oxidative phosphorylation, which takes place in the mitochondria of the cell; glycolysis, which releases ATP when glucose is broken down into lactate; and the citric acid cycle. Thus, energy can be derived from various dietary components, but this common energy compound allows energy from any source to be used for a host of metabolic purposes.

Energy consumed in the diet undergoes digestion in the intestinal tract. Proteins are broken down by peptidases into free amino acids as well as shorter-length peptides, which are then absorbed across the intestinal barrier into the blood stream. The first priority for absorbed amino acids is synthesis of proteins, such as muscle and enzymes. However, the balance of amino acids in the diet rarely if ever matches the exact proportions of amino acids required for protein synthesis, so excess amino acids will be catabolized to release ammonia, along with a carbon skeleton that is available for further metabolism. Some amino acids, the so-called gluconeogenic amino acids, are degraded to pyruvate or TCA cycle intermediates, and can either be used to generate free glucose or converted to glycogen or fatty acids for storage. Other amino acids, called ketogenic amino acids, cannot be used for glucose synthesis, but rather are degraded to acetyl-CoA or acetoacetate and either used for energy directly in the TCA cycle, or converted to ketone bodies or fatty acids. Irrespective of the metabolism of specific amino acids, they can be – and are – all used as a source of energy for the body. It can therefore be seen that, while it is not their primary role in the body, amino acids represent a significant source of energy to the pig when present in excess of that needed for protein synthesis.

Carbohydrates in the body have a primary role of supplying energy to the pig. Some carbohydrates, such as starch and lactose, can be digested in whole or in part by enzymes present in the intestinal tract or on the surface of the small intestine of the pig. These enzymes break carbohydrate chains into single sugars, such as glucose (from starch, sucrose and lactose), galactose (from lactose) and fructose (from sucrose), which can then be absorbed and delivered via the blood stream to the tissues requiring energy. Some tissues are obligate glucose users, which means they can only use glucose for energy, while other tissues can utilize fatty acids or other energy sources. When sugars are present in excess of that needed for immediate use as fuel, they can be stored in small quantities as glycogen. Since this storage form has limited capacity, most of the excess intake of sugars is converted into triglycerides, with the formation of glycerol and fatty acids as intermediate steps.

Other carbohydrates cannot be broken down by enzymes produced in the intestine of the pig and are therefore referred to as fibre. The only way they can contribute to the energy balance of the pig is via fermentation, which occurs primarily, but not exclusively, in the caecum and the large intestine. The products of fermentation are short-chained fatty acids, namely acetic acid, propionic acid, butyric acid and lactic acid. These fatty acids can be used by many tissues within the body as fuel but it is believed that most are used by the intestinal tissues themselves, and little if any is available for use by the periphery. However, fermentation is not a very efficient process, so from an efficiency perspective, nutritionists prefer products which are absorbed as sugars rather than fermented and absorbed as short-chained fatty acids. This explains the interest in the use of exogenous enzymes in the diet of the pig. If these enzymes can move digestion from the large intestine to the small intestine, then more efficient use of dietary energy is possible.

Fats are handled somewhat differently from amino acids, glucose or fibre. Most fat in the diet exists as triglycerides, which consist of three fatty acid chains bound to a glycerol 'backbone'. Triglycerides cannot be absorbed as such, but must first be broken down by the enzyme lipase to form monoglycerides and diglycerides as well as free fatty acids; this process can only occur under certain conditions in the intestine, requiring emulsification by bile salts, which, like lipase, are released from the pancreas. This combination of mono-, di- and triglycerides plus free fatty acids form mixed micelles in the presence of bile salts; these mixed micelles are then absorbed by the enterocytes of the small intestine. Once inside these intestinal cells, triglycerides are reformed and enter the lymph system for transportation to the blood stream. These chylomicrons may be absorbed into fat cells throughout the body, following digestion by lipoprotein lipase, which is concentrated in adipose tissue but is present in smaller concentrations in other tissues as well. Thus, fatty acids can be stored in the body in adipose tissue or can be utilized as an energy source via β-oxidation.

In a typical North American diet formulated for grow-finish pigs, containing 450 g/kg corn, 300 g/kg soya bean meal, 200 g/kg corn distillers dried grains with solubles and 30 g/kg choice white grease (remaining 20 g/kg represents vitamins and minerals), approximately 39% of NE will be derived from starch, 23% from protein, 23% from fat and 15% from fibre.

4 Energy systems: overview, gross energy and digestible energy (DE)

4.1 Overview

In the field of nutrition, there is always a need to quantify the amount of energy or nutrients in a diet or a feed ingredient. There is also always a need to quantify requirements. This is the role of systems. Without such systems – or something in its place – it would be impossible to quantify the energy content of ingredients or to define requirements. It would also be impossible to formulate diets. This is a common theme throughout the field of nutrition.

In the case of amino acids, the system might be standardized ileal digestible amino acids, although sometimes apparent ileal digestible values may be used (NRC, 2012). In the case of vitamins, it will be based on international units or as a concentration (mg/kg, or µg/kg), and in the case of minerals, it may be total concentration or, in some instances, digestible concentration, such as standardized total tract digestible phosphorus (Almeida et al., 2010; NRC, 2012).

Atwater (1910) provided an overview of the early concept of energy or 'fuel value' of dietary components for humans, and suggested that the fuel value of protein and carbohydrates was 4 calories per gram, while that of fat was 9 calories per gram. Fuel value was defined as 'the number of calories of heat which one pound of a given material would yield upon combustion, allowance being made for the nitrogenous products of metabolism of protein compounds which are not consumed in the body' (Atwater and Wood, 1896). These caloric values remain in use today on human food labels provided to consumers in the United States. As innovative as this approach was at the time, it is surprising that these 'Atwater values' are still in use, despite more than a century of research which has revealed the wide range in digestibility of these components across a diversity of foods and the variation in energy concentration within these broad categories. This approach in human nutrition has the advantages of simplicity and ease of application, but is inaccurate as it relates to specific foods.

In the world of pig nutrition, numerous energy systems are being used around the world, but considered objectively, none are perfect. All are more robust than the Atwater system, because they assign values to individual ingredients based on their specific degree of digestibility and, in some instances, as explained below, on the net available energy after accounting for the cost of utilization as well as digestibility. The problem with developing a universally accepted energy system is the fact that energy is derived from different sources, each of which is utilized by the pig with a differing degree of efficiency, combined with the fact that energy is used differently according to its metabolic fate. For this reason, some nutritionists believe that the ultimate energy system will not be a 'system' at all, but will involve a modelling approach that can handle the breadth and width of the complexity of the subject (Emmans, 1995; Birkett and de Lange, 2001b; NRC, 2012). There are beginnings of such detailed energy models being used commercially, but at the moment, the pig industry is dependent on the more conventional systems approach. Models will more accurately quantify the energy being used by the pig under specific circumstances of growth, gestation or lactation, adjusted for the impact of environmental, health and social factors.

Finally, irrespective of which energy system is used, it is essential to utilize data that have been determined using similar methodologies. For example, NE can be determined in one of three ways: serial slaughter, indirect calorimetry or regression. The serial slaughter method involves a feeding trial in which some pigs are harvested at the beginning of a study while the remaining animals are fed the test diets and harvested at the end. The quantity of energy retained in the carcass over the experimental period is calculated as retained energy. Adding an estimate for the maintenance requirement to the retained energy will provide the NE content of the diet (Kil et al., 2011).

Alternatively, and more commonly, pigs are placed in respiratory chambers in which oxygen consumption and carbon dioxide production can be accurately measured. This information can in turn be used to estimate fasting heat production, which when added to retained energy is equivalent to NE. Retained energy is calculated as ME intake less heat production (Noblet et al., 1994a). The regression method will be explained below. The three methods may give very different NE values for a given ingredient or mixed feed, so it is very important that when choosing an energy system, values are derived using the same methods in all instances (NRC, 1998).

4.2 Gross energy

The classical approach to defining the available or effective quantity of energy content of ingredients and diets involves gross energy (GE), DE, ME or NE (Fig. 1; Patience, 2012). However, GE is never utilized in pig diet formulation because it ignores losses due to digestibility and metabolizability, and because it also fails to adjust for the efficiency with which differing sources of energy are used by the pig.

GE can be determined by bomb calorimetry, which involves the complete combustion at 25°C and standard pressure; the consequent rise in temperature is recorded and converted to an energy value, assuming a 1°C rise in temperature of 1 gram of water equals 1 calorie (Hargrove, 2006). Alternatively, one can determine the constituents of an ingredient or feed, namely carbohydrates, protein and fat, and using the mean enthalpy of those

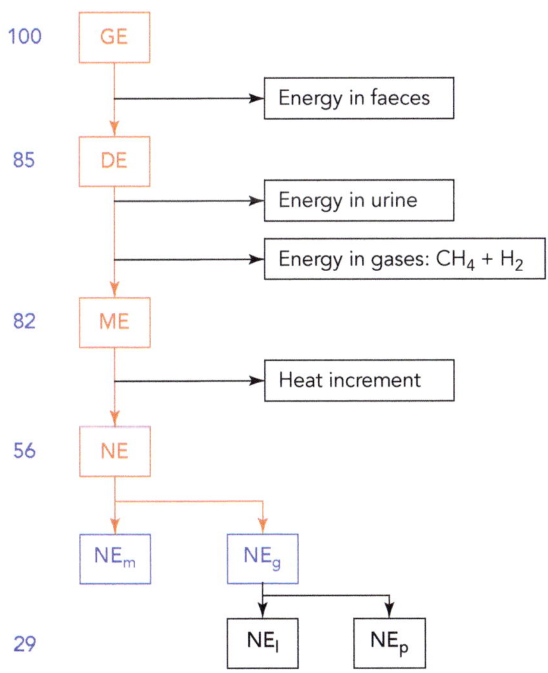

Figure 1 Representation of the classical energy systems and how they are calculated. The numbers on the left represent 'typical' percentages of gross energy included in each energy system.

constituents, estimate GE. Oresanya et al (2008) applied both methods to pig carcass analyses, and the resulting GE values differed by less than 2%.

4.3 Digestible energy

DE is greatly preferred over GE because it adjusts for losses of energy via the faeces; this system is broadly used and appreciated for its simplicity. Typically, the determination of DE is carried out in metabolism crates which permit control of animal movement and feed intake. Two approaches are possible. The total collection method involves gathering of all faecal material over a specified time period, and then comparing the energy consumed in the diet with that lost in the faeces. The alternative is the marker method, which involves the use of an indigestible marker and calculation of apparent total tract digestibility (ATTD) according to the following equation (Oresanya et al., 2008):

ATTD, % = [100 – [100 × (% marker in feed/% marker in faeces) × (concentration of energy in faeces/concentration of energy in feed)]].

The ATTD of energy is then multiplied by the GE of the diet to determine the DE concentration of the diet; this equation can be applied as either fed basis or dry matter basis, as long as one is consistent throughout. Commonly used markers include chromic oxide, titanium dioxide and acid-insoluble ash (AIA; Kavanagh et al., 2001). AIA can be a great choice of marker when the ingredients in the diet contain sufficient naturally occurring AIA to serve as a marker – preferably >0.35%. This removes the need to add an exogenous marker. This achieves considerable cost savings, and the AIA assay is not difficult.

The marker method can be applied under field conditions to permit determination of the DE content of a diet used commercially. However, it must be noted that values obtained under such conditions may differ somewhat from those determined when feed intake is restricted. Indeed, the topic of selection of methodology to undertake the determination of DE is somewhat controversial, as results will be influenced by the housing system (Bakker and Jongbloed, 1994; Holloway and Patience, 2014), the specific marker employed (Kavanagh et al., 2001), the collection method (Li et al., 2016), the age of the pig (Shi and Noblet, 1993) and the composition of the basal diet (Li et al., 2016).

The DE content of a diet or ingredient can also be estimated from their chemical composition. Following are example equations which are available for this purpose (Noblet and Perez, 1993):

DE (kcal/kg DM) = 1,161 + (0.749*GE) – (4.3*Ash) – (4.1*NDF)

DE (kcal/kg DM) = 4,168 – (9.1*Ash) + (1.9*CP) + (3.9*EE) – (3.6*NDF)

where NDF is neutral detergent fibre, CP is crude protein and EE is ether extract, all expressed as g/kg DM.

Also, DE can also be estimated using near-infrared spectroscopy (NIRS). For this purpose, DE is determined using classical digestibility experiments, and samples are then scanned using an NIRS instrument (van Barneveld et al., 1999). When sufficient samples have been determined – probably in the range of 75 to 100 for each ingredient or diet type – a satisfactory prediction equation can be developed and applied to future samples of these same ingredients or feed types (Aufrère et al., 1996). Over time, additional samples should be evaluated using classical methods to ensure that there has been no 'drift' and that the NIRS prediction equations are still valid. This process of recalibration is particularly important with each new crop year, when changes are likely to be greatest.

As mentioned above, DE is determined by measuring the quantity of energy excreted from the body in the faeces. It is therefore assumed that if energy does not appear in the faeces, it must have been absorbed and therefore was 'digested'. However, this assumption ignores another preabsorptive loss of energy from the intestinal tract, namely gases emitted – primarily methane and hydrogen – resulting from fermentation processes. However, by convention, these are not included in DE, but rather are considered in ME and will be discussed below.

DE represents about 80 to 85% of GE in typical commercial diets, but can vary widely among individual ingredients, from 50% to 98%. If all other conditions remain the same, this ratio tends to narrow as pigs grow from birth through to adulthood.

5 Energy systems: further options

5.1 Metabolizable energy

ME is calculated by deducting urinary and fermentative gaseous losses of energy from DE. Loss of energy via the urine is correlated with amino acids, which are not incorporated into body protein (lean plus enzymes) and are therefore metabolized, with the associated amino group excreted as ammonia or urea in the urine.

Urinary gaseous losses represent the consequence of fermentative processes and are related to the fermentable fibre content of the diet. Gaseous losses are often ignored in the determination of ME due to the assumption that such losses are relatively minor in magnitude. Noblet et al. (1994b) reported that methane losses amount to only about 0.4% of DE intake in growing pigs. However, gaseous losses are increased in diets containing higher levels of fermentable fibre, to as high as 3% of DE (Shi and Noblet, 1993). There is another reason why gaseous losses are rarely measured; it is very expensive, and requires highly specialized equipment, which, due to its expense, is available to only a handful of researchers and no one in the private sector.

ME can be, and often is, estimated from DE using prediction equations, due to the cost and time required to collect urine and measure its GE content. Since the major source of energy in the urine is excreted urea, LeGoff and Noblet (2001) developed the following equation to estimate ME from DE:

$$ME/DE = 100.3 - (0.021*CP)$$

where CP is the crude protein content of the diet, expressed as g/kg of DM. Other approaches have been used. Rather than the direct determination of the GE content of urine, it can be estimated from the nitrogen content of urine according to the following equation (Noblet and van Milgen, 2004):

$$GE_{urine}\ (kcal/kg) = 45.9 + 7.4*N$$

where N is the nitrogen content of urine expressed as g/kg. ME can thus be calculated by subtracting the energy lost via the urine, from the DE content of the diet.

ME is widely used by the pig industry in North and South America, and is considered to be slightly more precise than DE. ME is typically 94 to 97% of the DE content of most practical diets. This range widens, of course, when considering certain ingredients with high crude protein content, such as soya bean meal or canola meal, where the ratios are 91 and 92%, respectively (NRC, 2012).

5.2 Net energy

NE is gaining increased attention in North America and is used in a number of European countries; in addition to the faecal and urinary losses addressed by DE and ME, respectively, NE also accounts for the varying heat increment among ingredients, which includes energy used for ingestion, digestion and the metabolic utilization of ME (Noblet and van Milgen, 2013). Because NE quantifies the energy available to the pig for maintenance and productive purposes, it is generally considered to be the best of the currently available energy systems (Noblet and van Milgen, 2004; Acosta et al., 2016). However, it is not the perfect energy system for a number of reasons. First, NE is difficult and expensive to determine; as explained above, indirect calorimetry is most commonly used, but few institutions have the required equipment and the procedure is slow and laborious, and therefore expensive. Furthermore, the conditions under which NE is determined, in a calorimetry chamber, is far removed from the conditions under which pigs are housed on farms. It is not known how much error is involved, but it is a question of interest. Boyd et al. (2015) suggested that large-scale growth studies under commercial conditions can be used to validate NE values of ingredients. Indeed, this approach should work equally well for DE and ME.

The second concern with NE is related to the fact that NE is used with differing efficiencies, depending on its fate, so even a more precise measurement of energy content of the diet may not provide greater accuracy in predicting animal performance. If maintenance varies among genotypes, the differences are likely to be small, but the ratio of energy directed to lipid accretion versus protein accretion can be quite large. Furthermore, maintenance energy may differ widely depending on the health status of the pigs or the physical (thermal) environment in which they live. This is important, because the efficiency with which dietary energy is used for maintenance, protein accretion and lipid accretion varies to a considerable extent (Patience, 2012). However, having noted this concern, NE more closely quantifies the energy available for maintenance and gain than does DE or ME, and for this reason, is being more widely adopted. Indeed, these criticisms of NE also apply to DE and ME. The efficiency with which DE or ME are used for maintenance, protein accretion and lipid accretion will also differ. The fundamental advantage of NE, then, is to get metabolically closer to the ultimate utilization of energy by the pig, after compensating for issues largely related to the ingredient composition of diets: digestibility and metabolizability.

NE can be estimated from DE or ME using prediction equations based on the chemical composition of the feed (Noblet et al., 1994):

$$NE = (0.700 \times DE) + (1.61 \times EE) + (0.48 \times starch) - (0.91 \times CP) - (0.87 \times ADF)$$

$$NE = (0.726 \times ME) + (1.33 \times EE) + (0.39 \times starch) - (0.62 \times CP) - (0.83 \times ADF)$$

Table 1 Relative energy values of common feed ingredients. The energy value of corn was arbitrarily set at 100 as the basis for these values

Ingredient	Digestible energy	Metabolizable energy	Net energy
Corn	100	100	100
Barley	91	91	87
Wheat	96	95	93
Wheat middlings	89	87	79
Oats	76	75	71
Oat groats	107	106	102
Field peas	102	99	91
Lentils	103	99	91
Sunflower meal (dehulled, extracted)	82	76	55
Canola meal (extracted)	95	89	71
Soybean meal (dehulled, extracted)	105	97	78
Choice white grease	240	239	268
Corn oil	254	253	283
Tallow, beef	232	231	258

Source: NRC (2012).

where EE is ether extract, CP is crude protein and ADF is acid detergent fibre. All dietary constituents are expressed as g/kg, and NE and DE are expressed as kcal/kg.

The ratio of NE to ME varies widely, due to the impact of ingredient composition on heat increment. It will be relatively narrow for ingredients low in fibre and protein, such as corn (79%) and much wider for ingredients high in protein such as soya bean meal (63%) or high in both protein and fibre such as canola meal (66%). Because this ratio varies widely among ingredients, NE is considered more precise than ME or DE. Theoretically, the benefit of utilizing NE is modest in terms of predicting animal performance – perhaps in the range of 1 to 2%. The greatest benefit of using NE in commercial practice is the more effective relative pricing of ingredients that vary in their contribution to heat increment, and thus their value to the pig as an energy source. Due to the widely varying levels of protein and complex carbohydrates across ingredients, the relative value of ingredients will vary, depending on the energy system used. This is illustrated in Table 1, which is based on *Nutrient Requirements of Swine* (NRC, 2012) energy values.

5.3 Other energy systems

The Danes have developed their own energy system, known as the PPE system (Boisen, 2007; Tybirk et al., 2013). This system is based on physiologically available energy (ATP), which is obtainable following the oxidation of diet constituents at the cellular level. PPE is determined in three steps: chemical analysis of dietary constituents, namely water, ash, crude protein and crude fat, followed by determination of *in vitro* digestibility at the ileal and faecal levels, and, finally, followed by the estimation of the energy values from the above information (Tybirk et al., 2013).

It is beyond the scope of this chapter to discuss the various energy systems in detail, but readers are directed to the following recent reviews for additional information (Patience, 2012; Kil et al., 2013; Szabo and Halas, 2013; Velayudhan et al., 2015).

5.4 Another option: modelling

Perhaps the ultimate answer to achieving the highest level of predictability between dietary energy content and animal performance is the use of modelling. As previously stated above, energy is a particularly difficult dietary component to study due to the fact that energy comes from different feed constituents (fibre, protein, lipid and starch), each with different post-absorptive utilization. Further, such efficiency is driven not only by source but also by metabolic utilization (maintenance vs. lipid accretion). It is therefore unreasonable to expect a classical energy system to accurately predict animal performance. The fact that energy is a very significant component of the total cost of pork production provides a strong motivation to achieve a higher degree of precision.

However, modelling of energy utilization has received relatively little attention (Kielanoswki, 1966; Birkett and de Lange, 2001a,b). It is a highly complex subject, and progress in the development of a model is likely to be slow and painstaking.

6 Evaluating energy sources, determining energy requirements and improving energy availability

6.1 Energy sources

Energy can be supplied in the diet by starch and other simple carbohydrates such as lactose, by more complex carbohydrates known as fibre, by individual amino acids within protein and by lipids. The quantity of energy supplied by a dietary constituent will be a function of the ratio of carbon plus hydrogen to oxygen [(C+H):O]. This explains, in part, the higher energy provided by fatty acids, as a large portion of their structure consists of carbon and hydrogen, relative to oxygen, compared to, for example, glucose or amino acids.

The GE of amino acids varies by a factor of 2.3, so the GE of a protein will be dependent on its amino acid composition. Phenylalanine contains the greatest quantity of GE at 6.74 Mcal/kg, followed by isoleucine, leucine and tryptophan with 6.60 Mcal GE/kg. At the low end, aspartic acid, glycine and serine each contain 2.89, 3.08 and 3.30 Mcal GE/kg, respectively. In terms of the so-called most limiting amino acids in most practical diets, lysine contains 5.62 Mcal GE/kg, methionine contains 4.45 Mcal GE/kg, threonine contains 4.11 Mcal GE/kg and tryptophan contains 6.57 Mcal GE/kg. On average, a typical protein contains 5.66 Mcal GE/kg, but of course, this depends on the proportion of individual amino acids present (Boisen and Verstegen, 2000).

The GE of carbohydrates varies from 4.18 Mcal/kg for starch and cellulose to 3.73 Mcal/kg for glucose. The GE of short-chained fatty acids ranges from 5.95 Mcal/kg for butyric acid to 3.49 Mcal/kg for acetic acid. Finally, the GE of longer chain fatty acids ranges from 9.54 Mcal/kg for oleic and stearic acid to 7.74 Mcal/kg for caprylic acid. The mean GE for fats is 9.30 Mcal/kg, about 1.64 times that of protein and 2.2 times that of glucose and cellulose

Table 2 Marginal energetic efficiency of using digestible nutrient components for generating ATP or depositing lipid

	Generating ATP		Depositing lipid	
Source	Efficiency	% relative to fatty acids	Efficiency	% relative to fatty acids
Fatty acids	66	100	90	100
Glucose	68	103	74	82
Amino acids	58	88	53	59
Crude fat	50	76	62	69

Source: Black (1995), as adapted by Birkett and de Lange (2001a).

(Boisen and Verstegen, 2000). Again, the actual GE of a fat will depend on its fatty acid profile.

The digestibility of dietary components is also an important variable. While the GE of starch and cellulose is the same, they vary widely in their DE, due to the relative ease with which starch can be broken down into glucose compared to the absence of not only necessary cellulytic enzymes, but also a microbiota that can assimilate cellulose. The third factor is the efficiency with which dietary components can be used by the pig to generate ATP, for the purpose of maintenance and protein accretion or to deposit lipid in the body. The differences in such efficiencies are clearly summarized in Table 2.

From a practical perspective, physical bulkiness will also influence the quantity of energy that ingredients can supply to the pig. Concentrated sources of energy, such as starch or lipids, can be consumed in relatively large quantities without restriction due to the volumetric capacity of the GI tract. On the other hand, bulky sources, such as highly fibrous ingredients, cannot meet the pig's need for the quantity of energy required for a high growth rate, or high rate of lactation, due to physical gut capacity.

The energy value of fats typically increases with an increase in the ratio of unsaturated fatty acids to saturated fatty acids (Wiseman et al., 1998). The change in energy content is greatest when the ratio increases from 1.25:1 to 2.25:1, and is more marked in younger than older pigs.

6.2 Energy requirements

Few tables of nutrient requirements provide a 'requirement' for energy. There are many reasons for this. One is regional differences in energy levels that result from the constraints of the local economy. Another is a lack of knowledge of the energy requirements of specific genotypes of pigs arising from inadequate information on the relative rates of protein and lipid deposition. The response of the pig to dietary energy concentration represents a continuum of rate and efficiency of growth, as well as carcass composition, specifically the ratio of protein to lipid. For example, the requirement for an amino acid is typically defined by the concentration in the diet, or the daily intake, that maximizes rate and efficiency of gain. Levels of the amino acid below this concentration will result in slower and less efficient growth and a lower ratio of protein to lipid in the carcass. Feeding above the requirement offers no change in any of these outcomes. In the case of energy, there is no concentration in the diet that maximizes growth rate, feed efficiency and carcass lean content. Feeding less energy in the diet will reduce growth rate and feed efficiency but feeding higher levels will increase growth rate – up to the maximum capacity of the pig. Feed efficiency will continue to improve as energy concentration is further increased (Beaulieu et al., 2009). As a result, the 'requirement' for energy is less well defined than it is for nutrients.

In most instances, nutritionists set the level of energy in the diet based on economics. Higher energy levels often lead to faster growth, and in the North American pork industry production model, rapid growth is a very high priority. In other parts of the world, growth rate is secondary to carcass fat content, leading to perhaps lower energy levels in the diet, or even limit feeding of pigs, especially barrows, to reduce lipid accretion in the body. Nonetheless, producers located in corn-growing areas will typically feed higher energy diets than those in cereal grain-growing areas. Because energy is the most expensive component of the diet, selecting the optimal energy level in the diet is critical to the overall achievement of animal performance targets, balanced against financial goals.

A pig's use of energy can be defined by the following equation:

$$ME = ME_{M} + ME_{PD} + ME_{LD}$$

where ME_M represents the ME required for processes associated with maintenance, ME_{PD} represents the ME required for protein deposition and ME_{LD} represents the ME required for lipid deposition. In this instance, maintenance will include not only fasting heat production, but all non-productive functions, including the energy required for thermal homeostasis, immunological response to disease challenges, activity and responses to social stressors. In the case of the nursing sow, the energy requirement is the sum of that required for maintenance and lactation and, in some instances, lean gain if the sow has not yet reached mature body weight. In the gestating sow, the energy requirement is the sum of that required for maintenance and the developing foetal and reproductive tissue, with an allowance for growth in the instance of a gilt. In the case of sows, there is the energy needed for the replenishment of body reserves lost during the previous lactation.

The partitioning of energy across the various functions can be estimated arithmetically. Using a 50 kg pig growing at 850 g/d as an example, the maintenance energy (ME, kcal/d) can be estimated as 197 X $BW^{0.060}$ (NRC, 2012), or 2,060 kcal ME/d. If a diet contains 3,200 kcal ME/kg, then 0.64 kg of feed per day would be required to meet the pig's maintenance energy requirement. This estimate would need to be adjusted upward if the pig is exposed to thermal stress, social stress or disease. Assuming that the pig is gaining 135 g protein per day, the ME required for this purpose would be 1,354 kcal/d, assuming the efficiency with which diet ME is used for protein accretion is 10.03 kcal/g (Patience, 2012). The remainder of the daily ME intake would be used for lipid accretion; in this example, a 50 kg pig would reasonably be expected to consume 6,240 kcal ME/d. Thus, the pig would have 2,826 kcal ME/d available for lipid accretion. Assuming an efficiency of 11.65 kcal ME/g of lipid deposited (Patience, 2012), the lipid accretion rate would be 243 g/d. This results in a lipid:protein ratio for a gain of 1.8, which is very reasonable. It also results in a feed conversion ratio of 2.30 kg feed per kg body weight gain, which is also reasonable for a healthy pig. Of course, these estimates represent averages. Individual pigs and groups of pigs may vary considerably based on genotype, diet, season, health and social factors.

6.3 Processing diets to improve energy availability

Most ingredients are processed before being included in pig or poultry diets. For example, grains will be ground to reduce particle size and thus improve digestibility. Often, diets are also pelleted to increase availability of energy. In specialized circumstances, more expensive options are employed, including expansion and extrusion (Stark, 2012).

The addition of digestive enzymes to swine diets is also practised to improve the digestibility of energy. Examples include carbohydrases (xylanases, β-glucanases, cellulases) and proteases. These products tend to be more effective in the diet of young pigs, but further development offers the potential for expanded use. Any technology that can help the pig industry utilize highly fibrous ingredients more effectively will be welcomed, provided the benefit is economically sound.

7 Summary

In summary, meeting the energy needs of the pig is a highly complex topic. Energy itself is one of the most complicated topics in nutrition, because it is provided by a diverse array of components in each ingredient: protein, lipid, starch and fibre. Each component is processed in a somewhat different fashion, and the efficiency with which each is metabolized depends on its post-absorptive fate. Yet, energy is by far the most expensive component of the diet and as such, warrants a high level of research to improve our understanding and achieve greater success in its use. Indeed, it is clear that the pork producer – individual or regional – which acquires dietary energy at the lowest cost, and uses that energy most efficiently, will be the most competitive in the future, all other factors being equal.

Future research is likely to focus on enhancing our understanding of how to best supply energy to the pig and how the pig uses that energy for maintenance and productive purposes. This will include continued research on the most effective way to incorporate highly fibrous – but often less expensive – ingredients in the diet of the pig, because agriculture is in competition with other industries for basal ingredients such as corn, sorghum, wheat and barley. There will also be continued research on dietary fats as a way to enhance pig performance, while also avoiding adverse effects on the quality of the carcass or pork. Fats are high in energy, but also tend to be expensive. Fats currently represent an effective way to counter the energy reducing effects of fibre in the diet, but this will only be successful if fats are competitively priced.

Research will also focus on how excess amino acids in the diet are utilized for energy. Because starch is valued in various industrial processes, there will likely be increased research on the minimum glucose requirement of the pig. This seems far-fetched at this time, but starch could be a highly prized commodity in the not-too-distant future.

Finally, future research will seek to understand how the pig responds to energy, in terms of the rate and efficiency of gain, the impact on the carcass and the impact on the environment. Modelling is likely to gain in acceptance and adoption as a means to better understand in a holistic manner the response of the pig to energy, a vehicle to the development of the most effective feeding programmes.

There are no textbooks on energy nutrition of the pig, and often, books on nutrition devote only small portions of the total volume to the subject. Interested readers are encouraged to seek out these chapters in books, most of which are listed below. They are also invited to study review articles on the topic; again, excellent examples are listed in the references section of this chapter.

8 References

Acosta, J., J. F. Patience and R. D. Boyd. 2016. Comparison of growth and efficiency of dietary energy utilization by growing pigs offered feeding programs based on the metabolizable energy or the net energy system. *J. Anim. Sci.* 94:1520–30.

Atwater, W. O. 1910. *Principles of Nutrition and Nutritive Value of Food.* Farmers' Bulletin No. 142, U.S. Department of Agriculture, Washington, DC.

Atwater, W. O., and C. D. Woods. 1896. *The Chemical Composition of American Food Materials.* Experiment Stations Bulletin No. 28, U.S. Department of Agriculture, Washington, DC.

Aufrère, J., D. Graviou, C. Demarquilly, J. M. Perez and J. Andrieu. 1996. Near infrared reflectance spectroscopy to predict energy value of compound feeds for swine and ruminants. *Anim. Feed Sci. Technol.* 62:77–90.

Bakker, G. C.M., and A. G. Jongloed. 1994. The effect of housing system on apparent digestibility in pigs using the classical and marker (chromic oxide, acid insoluble ash) techniques in relation to dietary composition. *J. Sci. Food Agric.* 64:107–15.

Baldwin, R. L., and A. C. Bywater. 1984. Nutritional energetics of animals. *Ann. Rev. Nutr.* 4:101–14.

Beaulieu, A. D., N. H. Williams and J. F. Patience. 2009. Response to dietary digestible energy concentration in growing pigs fed cereal-grain based diets. *J. Anim. Sci.* 87:965–76.

Birkett, S., and C. F.M. deLange. 2001a. A computational framework for a nutrient flow representation of energy utilization by growing monogastric animals. *Br. J. Nutr.* 86:661–74.

Birkett, S., and C. F. M. deLange. 2001b. Limitations of conventional models and a conceptual framework for a nutrient flow representation of energy utilization by animals. *Br. J. Nutr.* 86:647–59.

Black, J. L. 1995. Modelling energy metabolism in the pig – critical evaluation of a simple reference model. In P. J. Moughan, M. W. A. Verstegen and M. I. Visser-Reyneveld (eds), *Modelling Growth in the Pig*, EAAP Publication no. 78. Wageningen Press, Wageningen, The Netherlands, pp. 87–102.

Boisen, S. 2007. New concept for practical feed evaluation systems. DJF Animal Science No. 79. Research Centre Foulum, Denmark.

Boisen, S., and M. W. A. Verstegen. 2000. Developments in the measurement of the energy content of feed and energy utilization by the pig. In: P. J. Moughan, M. W. A. Verstegen and M. I. Visser-Reyneveld (eds), *Feed Evaluation – Principles and Practice*. Wageningen Academic Press, Wageningen, The Netherlands, pp. 57–76.

Botham, K. M., and P. A. Mayes. 2006. Bioenergetics: The role of ATP. In R. K. Murray, D. K. Granner and V. W. Rodwell (eds), *Harper's Illustrated Biochemistry*. Lange Medical Books/McGraw-Hill, New York, NY, pp. 88–93.

Boyd, R. D., C. E. Zier-Rush, M. McGrath, P. Palan, J. Picou and E. van Heugten. 2015. Calibration of net energy for fat by growth assay in early and late phases of growth in pigs. *J. Anim. Sci.* 93(Suppl. 2):73.

Emmans, G. C. 1995. Models of the relationship between animal performance and dietary energy. In: P. J. Moughan, M. W. A. Verstegen and M. I. Visser-Reyneveld (eds), *Modelling Growth in the Pig*. Wageningen Academic Press, Wageningen, The Netherlands.

Hargrove, J. L. 2006. History of the calorie in nutrition. *J. Nutr.* 136:2957–61.

Holloway, C. L., and J. F. Patience. 2014. Determination of the impact of housing system on the determination of apparent total tract digestibility of energy and dry matter. *J. Anim. Sci.* 92(Suppl. 2):63.

Kavanagh, S., P. B. Lynch, F. O'Mara and P. J. Caffrey. 2001. A comparison of the total collection and marker technique for the measurement of apparent digestibility of diets for pigs. *Anim. Feed Sci. Technol.* 89:49–58.
Kielanowski, J. 1966. Conversion of energy and the chemical composition of gain in bacon pigs. *Anim. Prod.* 8:121–8.
Kil, D. Y., R. B. Hinson, A. D. Beaulieu, F. Ji, L. L. Stewart, G. L. Allee, J. F. Patience, J. E. Pettigrew and H. H. Stein. 2011. Net energy of soybean oil and choice white grease in diets fed to growing and finishing pigs. *J. Anim. Sci.* 89:448–59.
Kil, D. Y., B. G. Kim and H. H. Stein. 2013. Feed energy evaluation for growing pigs.. *Asian Australia J. Anim. Sci.* 26:1205–17.
Kleiber, M. 1975. *The Fire of Life: An Introduction to Animal Energetics.* Krieger Publ. Co., Malabar, FL.
Le Goff, G., and J. Noblet. 2001. Comparative total tract digestibility of dietary energy and nutrients in growing pigs and adult sows. *J. Anim. Sci.* 79:2418–27.
Li, Y. S., H. Tran, J. W. Bundy, T. E. Burkey, B. J. Kerr, M. K. Nielsen and P. S. Miller. 2016. Evaluation of collecting method and diet effects on apparent digestibility and energy values of swine diets. *J. Anim. Sci.* 94:2415–24.
Noblet, J., H. Fortune, X. S. Shi and S. Dubois. 1994a. Effect of body weight on net energy value of feeds for growing pigs. *J. Anim. Sci.* 72:648–57.
Noblet, J., X. S. Shi and S. Dubois. 1994b. Prediction of net energy value of feeds for growing pigs. *J. Anim. Sci.* 72:344–54.
Noblet, J., and J. M. Perez. 1993. Prediction of the digestibility of nutrients and energy values of pig diets from chemical analysis. *J. Anim. Sci.* 71:3389–98.
Noblet, J., and J. van Milgen. 2004. Energy value of pig feeds: Effect of pig body weight and energy evaluation system. *J. Anim. Sci.* 82(E. Suppl.):E229–38.
Noblet, J., and J. van Milgen. 2013. Energy and energy metabolism in swine. In: L. Chiba, (ed.), *Sustainable Swine Nutrition.* John Wiley and Sons, Ames, IA, pp. 23–57.
NRC. 1998. *Nutrient Requirements of Swine.* 10th rev. ed. National Academies Press, Washington, DC.
NRC. 2012. *Nutrient Requirements of Swine.* 11th rev. ed. National Academies Press, Washington, DC.
Oresanya, T. F., A. D. Beaulieu and J. F. Patience. 2008. Investigations of energy metabolism in weanling barrows: The interaction of dietary energy concentration and daily feed (energy) intake. *J. Anim. Sci.* 86:348–63.
Patience, J. F. 2012. The influence of dietary energy on feed efficiency in grow-finish swine. In: J. F. Patience (ed.), *Feed Efficiency in Swine.* Wageningen Academic Press, Wageningen, The Netherlands, pp. 101–29.
Shi, X. S., and J. Noblet. 1993. Contribution of the hindgut to digestion of diets in growing pigs and adult sows: effect of diet composition. *Livest. Prod. Sci.* 34:237–52.
Stark, C. R. 2012. Feed processing to maximize feed efficiency. In: J. F. Patience (ed.), *Feed Efficiency in Swine.* Wageningen Academic Press, Wageningen, The Netherlands, pp. 131–53.
Szabo, C., and V. Halas. 2013. Shortcomings of the energy evaluation systems in pigs: A review. *Agric. Conspec. Sci.* 78:153–8.
Tybirk, P., N. Morten Sloth and L. Jorgensen. 2013. *Nutrient requirement standards.* Pig Research Centre, Copenhagen, Denmark. 10pp.
Van Barneveld, R. J., J. D. Nuttall, P. C. Flinn and B. G. Osborne. 1999. Near infrared reflectance measurement of the digestible energy content of cereals for growing pigs. *J. Near Infrared Spectrosc.* 7:1–7.
Velayudhan, D. E., I. H. Kim and C. M. Nyachoti. 2015. Characterization of dietary energy in swine feed and feed ingredients: A review of recent research results. Asian Australas. *J. Anim. Sci.* 28:1–13.
Wiseman, J., J. Powles and F. Salvador. 1998. Comparison between pigs and poultry in the prediction of the dietary energy value of fats. *Anim. Feed Sci. Technol.* 71:1–9.

Chapter 8

Meeting amino acid requirements in pig nutrition

Sung Woo Kim, North Carolina State University, USA

1 Introduction

Protein is one of the major components of the pig's body. In terms of dry matter, protein makes up the largest percentage of the pig's body at birth; however, the protein accretion slows down and the fat accretion accelerates as the pig grows. While pigs can store fat in their cells as triacylglycerides, protein cannot be stored in the body. The amount of protein in the pig's body is the net balance between protein synthesis and protein degradation. Protein turnover may be fast, as short as few seconds, or slow, as long as several days, depending on the types of protein in the body. The pig gains body protein when the rate of protein synthesis exceeds the rate of protein degradation, but loses body protein in the opposite scenario.

Given excess starch and fat in the diet, the fat in pigs' bodies increases; dietary starch provides glucose for fatty acid synthesis in the liver and dietary fat provides fatty acids to the liver, providing the necessary materials to build triacylglycerides, which can then be stored as depot fat in adipocytes. However, excess dietary protein does not lead to increased body protein. Instead, it is degraded into amino acids and then further catabolized to glucose or ketone bodies, which are then used as energy sources. It is therefore not necessary or helpful to provide excess dietary protein, because it will not increase body protein (lean gain).

Protein is composed of 20 amino acids. Among these, nine amino acids are known as essential (or indispensable), two as conditionally essential and the remaining nine

http://dx.doi.org/10.19103/AS.2017.0013.08

Table 1 Requirements of selected essential amino acids in pigs by growth stages[1]

	Body weight, kg						
SID AA[2], g/kg	5–7	7–11	11–25	25–50	50–75	75–100	100–135
Lysine	15.0	13.5	12.3	9.8	8.5	7.3	6.1
Threonine	8.8	7.9	7.3	5.9	5.1	4.4	3.7
Methionine	4.3	3.9	3.6	2.8	2.4	2.1	1.8
Tryptophan	2.5	2.2	2.0	1.7	1.5	1.3	1.1
Valine	9.5	8.6	7.8	6.4	5.5	4.8	4.1
Leucine	15.0	13.5	2.3	9.9	8.5	7.4	6.2
Isoleucine	7.7	6.9	6.3	5.1	4.5	3.9	3.3

[1] Adapted from NRC (2012).
[2] Standardized ileal digestible amino acids.

as non-essential (dispensable) amino acids. If a diet is deficient in one amino acid, the protein synthesis will be limited. There is, therefore, no overall, undifferentiated protein requirement for pigs; rather, it is important to know the pig's requirements for individual amino acids. Table 1 shows the amino acid requirements of pigs in their different growth stages, as suggested by the National Research Council (NRC) in the United States (NRC, 2012).

Feedstuffs are made up of ingredients which are mixed to provide the nutrients essential for the growth of pigs. The feedstuffs used in pig diets are often concentrates, which include energy feeds (IFN feedstuff category #4) and protein supplements (IFN feedstuffs category #5). Protein supplements are feedstuffs with a minimum of 200 g/kg crude protein and a maximum of 180 g/kg crude fibre. These are used to provide protein in pig diets. Each protein supplement will have its own, unique amino acid composition. A single protein supplement is unlikely to provide an adequate amount of all amino acids, and so multiple protein supplements are often used together to meet the requirements of pigs.

In this chapter, we review the principles and practical aspects of meeting the amino acid requirements of pigs, with information organized according to the growth stage and physiological status. The stages of growth include nursing, nursery, grower and finisher, based on the body weight of pigs as suggested by the NRC in their publication (NRC, 2012). It is advised that the research data cited in this chapter may not cover the entire spectrum of literature.

2 Gestating sows

2.1 Principles

In the nutritional management of gestating sows, two of the most important factors worth considering are the growth of foetuses and the growth of mammary glands. The growths of both foetuses and mammary glands are complex processes. In this section, we focus only on the nutritional aspect of understanding foetal and mammary gland growths, as this relates to providing for amino acid requirements during gestation.

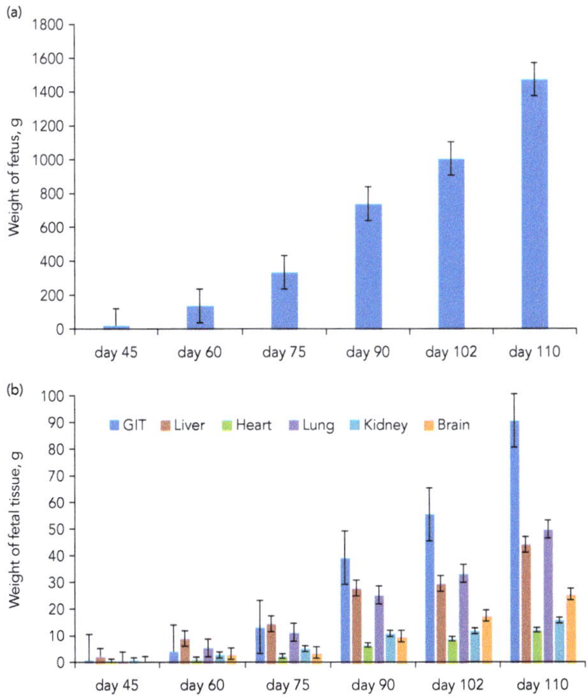

Figure 1 (a) Weight of a porcine foetus at different days of gestation (McPherson et al., 2004). (b) Weights of porcine foetal tissues at different days of gestation (McPherson et al., 2004).

McPherson et al. (2004) and Ji et al. (2006) investigated the growth of the foetus and foetal tissues during the course of gestation in pigs (Fig. 1). There are also a few other, earlier papers which have investigated the foetal growth (Ullrey et al., 1965; Fall et al., 2003). It has been well established that the weight gain of foetal tissues is minimal in early gestation and increases from mid-gestation. Kim et al. (2009) showed that the foetal tissue growth increases significantly from d 70 of gestation (Fig. 1). Up until d 70, the protein accretion in a foetus is fairly small, and in the study, each foetus gained only 0.25 g protein per day until this point. However, the foetus starts to gain protein rapidly from d 70 of gestation onwards. Each foetus gained 4.63 g protein per day from d 70 of gestation until farrowing (Ji et al., 2005). This is almost a 19-fold increase in the protein accretion, which could be a significant amount if a sow has 14 or more foetuses (Kim et al., 2013).

Ji et al. (2006) and Kim et al. (1999a) investigated the growth of mammary glands during the course of gestation in pigs (Fig. 2). There are also a few other papers which have investigated the mammary gland growth during gestation (Kensinger et al., 1982; Sørensen et al., 2002). It has been shown that the weight gain of mammary parenchymal tissues is minimal in early gestation and increases from mid-gestation. Kim et al. (2009) showed that mammary parenchymal tissues grow mainly from d 70 of gestation onwards (Fig. 2), with major compositional changes involving the loss of adipocytes and gain of mammary cells (Fig. 3). The protein accretion in a mammary gland is also fairly small until d 70 of gestation. In the study, up until this point, each

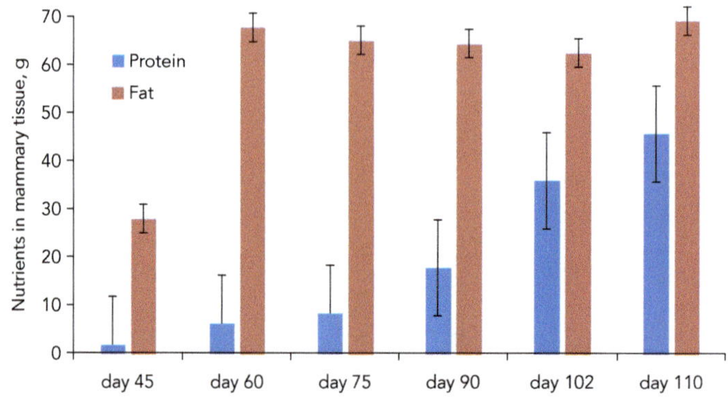

Figure 2 Amount of protein and fat in porcine mammary tissues at different days of gestation (Ji et al., 2006).

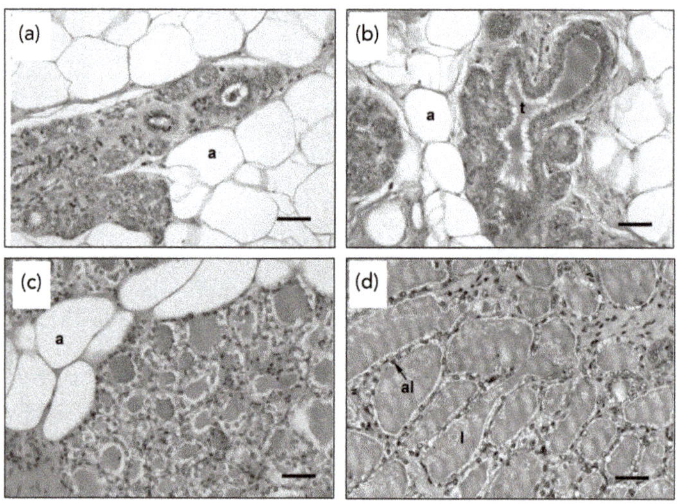

Figure 3 Histomorphology of mammary gland tissues from gilts during gestation. (a) day 45 of gestation; (b) day 60 of gestation; (c) day 90 of gestation and (d) day 112 of gestation. Thick black bars indicate 50 μm. Photos are adopted from Ji et al. (2006).

mammary gland gained only 0.14 g protein per day. After d 70 and up until farrowing, each mammary gland gained 3.41 g protein per day. This is almost a 24-fold increase in the protein accretion, which could be a significant amount if a sow has 16 or more mammary glands (Kim et al., 2013).

Young sows continue to grow during gestation, whereas the growth is minimal for multiparous sows. Ji et al. (2005) investigated the growth of maternal tissues during the course of gestation in pigs (Fig. 4). Within a sow, the rate of protein gain in maternal

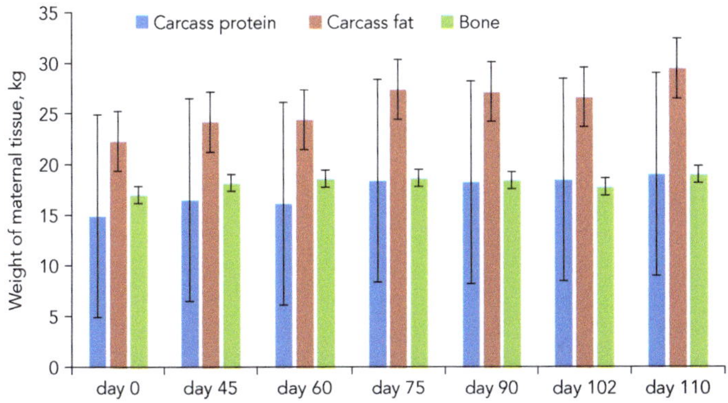

Figure 4 Carcass protein, carcass fat and bone in sows at different days of gestation (Ji et al., 2005).

tissues seems to be constant during gestation. However, these rates differ between young sows and multiparous sows.

Therefore, the nutritional management of gestating sows needs to consider how nutrient requirements vary depending on the stage of gestation and the number of foetuses and mammary glands.

2.2 Protein sources for gestating sows

Gestating sows are mature pigs which are capable of digesting proteins from various plants and animal-originated feedstuffs. Legume seed meals are the most popular protein supplements used in feeding gestating sows, including soybean meal (5-04-612), canola meal (5-06-145) and flaxseed meal (5-02-048). Coproducts from wet milling and brewing processes have also been used effectively, including distillers' dried grains with solubles (5-02-843) and potato protein concentrate (5-25-392). For gestating sows, the protein and amino acid digestibility of these feedstuffs is fairly high, compared with their digestibility by growing pigs (Stein et al., 2001).

2.3 Amino acid requirements for gestating sows

Due to the changes in the growth rates of foetuses and mammary glands during the course of gestation, the amino acid requirement and ideal amino acid pattern for a gestating sow differ according to the stage of gestation and the number of foetuses and mammary glands.

Kim et al. (2009) showed the changes in the amino acid accretion in a foetus (Table 2) during different stages of gestation. Lysine, an example of an essential amino acid, accreted with a rate of 19 mg per day until d 70 of gestation, when it increased to 283 mg per day (a 15-fold increase). In the case of tryptophan, the accretion increased from 3 to 56 mg per day (a 19-fold increase). It therefore appears that a sow must provide a significantly increased amount of amino acids to support the growth of foetuses from d 70 of gestation onwards. Lysine-based amino acid ratios accreted in foetal tissues also differ according to the stage of gestation (Table 2). Ratios among lysine, threonine, methionine, tryptophan

Table 2 Amino acid accretion and its relative ratio to lysine in foetal and mammary parenchymal tissues of sows during gestation

	Foetal tissue		Mammary parenchymal tissue	
Gain, g/d	Day 0–70	Day 70 to farrowing	Day 0–70	Day 70 to farrowing
Protein	0.25	4.63	0.14	3.41
Lysine	0.019	0.283	0.011	0.256
Threonine	0.010	0.162	0.006	0.145
Methionine	0.006	0.092	0.003	0.068
Tryptophan	0.003	0.056	0.002	0.040
Valine	0.013	0.211	0.008	0.194
Leucine	0.020	0.332	0.012	0.286
Isoleucine	0.009	0.142	0.006	0.141
Ratio to lysine				
Lysine	1.00	1.00	1.00	1.00
Threonine	0.53	0.57	0.55	0.57
Methionine	0.32	0.33	0.27	0.27
Tryptophan	0.16	0.20	0.18	0.16
Valine	0.68	0.75	0.73	0.76
Leucine	1.05	1.17	1.09	1.12
Isoleucine	0.47	0.50	0.55	0.55

Table 3 Amino acid requirements of gestating sows with 160 kg body weight at breeding, 50 kg weight gain during gestation, 12 foetuses and 14 mammary glands

g/d	Lys	Thr	Met	Trp	Val	Leu	Ile
Day 0–70	5.42	4.59	1.39	0.85	3.52	4.72	3.25
Day 70 to farrowing	13.74	9.86	3.77	2.28	9.30	13.35	7.67

and valine are 100:53:32:16:68 up to day 70 of gestation, and 100:57:33:20:75 from d 70 onward.

Kim et al. (2009) also showed the changes in the amino acid accretion in mammary glands (Table 2) during different stages of gestation. Lysine accreted with a rate of 11 mg per day until d 70 of gestation, and then increased to 256 mg per day from d 70 (a 23-fold increase). In the case of tryptophan, the accretion was increased from 2 to 40 mg per day (a 20-fold increase). It therefore seems that a sow must provide a significantly increased amount of amino acids to support the growth of mammary glands from d 70 of gestation. However, lysine-based amino acid ratios accreted in foetal

Table 4 Amino acid requirements and lysine-based amino acid ratio for sows with varied numbers of foetuses and mammary glands during gestation (day 70 to farrowing)[1]

	Number of foetuses[2]			Number of mammary glands[3]			Body weight at breeding[4]		
g/d	9	12	15	14	16	18	160	180	200
Protein									
Lysine	13.7	13.7	13.8	13.4	13.7	14.1	13.7	13.9	14.0
Threonine	9.9	9.9	9.9	9.7	9.9	10.1	9.9	10.1	10.3
Methionine	3.7	3.8	3.9	3.7	3.8	3.9	3.8	3.8	3.9
Tryptophan	2.2	2.3	2.4	2.2	2.3	2.4	2.3	2.3	2.4
Valine	9.1	9.3	9.5	9.0	9.3	9.6	9.3	9.4	9.5
Leucine	13.0	13.4	13.7	12.9	13.4	13.8	13.4	13.5	13.6
Isoleucine	7.7	7.7	7.7	7.5	7.7	7.9	7.7	7.8	7.9
Ratio to lysine									
Lysine	1.00	1.00	1.00	1.00	1.00	1.00	1.00	1.00	1.00
Threonine	0.72	0.72	0.71	0.72	0.72	0.71	0.72	0.87	0.73
Methionine	0.27	0.27	0.28	0.27	0.27	0.27	0.27	0.26	0.27
Tryptophan	0.16	0.17	0.17	0.17	0.17	0.17	0.17	0.16	0.17
Valine	0.66	0.68	0.69	0.67	0.68	0.68	0.68	0.65	0.68
Leucine	0.95	0.97	0.99	0.96	0.97	0.98	0.97	0.87	0.97
Isoleucine	0.56	0.56	0.56	0.56	0.56	0.56	0.56	0.60	0.56

[1] Expected body weight gain of sows (including conceptus) during gestation is 50 kg.
[2] When a sow with 160 kg body weight at breeding has 16 mammary glands with various numbers of foetuses.
[3] When a sow with 160 kg body weight at breeding has 12 foetuses with various numbers of mammary glands.
[4] When a sow has 12 foetuses and 16 mammary glands with various body weights at breeding.

tissues also differ by the stage of gestation (Table 5). Ratios among lysine, threonine, methionine, tryptophan and valine are 100:55:27:18:73 up until d 70 of gestation, and 100:57:27:16:76 from d 70.

Amino acid requirements for foetal growth, mammary gland growth and maternal gain (in the case of young sows) can be combined to show the overall amino acid requirement for gestating sows, based on the amino acid accretion in each component (Table 3). The required amounts of amino acids differ between early and late gestation (i.e. before and after d 70 of gestation), mainly due to the differences in the amino acid accretion in foetuses and mammary glands. Lysine-based amino acid ratios required by gestating sows also differ between early and late gestation, mainly due to the differences in amino acid ratios accreted in foetuses and mammary glands. Therefore, determining the amino acid requirements for gestating sows should be based on the stage of gestation, the number of foetuses and the number of mammary glands (Table 4).

Table 5 Protein content and amino acid composition in colostrum and milk

	Colostrum[1]	Milk[2]
DM, g/kg	243 ± 8	166 ± 4
Crude fat, g/kg DM	219 ± 21	416 ± 51
Crude protein, g/kg DM	497 ± 23	286 ± 2
Immunoglobulin G, g/L	8.96 ± 0.15	4.83 ± 0.13
Amino acid, g/kg DM		
Ala	20.3 ± 0.9	11.6 ± 0.3
Arg	21.2 ± 0.9	9.0 ± 0.2
Asp	38.1 ± 3.1	32.4 ± 0.8
Cys	7.3 ± 0.1	4.5 ± 0.1
Gly	19.6 ± 0.9	6.8 ± 0.2
Glu	78.0 ± 5.0	58.7 ± 1.4
His	11.7 ± 0.6	5.7 ± 0.1
Ile	18.6 ± 1.3	13.9 ± 0.4
Leu	44.6 ± 2.4	28.0 ± 0.8
Lys	65.9 ± 2.6	25.2 ± 0.7
Met	10.5 ± 0.6	6.4 ± 0.1
Phe	20.3 ± 1.5	12.1 ± 0.3
Pro	56.3 ± 0.9	34.9 ± 0.7
Ser	27.4 ± 1.2	14.7 ± 0.3
Thr	26.7 ± 1.6	13.8 ± 0.3
Trp	6.6 ± 0.1	4.1 ± 0.1
Tyr	18.5 ± 1.0	11.8 ± 0.3
Val	32.5 ± 2.0	15.2 ± 0.4

[1] Colostrum was collected within 12 h postpartum (Adopted from Lin et al. (2009) and Shen et al. (2011)).
[2] Milk was collected on d 17–21 of lactation (Adopted from Mateo et al. (2008) and Shen et al. (2011)).

3 Lactating sows

3.1 Principles

In the nutritional management of lactating sows, the three most important factors to consider are milk production, piglet growth and growth of lactating mammary glands. These three components are closely related; suckling stimuli from piglets to mammary glands are essential for the galactopoiesis and mammary cell proliferation (Brooks and Burke, 1998; Kim et al., 1999b), the number of mammary cells determines the milk production (Kim et al., 2000) and the increased milk production enhances the growth of nursing pigs.

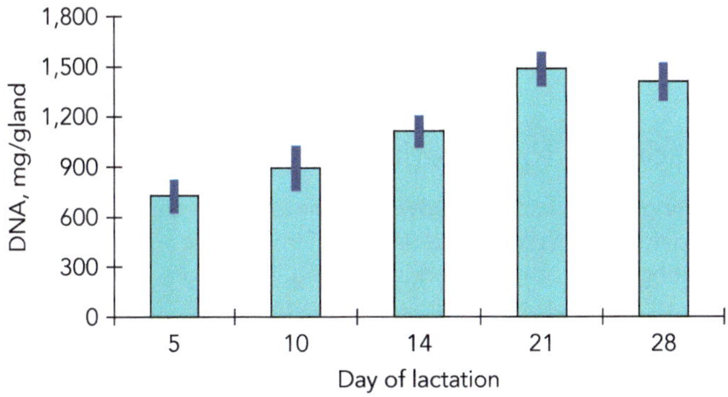

Figure 5 Total DNA in mammary parenchymal tissues at different days of lactation (adopted from Kim et al., 1999a).

Sows provide significant amounts of amino acids to support milk production, regardless of dietary intake. When a sow's amino acid intake is limited, she mobilizes her body protein to provide amino acids for milk production, which causes weight loss during lactation. Limited feed intake (appetite) often becomes a challenge for the successful nutritional management of lactating sows, because the genetic selection for pigs has focused on high lean growth, which is not necessarily correlated with increased feed intake or appetite (Kanis, 1990). The nutritional management of lactating sows, therefore, must focus on providing sufficient nutrients with nutrient-concentrated diets in order to balance the requirements of milk production and mammary gland growth.

There is extensive research describing the growth of lactating mammary glands and the associated nutritional requirements (Kim et al., 1999a,b,c, 2000). Supporting the growth of lactating mammary glands is important, because the milk production is determined by the number of mammary cells. Mammary glands start to grow upon conception, with significant growth from d 70 of gestation. Upon farrowing, mammary cells are fully functional, thereby synthesizing and secreting colostrum and then changing the composition to mature milk within a few days. The number of mammary cells continues to increase during lactation (Kim et al., 1999a), allowing an increase in milk production up to week 4 of lactation. It has been shown that mammary glands continue to grow until week 4 of lactation, with associated increases in mammary cell numbers (Fig. 5). The rates of protein gain in lactating mammary glands seem to be constant until week 4 of lactation, and then decelerate if lactation continues. Those mammary glands not suckled by piglets start to involute, and rapidly lose the majority of their proteins within five days of farrowing (Kim et al., 2001a). Upon weaning, the majority of proteins in the mammary glands are lost within two days (Ford at al., 2003).

The protein content and amino acid composition of colostrum (Lin et al., 2009; Shen et al., 2011) and milk (Mateo et al., 2008; Shen et al., 2011) are shown in Table 5. Milk protein and carbohydrate contents are largely unaffected by factors such as breed, nutritional status and parity of sows (Klaver et al., 1981; Klobasa et al., 1987; Farmer et al., 2001; Hurley, 2015), whereas milk fat content can be affected by dietary fat supplementation (Tilton et al., 1999). The milk yield increases until week 4 of lactation, but without major changes in the protein content and amino acid composition. In feeding lactating sows, the dietary nutrient composition should aim to minimize the negative balance between nutrient supply and nutrient use. A negative nutrient balance will lead to maternal tissue mobilization to provide deficient nutrients for the milk production and mammary gland growth.

3.2 Protein sources for lactating sows

Similar to gestating sows, lactating sows are mature and capable of digesting proteins from various feedstuffs from plants and animals. Similar types of protein supplements used in feeding gestating sows can be used in feeding lactating sows. The protein and amino acid digestibility of major protein supplements is fairly high for lactating sows, compared with the digestibility by growing pigs (Stein et al., 2001).

3.3 Amino acid requirements for lactating sows

Lactating sows require amino acids to synthesize milk protein and to increase the number of mammary cells to support increasing milk production. When a feed or feed intake does not provide a sufficient amount of amino acids, a sow mobilizes her body protein to provide amino acids for milk production and mammary gland growth. Insufficient feed intake or low appetite often becomes a limiting factor in amino acid supply for milk production and mammary gland growth during lactation. In formulating a diet for lactating sows, amino acid contents should be adjusted based on expected feed intake and milk production. When a sow has a low feed intake, the amino acid content of her feed should be increased compared with a sow possessing a high feed intake.

The amino acid compositions of milk, mammary parenchymal tissues and maternal tissues are different (Table 6). The dietary amino acid ratio should aim to balance the differences between the amino acid composition supplied by maternal tissue mobilization and the amino acid composition required for the milk production and mammary gland growth. Therefore, the dietary amino acid ratio should be affected by these three factors (i.e. tissue mobilization, milk production and mammary gland growth). However, the milk

Table 6 Amino acid patterns in milk, mammary glands and plasma in lactating sows[1]

Ratio to lysine	Milk[2]	Mammary parenchymal tissue[3]	Retained in mammary glands[4]	Maternal tissue mobilization[5]
Lysine	1.00	1.00	1.00	1.00
Threonine	0.55	0.57	1.50	0.39
Methionine	0.26	0.27	0.53	0.28
Tryptophan	0.16	0.15	0.25	0.16
Valine	0.60	0.76	2.16	0.75
Leucine	1.11	1.13	4.03	0.98
Isoleucine	0.55	0.55	2.31	0.58

[1] Table is adapted and modified from Kim and Easter (2003).
[2] Data are from Mateo et al. (2008).
[3] Data are from Kim et al. (1999a).
[4] Data are from Trottier et al. (1997).
[5] Data are from Kim et al. (2001b).

Table 7 Ideal amino acid patterns and the order of limiting amino acids for lactating sows. Data are adapted from Kim et al. (2001a)

Estimated 21-d weight loss (kg)[1]	75–80	33–45	12–15	6–8	0
Level of tissue mobilization (%)[2]	50	40	20	5	0
Ideal AA pattern (% of Lys)					
Lys	100	100	100	100	100
Thr	75	69	63	60	59
Val	78	78	78	77	77
Leu	128	123	118	115	115
Ile	60	59	59	59	59
Arg	22	38	59	69	72
Order of limiting amino acids[3]					
First limiting amino acid	Thr	Lys	Lys	Lys	Lys
Second limiting amino acid	Lys	Thr	Thr	Val	Val
Third limiting amino acid	Val	Val	Val	Thr	Thr

[1] These values refer to weight loss of sows during lactation estimated based on the amount of protein loss and tissue composition measured by Kim et al. (2001a).
[2] These values refer to amino acid % in milk output derived from tissue protein catabolism in the sow.
[3] This assumes that a typical corn–soybean meal diet is fed during lactation.

production and mammary gland growth are unaffected by dietary changes because sows ensure that the nutritional requirements for the milk production are met by mobilizing their body tissues when the intake is not sufficient. Consequently, the dietary amino acid ratio should be largely determined by the amount of maternal tissue mobilization that is related to voluntary feed intake or appetite. The dietary amino acid ratio required for a sow will differ depending on her feed intake (which, in turn, affects the maternal tissue loss). It has been shown that a sow with severe tissue mobilization (due to low feed intake in the case of young sows) will have a greater requirement for threonine, whereas a sow with minimal tissue mobilization (due to sufficient feed intake in the case of multiparous sows) will have a greater requirement of valine (Table 7). Therefore, the nutritional management of lactating sows could be based on their parities in order to provide the ideal amino acid ratio needed to maximize the milk production and mammary gland growth.

4 Nursing pigs

4.1 Principles

The growth of nursing pigs largely depends on their milk intake from suckling mammary glands. It has been shown that the milk production is not constant among lactating mammary glands regardless of anatomical location, due to their different sizes (Fig. 6). Pigs suckling five pairs of anterior mammary glands grow faster than pigs suckling mammary glands in other locations (Fig. 7). When the milk production is not sufficient from a sow,

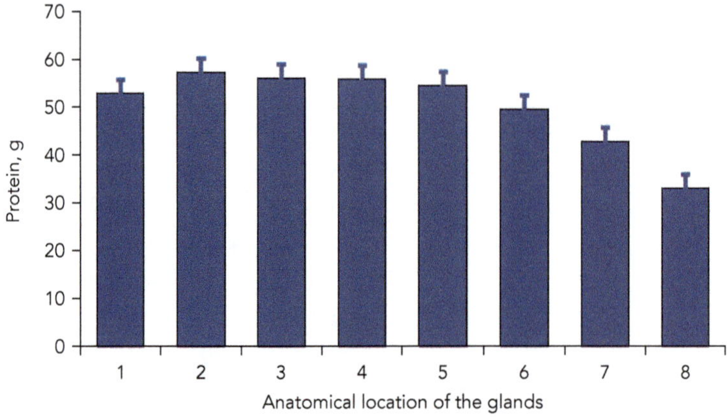

Figure 6 Protein content (g) in lactating mammary glands by their anatomical location.

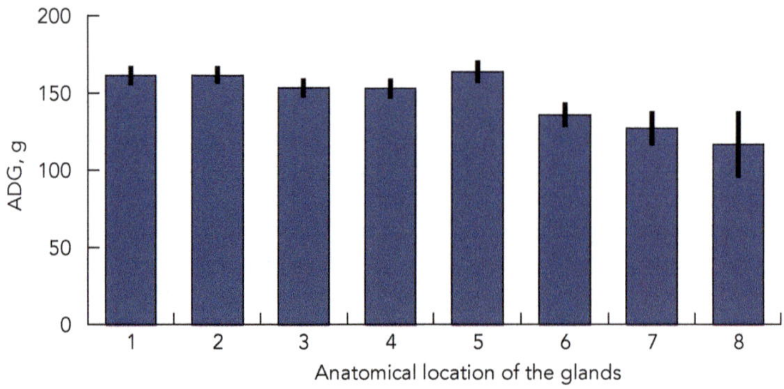

Figure 7 Average daily gain (g) of suckling piglets by the anatomical location of mammary glands.

creep feeding can be practised, which provides additional nutrients and also enhances appetite for a solid diet, helping post-weaning adaptation.

4.2 Protein sources for nursing pigs

Newborn pigs receive nutrients through colostrum and milk. Protein contents in porcine colostrum and milk are 12.1 and 4.7%, respectively (Lin et al., 2009; Shen et al., 2011). These values are high compared with those from other animal species. Proteins in porcine milk include casein (600 g/kg) and whey protein (400 g/kg). This proportion is different from that of bovine milk (casein 800 g/kg and whey protein 200 g/kg). Casein is composed of α-casein (480 g/kg), β-casein (360 g/kg), κ-casein (130 g/kg) and γ-casein (30 g/kg), whereas whey protein is composed of β-lactoglobulin (450 g/kg), α-lactalbumin (170 g/kg), proteose-peptone (110 g/kg), immunoglobulins (110 g/kg), membrane proteins (90 g/kg), serum albumin (60 g/kg), lactoferrin (10 g/kg) and lactoperoxidase (10 g/kg) (Fox et al., 2007). Both casein and whey proteins are highly digestible by newborn pigs.

Table 8 Ideal amino acid pattern for young pigs and amino acid ratios in porcine milk and protein supplements

Ratio to lysine	Ideal protein for young pigs	Porcine milk[2]	Fishmeal	Plasma protein	Soybean meal
Lysine	1.00	1.00	1.00	1.00	1.00
Threonine	0.59	0.55	0.53	0.60	0.60
Methionine	0.29	0.26	0.38	0.11	0.23
Tryptophan	0.16	0.16	0.12	0.22	0.23
Valine	0.64	0.60	0.65	0.70	0.74
Leucine	1.00	1.11	0.95	1.07	1.21
Isoleucine	0.51	0.55	0.54	0.38	0.72

Lin et al. (2009) investigated the total tract amino acid digestibility of porcine colostrum by newborn pigs. The digestibility of the amino acids in colostrum was over 98% when consumed by newborn pigs. Mavromichalis et al. (2001) investigated the ileal amino acid digestibility of porcine milk by nursing pigs. The ileal digestibility of amino acids in milk was fairly high at near 90%. Colostrum and milk provide highly digestible essential proteins for nursing pigs.

4.3 Amino acid requirements for nursing pigs

Amino acid compositions in milk and other protein supplements are shown in Table 8. It is interesting to observe that the amino acid composition in porcine milk is fairly similar to the ideal amino acid pattern for young pigs suggested by NRC (2012), compared with other protein supplements.

5 Nursery pigs

5.1 Principles

In the modern pig production, pigs are weaned early, typically from two to four weeks of age, whereas sows would naturally wean their litters at 10–12 weeks of age. Weaning age may differ depending on the region and management programme. For example, it is typical to wean pigs at two to three weeks of age in the United States and South America, whereas weaning at three to four weeks of age is more common in the EU and China.

Newly weaned pigs are vulnerable to environmental stressors, often leading to growth depression. A solid diet given to pigs upon weaning could also be a major stressor; it may cause intestinal challenges including physical damage, allergic reactions, inflammatory responses and oxidative damage to intestinal mucosal layers. After that, these reduce feed intake, cause diarrhoea and finally reduce growth.

The gastrointestinal tract is not fully mature at two to four weeks of age. Pigs at this age will have limited HCl secretion in the stomach and limited secretion of amylolytic enzymes, whereas lactase will be highly active. The nutritional management of nursery pigs should, therefore, aim for a smooth transition from sows' milk to a solid diet by phase feeding, starting with feedstuffs that contain substrates that are highly digestible, and then gradually replacing these with plant-based feedstuffs.

5.2 Protein sources for nursery pigs

In feeding newly weaned nursery pigs, a widely accepted practice is the use of protein supplements obtained from animal tissues and animal products such as skim milk powder (IFN 5-01-175), casein (5-01-162), whey protein concentrate (5-06-836), fishmeal (5-01-977), spray dried egg, blood cells, blood meal (5-26-005), blood plasma, meat- and bonemeal (5-00-388), meat meal (5-00-385), poultry by-product (5-03-800) and poultry meal (5-03-798). The use of these protein supplements will depend on the region, due to accessibility (or cost) and regulation. For example, the use of the brain and spinal cord of cattle older than 20 months of age is prohibited in the United States. The poultry meal is more accessible in southern parts of the United States, while the fishmeal is more accessible in Southeast Asian countries. Skim milk powder, casein and whey protein concentrate are typically expensive because they are consumed as human food, and this limits their use in nursery diets. However, off-the-shelf dairy foods such as cheese powder can be collected and used as protein supplements for nursery pigs. Proteins from dairy products contain amino acids at close to the ideal ratio for nursery pigs (Table 8). Blood plasma has also been used in diets for newly weaned pigs because of its high nutrient digestibility and the health benefits associated with the functional proteins included in blood plasma. Despite its high cost, blood plasma has been used to make up 3–8% of diets for newly weaned pig because of its nutritional and functional benefits.

Protein supplements obtained from plants are also in use, for example, legume seed meals, and coproducts from wet milling and brewing processes. However, the use of protein supplements from plants has a certain maximum allowance, due to antinutritional compounds found in these feedstuffs. For example, the dehulled soybean meal (5-04-612) is one of the most popular protein supplements in feeding pigs, but its use is limited to feeding newly weaned nursery pigs because it contains multiple antinutritional compounds such as trypsin inhibitors, allergenic proteins (i.e. glycinin and β-conglycinin) and flatulence-producing compounds (i.e. raffinose family oligosaccharides). For newly weaned pigs with a body weight of 5–7 kg, the soybean meal may be included in the diet, but it typically makes up less than 20% of the feed. However, the use of soybean meal in newly weaned pig diets can be increased if these antinutritional compounds can be reduced by processing methods, such as fermentation or enzymatic hydrolysis. The enzyme-treated soybean meal and the fermented soybean meal can be used with reduced concerns about antinutritional compounds (Hong et al., 2004; Kim et al., 2010). Typically, these processed soybean meals are used to make up 3–10% of the diet, without reducing the amount of conventional soybean meal in the diet.

Supplemental amino acids have been used extensively in nursery diets to provide the required amino acid ratios. There are at least five to seven commercially available supplemental amino acids which include lysine, methionine, threonine, tryptophan, valine, leucine and isoleucine. The use of supplemental amino acids can reduce the requirement for protein supplements, which are either expensive or contain antinutritional compounds. Diets for nursery pigs are often supplemented with lysine, methionine, threonine and tryptophan. The amount of supplemental lysine determines the amount and number of other supplemental amino acids in the diet. Supplemental L-lysine HCl can be used at levels as high as 5 g/kg in the diet without affecting the growth performance of nursery pigs. The use of supplemental amino acids allows the nutritionists to formulate diets with ideally balanced amino acids and to reduce the crude protein intake. In the pig production without the use of antimicrobial growth promoters, it is important to formulate a nursery diet with a reduced amount of dietary crude protein in order to reduce the incidence of diarrhoea.

5.3 Amino acid requirements for nursery pigs

Amino acid requirements for nursery pigs have been extensively investigated. The requirements are based on the amino acid content that will be digestible and absorbable in the small intestine before reaching the large intestine. These are called standardized ileal digestible (SID) amino acids or true ileal digestible amino acids. For nursery pigs, feed should meet or slightly exceed the digestible essential amino acids' requirements based on body weight. There are several publications suggesting amino acid requirements. An example from the NRC is shown in Table 1.

Meeting amino acid requirements in nursery diets requires the consideration of both the amount of and balance (ratios) between essential amino acids. When formulating nursery diets to meet amino acid requirements, protein supplements from animal sources can be introduced to the formula, providing proteins that are highly digestible and amino acids with a ratio similar to the requirement (Table 8). The amount of supplemental lysine should be determined first (and can be as high as 5 g/kg). This, in turn, determines the amount of protein supplements from plant sources that can be used in the diet. Finally, other supplemental amino acids (typically methionine and threonine, and sometimes tryptophan and valine) can be added to meet the requirements. The use of supplemental amino acids is important in formulating a diet with an ideal amino acid ratio for nursery pigs.

6 Growing and finishing pigs

6.1 Principles

At this growth stage, pigs eat well and grow fast. At some points, the average daily weight gain can be as high as 1.3 kg during this stage. The nutritional management of growing and finishing pigs primarily focuses on high lean gain, affordable feed cost, enhancing feed efficiency, minimizing nutrient waste to the environment, and producing safe and high-quality pork. Growing and finishing pigs can digest cereal grains and legume seed meals without problems, and so plant-based feedstuffs are usually used in formulating diets in order to optimize feed costs. Costly feedstuffs from animal sources can be excluded at this time.

6.2 Protein sources for growing and finishing pigs

Legume seed meals such as soybean meal, canola meal and flaxseed meal are extensively used as protein supplements in diets for growing and finishing pigs. Coproducts from wet milling and brewing processes such as distillers' dried grain with solubles (DDGS; 5-02-843), brewers' grains (5-00-516), corn gluten meal (5-02-900) and potato protein concentrate (5-25-392) are also extensively used depending on the region. Protein supplements from these coproducts can be used to partly replace conventional protein supplements, without negative impacts on the growth performance (Cromwell et al., 2011; Yaakugh et al., 1994; Tusnio et al., 2011; Schedle et al., 2016). Supplemental amino acids are used to reduce the requirement for protein supplements. This helps to reduce dietary crude protein, and thus to reduce nitrogen excretion in faeces. Supplemental L-Lys HCl can be used up to a level of 5 g/kg without having negative impacts on the growth performance, as long as essential amino acid requirements are met by supplementing methionine, threonine and occasionally tryptophan and valine as needed (Jones et al., 2014; Tuitoek et al., 1997; Lee et al., 2001; Toledo et al., 2014).

6.3 Amino acid requirements for growing and finishing pigs

When formulating diets for growing and finishing pigs, it is important to ensure that the amino acid balance meets the requirements in order to reduce the catabolism of dietary amino acids and thus reduce the nitrogen excretion through faeces. Amino acid requirements and the ideal balance relative to lysine have been extensively investigated and recorded in publications such as ARC (1981), NRC (2012) and Rostagno et al. (2005). The use of typical protein supplements alone cannot meet the ideal amino acid balance needed by growing and finishing pigs, and thus supplemental amino acids have been used extensively to meet the requirements. The use of lysine, the first limiting amino acid, can reduce the crude protein content by approximately 30 g/kg, and the use of three major limiting amino acids (lysine, threonine and tryptophan or methionine) can reduce the crude protein content by approximately 50–60 g/kg.

Amino acid requirements decline as pigs mature and get closer to market weight. Depending on the region (including the United States, Canada, Japan and Korea), ractopamine may be used during the last three weeks before marketing in order to enhance protein gain. When ractopamine is supplemented in finishing diets, the lysine content is raised up to 9 g/kg SID Lys in order to meet the amino acid requirements for the increased protein synthesis. However, except for the case of ractopamine use, the dietary amino acid content is decreased.

7 Summary

This chapter briefly reviewed the amino acid requirements of pigs and the major factors to consider when determining them, with an emphasis on the difference in requirements according to the growth stage and physiological status of the pigs. The amino acid levels and ratios required for foetal and mammary gland growths affect dietary amino acid requirements, particularly during late gestation in sows. The amino acid levels and ratios required for the milk production and mammary gland growth affect dietary amino acid requirements for lactating sows, and this is complicated by the issue of maternal tissue mobilization. Milk protein contains ideal amino acid ratios for the growth of nursing piglets. In feeding growing pigs, dietary amino acids should meet the requirements but also minimize excess provision to reduce amino acid catabolism. The use of supplemental amino acids allows the nutritionist to formulate feed which helps to reduce nitrogen excretion to the environment. With an increased number of supplemental amino acids (Lys, Met, Thr, Trp and Val) available in swine feed, nutritionists can formulate feeds with ideal amino acid patterns for each growth stage and physiological status, and greatly reduce the use of protein supplements.

8 Where to look for further information

Based on what we know and studied so far, it seems that further investigation is warranted on the subject of how much the crude protein content of feed can be reduced by increasing the use of supplemental amino acids without affecting production efficiency. Potential improvement in the genetic potential for lean gain and prolificacy could also challenge the nutritionists to revise their understanding of amino acid requirements and note any potential changes in their required ratios.

9 References

Anon. 2002. 'Food contact regulations', *US Federal Register*, 70(8):30–50, http://usgov/federalregister/2002/70/(accessed 4 January 2015).

ARC. 1981. The Nutrient Requirements of Pigs. Commonwealth Agricultural Bureaux, Slough, UK.

Brooks, P. H., and J. Burke. 1998. Behaviour of sows and piglets during lactation. In (M. W. A. Verstegen, P. J. Moughan, and J. W. Scharma Eds): The Lactating Sow. Wageningen Press, Wageningen, pp. 301–38.

Cromwell, G. L., M. J. Azain, O. Adeola, S. K. Baidoo, S. D. Carter, T. D. Crenshaw, S. W. Kim, D. C. Mahan, P. S. Miller, and M. C. Shannon. 2011. Corn distillers dried grains with solubles in diets for growing-finishing pigs – a cooperative study. *J. Anim. Sci.* 89:2801–11.

Fall, C. H. D., C. S. Yajnik, S. Rao, A. A. Davies, N. Brown, and H. J. W. Farrant. 2003. Micronutrients and fetal growth. *J. Nutr.* 133:1747S–1756S.

Farmer, C., Palin, M. F., Sorensen, M. T. and Robert, S., 2001. Lactational performance, nursing and maternal behavior of Upton-Meishan and large white sows. *Canadian Journal of Animal Science* 81: 487–93.

Ford, J. A., S. W. Kim, S. L. Rodriguez-Zas, and W. L. Hurley. 2003. Quantification of mammary gland tissue size and composition changes after weaning in sows. *J. Anim. Sci.* 2003. 81:2583–9

Fox, P. F., T. P. Guinee, T. M. Cogan, and P. L. H. McSweeney. 2017. Fundamentals of Cheese Science. 2nd Eds Pp. 771. Springer.

Hong, K. J., C. H. Lee, and S. W. Kim. 2004. Aspargillus Oryzae GB-107 fermentation improves nutritional quality of food soybeans and feed soybean meals. *J. Med. Food* 7:430–6.

Ji, F., G. Wu, J. R. Blanton Jr., and S. W. Kim. 2005. Changes in weight and composition in various tissues of pregnant gilts and their nutritional implication. *J. Anim. Sci.* 83:366–75.

Ji, F., W. L. Hurley, and S. W. Kim. 2006. Characterization of mammary gland development in pregnant gilts. *J. Anim. Sci.* 84:579–87.

Jones, C. K., M. D. Tokach, J. L. Usry, C. R. Neill, and J. F. Patience. 2014. Evaluating lysine requirements of nursery pigs fed low protein diets with different sources of nonessential amino acids. *J. Anim. Sci.* 92: 3460–70.

Kanis, E. 1990. Effect of food intake capacity on genotype by feeding regimen interactions in growing pigs. *Animal Production* 50:343–51.

Kensinger, R. S., R. J. Collier, F. W. Bazer, C. A. Ducsay, and H. N. Becker. 1982. Nucleic acid, metabolic and histological changes in gilt mammary tissue during pregnancy and lactogenesis. *J. Anim. Sci.* 54:1297–308.

Kim, S. W., W. L. Hurley, I. K. Han, and R. A. Easter. 1999. Changes in tissue composition associated with mammary gland growth during lactation in the sow. *J. Anim. Sci.* 77:2510–16.

Kim, S. W., I. Osaka, W. L. Hurley, and R. A. Easter. 1999b. Mammary gland growth as affected by litter size in lactating sows: impact on lysine requirement. *Journal of Animal Science*. 77:3316–21.
Kim, S. W., W. L. Hurley, I. K. Han, H. H. Stein, and R. A. Easter. 1999c. Effect of nutrient intake on mammary gland growth in lactating sows. *J. Anim. Sci.* 77:3304–15.
Kim, S. W., W. L. Hurley, I. K. Han, and R. A. Easter. 2000. Growth of nursing pigs related to the characteristics of nursed mammary glands. *Journal of Animal Science*. 78:1313–18.
Kim, S. W., R. A. Easter, and W. L. Hurley. 2001a. The regression of unsuckled mammary glands during lactation in sows: The influence of lactation stage, dietary nutrients, and litter size. *J. Anim. Sci.* 79:2659–68
Kim, S. W., D. H. Baker, and R. A. Easter. 2001b. Dynamic ideal protein and limiting amino acids for lactating sows: The impact of amino acid mobilization. *J. Anim. Sci.* 79:2356–66.
Kim, S. W., W. L. Hurley, G. Wu, and F. Ji. 2009. Ideal amino acid balance for sows during gestation and lactation. *J. Anim. Sci.* 87:E123–E132.
Kim, S. W., E. van Heugten, F. Ji, and R. D. Mateo. 2010. Fermented soybean meal as a vegetable protein source: I Effects on growth performance of nursery pigs.. *J. Anim. Sci.* 88:214–24 .
Kim, S. W., A. C. Weaver, Y. B. Shen, and Y. Zhao. 2013. Improving efficiency of sow productivity: nutrition and health. *J. Anim. Sci. Biotechnol.* 4:26.
Klaver, J., Van Kempen, G. J. M., De Lange, P. G. B., Verstegen, M. W. A. and Boer, H., 1981. Milk composition and daily yield of different milk components as affected by sow condition and lactation/feeding regimen. *Journal of Animal Science* 52: 1091–7.
Klobasa, F., Werhahn, E. and Butler, J. E., 1987. Composition of sow milk during lactation. *Journal of Animal Science* 64: 1458–66.
Lee, J. H., J. H. Kim, J. D. Kim, S. W. Kim, and I. K. Han. 2001. Effects of low crude protein diets supplemented with synthetic amino acids on performance, nutrient utilization and carcass characteristics in finishing pigs reared using a phase feeding regimen. *Asian-Aust. J. Anim. Sci.* 14:655–67.
Lin, C., D. C. Mahan, G. Wu, and S. W. Kim. 2009. Protein digestibility of colostrums by neonatal pigs. *Livestock Science* 121:182–6.
Mateo, R. D., G. Wu, H. K. Moon, J. A. Carroll, and S. W. Kim. 2008. Effects of dietary arginine supplementation during gestation and lactation on the performance of lactating primiparous sows and nursing piglets. *J. Anim. Sci.* 86:827–35.
Mavromichalis, I., Parr, T. M., Gabert, V. M., Baker, D. H., 2001. True ileal digestibility of amino acids in sow's milk for 7-day-old pigs. *J. Anim. Sci.* 79:707–13.
McPherson, R. L., F. Ji, G. Wu, J. R. Blanton, and S. W. Kim. 2004. Fetal growth and compositional changes of fetal tissues in pigs. *J. Anim. Sci.* 82:3014–20.
NRC. 2012. Nutrient requirements of swine. 11th rev. ed. Natl. Acad. Press, Washington, DC.
Rostagno, H. S., A. Teixeira, J. L. Donzele, P. C. Gomes, R. F. M. De Oliveira, D. C. Lopes, A. J. P. Ferreira, and S. L. Toledo Barreto. 2005. Brazilian Tables for Poultry and Swine: composition of feedstuffs and nutritional requirements. Universidade Federal de Viçosa, Departamento de Zootecnia, MG, Brazil
Schedle, K., A. Simmongiovanni, E. Corrent, and J. Bartelt. 2016. Effect of a protein reduced diet with and without high contents of industrial by-products, on zootechnical performance and slaughter parameters of fattening pigs. *Vet. Med. Zoot.* 73:113–17.
Shen, Y. B., J. A. Carroll, I. Yoon, R. D. Mateo and S. W. Kim. 2011. Effects of supplementing Saccharomyces cerevisiae fermentation product in sow diets on performance of sows and nursing piglets. *J. Anim. Sci.* 89:2462–71.
Smith, R. 2015. An introduction to plant breeding, Cambridge University Press.
Smith R. and B. Jones. 2010. Recent advances in plant breeding. *J. Agric. Science*, 22:1–20.
Sørensen, M. T., K. Sejrsen, and S. Purup. 2002. Mammary gland development in gilts. *Livest. Prod. Sci.* 75:143–8.

Stein, H. H., S. W. Kim, T. T. Nielsen, and R. A. Easter. 2001. Comparative standardized ileal protein and amino acid digestibilities by growing pigs and sows. *J. Anim. Sci.* 79:2113–22.
Tilton, S. L., Miller, P. S., Lewis, A. J., Reese, D. E., Ermer, R. M. 1999. Addition of fat to the diets of lactating sows: I Effects on milk production and composition and carcass composition of the litter at weaning.. *J. Anim. Sci.* 77, 2491–500 .
Toledo, J. B., A. C. Furlan, P. C. Pozza, J. Carraro, G. Moresco, S. L. Ferreira, and A. G. Gallego. 2014. Reduction of the crude protein content of diets supplemented with essential amino acids for piglets weighing 15 to 30 kilograms. *Revista Brasileira de Zootecnia* 43:301–9.
Trottier, N. L., C. F. Shipley, and R. A. Easter. 1997. Plasma amino acid uptake by the mammary gland of the lactating sow. *J. Anim. Sci.* 75:1266–1278 .
Tuitoek, K., L. G. Young, C. F. M. Lange, and B. J. Kerr, 1997. The effect of reducing excess dietary amino acids on growing-finishig pig performance: evaluation of the ideal protein concept. *J. Anim. Sci.* 75:1575–83.
Tusnio, A., B. Pastuszewska, E. Swiech, and M. Taciak. 2011. Response of young pigs to feeding potato protein and potato fibre - nutritional, physiological and biochemical parameters. *J. Anim. Feed Sci.* 20:361–78.
Yaakugh, I. D. I., T. S. B. Tegbe, S. A. S. Olorunju, and A. O. Aduku. 1994. Replacement value of brewers' dried grain for maize on performance of pigs. *J. Sci. Food Ag.* 66:465–71.

Chapter 9

Recent advances in understanding the role of vitamins in pig nutrition

Charlotte Lauridsen, Aarhus University, Denmark; and J. Jacques Matte, Agriculture and Agri-Food Canada, Canada

1 Introduction

There is, in general, a lack of scientific information on the requirements for vitamins in the modern intensive swine production, as official recommendations are primarily based on genetically outdated lines of pigs and production conditions which have changed dramatically during the last 30 years. Depending upon the source of information, either from Denmark or from Canada, growth and reproduction performance of modern lines of pigs would correspond to daily weight gains over 900 g in the finishing phase and to litter size of 13 (CCSI, 2015) to 16 liveborn/litter (VSP, 2015). Such a level of performance associated with reduced antibiotic usage pig production systems are factors influencing the vitamin requirement, and more information on the optimal vitamin nutrition in this aspect is needed. The level of scientific knowledge for each vitamin is also heterogeneous and empiricism is still present in this field, which is reflected by large variations of vitamin fortifications used in commercial conditions, as reported by previous and recent surveys (BASF, 1993; BASF, 2001; Flohr et al., 2016). Since the middle of the twentieth century, vitamin requirements for pigs have changed from prevention of deficiencies (practically absent nowadays) to optimization of performance for growth and reproduction. However, less information is available on vitamin requirements for optimal immunity and/or meat nutritional quality, although there is a major interest in optimizing these issues through

http://dx.doi.org/10.19103/AS.2017.0013.09

nutritional strategies in the pig production systems. The challenge for the future is to realign objectives from the performance criteria alone to other aspects such as stress and disease resistance.

Besides the decrease in neonatal piglet survival during the last two decades, which has become one of the major problems in modern pig production, pig production had to face several outbreaks of major diseases (e.g. PED, Circovirus and PRRSV). The rapid succession and wide spread of these different infections is a matter of concern both in economic terms and in terms of public good. With this macroscopic view on pig production, it is tempting to hypothesize that modern pig lines are just more fragile and susceptible to pathogens than older pig lines. In parallel, during the last 30 years, changes in modes of production have constantly limited the allowances of major and minor nutrients provided to the pig throughout its life from fertilization of ova to slaughter at market weight. For micronutrients such as vitamins, the development of hyperprolificacy has likely decreased the individual share of their transfer from the mother to embryos/foetus (Matte and Lauridsen, 2013). For fat-soluble vitamins in particular, this might be critical as they are generally poorly transferred to the foetus during gestation (Matte et al., 2014a,b). After birth, the post-natal transfer of fat-soluble vitamins occurs through colostrum and milk, but, given the large litter size with heterogeneous piglets and increased competition between littermates for colostrum (Theil et al., 2014), the partition of vitamins and other micronutrients from the sow to the offspring has also likely been reduced on an individual piglet basis. After weaning, piglets become completely independent from maternal transfer of vitamins and rely on dietary vitamin provision, which is primarily brought by the vitamin premix but may also be provided naturally by ingredients in the diet and/or endogenous synthesis. However, during the critical period after weaning from the sow, piglets may suffer from impaired capability for absorption and digestion, and the dietary composition of the weaning diet may reduce the bioavailability of some vitamins relative to milk.

During the last decades, growth performance has improved constantly towards global reduction of feed intake throughout the entire growing period. Fortification strategies, however, have not actually changed in commercial conditions during at least the last 20 years (BASF, 1993; Flohr et al., 2016). Therefore, it is plausible to assume that daily amounts of vitamins consumed per market pig for the whole starting, growing and finishing period have diminished by the same order of magnitude as feed intake during the last decades. Hence, there is lack of scientific information on the vitamin requirement for the fast-growing lean meat type of pigs according to their physiological responses. Furthermore, during the last decades, some feed regimens have proposed total withdrawal of the vitamin supplementations in premixes during part or for the whole finishing period (McGlone, 2000; Shaw et al., 2002), as an attractive practice to pork producers for reducing feed costs. Although such decrease of vitamin provision to market pigs can be applied without detrimental effect on growth performance (McGlone, 2000), it may have impact on vitamin deposition in carcass and pork products, and subsequently on meat quality of the pork. Despite some discrepancies in the literature probably related in part to experimental design, supplement withdrawal may decrease, for instance, concentrations of pork riboflavin and niacin (Shaw et al., 2002), thiamine and vitamin E (Edmonds and Arentson, 2001), vitamin A (Olivares et al., 2009) as well as vitamin B_{12} and folates (Giguère et al., 2012) . Such decreases in contents of these micronutrients in pork may affect not only meat quality parameters such as oxidative stability, colour and water-binding (Jensen et al., 1998), but also consumer perception for the nutritional value of this meat (Shaw et al., 2002) as pork is generally considered the most important among meat sources consumed by humans. In fact, conventional pork (without withdrawal of the vitamin supplementations) is recognized as an excellent source of several water-soluble vitamins providing a significant percentage (>65% for thiamine and niacin and >25% for riboflavin, vitamin B_6 and vitamin B_{12}) of the recommended daily allowances for humans (Sahlin and House, 2006). Besides the nutritive value of pork, the presence of antioxidative micronutrients might be important to promote oxidative stability. This is especially critical in the context of increasing interest in changing the fatty acid composition with higher proportion of omega-3 which can not only enhance health benefits of this meat for consumers but also challenge its oxidative stability.

Besides impacts on meat quality, another potential consequence of the above-mentioned global reduction of vitamin provisions during the whole life cycle of market pigs may be alteration of immune competence and risk of disease occurrence. Some vitamins play critical roles for this aspect of animal metabolism. This chapter aims to present recent scientific information related to the role of vitamins and their importance for oxidative mechanisms in relation to the development and competence of the immune system which are key contributors to optimal health status of pigs and to their ability to face pathogenic pressure during their life.

2 Supply of vitamins to pigs

2.1 The maternal dependence

In modern pig production, the duration of the post-weaning period until slaughter corresponds roughly to about half of the life duration of a pig (from fertilization of ova to slaughter). This means that for half of their life duration, pigs are entirely dependent on the transfer of vitamins from their mother (*in utero*, colostrum and milk). As mentioned above, this is particularly critical for fat-soluble vitamins. Indeed, vitamins D and E are not being transferred *in utero*, and although local transport of retinoids by retinol-binding protein has been demonstrated in the developing conceptus (Chew et al., 1993), vitamin A is transferred to a limited degree. Thus, as pigs are born almost deficient in vitamins D and E, and with almost no vitamin A depots, they have to rely entirely on the post-natal supply of these vitamins. Colostrum is an important source of especially vitamin E, which is more concentrated (times four) in colostrum than in milk (Lauridsen et al., 2002a). When individual colostrum intake is limited either because of the high number of littermates competing for colostrum or because of a limited synthesis by the sow, this may be a critical factor for vitamin E status in piglets. However, within a short period (four days) after birth, suckling piglets are capable of increasing their plasma vitamin E status by a factor as high as 83 (Lauridsen et al., 2002a). Among livestock, the piglet is born with the lowest vitamin D status, and dietary supplementation of vitamin D to the lactating sow seems not an efficient way of increasing the newborn piglets' vitamin D status (Lauridsen et al., 2010; Matte et al., 2016). Human studies have shown poor penetration of vitamin D and the metabolite 25-hydroxyvitamin D into milk (Kovacs, 2008), whereas, in pigs, no scientific information has been available until recently regarding the vitamin D content in sow colostrum and milk. These latest data indicate that vitamin D metabolites are also present in very small amounts in sow milk with concentrations approximately 10 times smaller than in blood serum (Matte et al., 2017). In fact, in sow milk, two forms, 25-hydroxyvitamin

D_3 (25(OH)D_3) and 24,25-dihydroxyvitamin D_3 (24,25$(OH)_2D_3$), are present, the latter accounting roughly for 55–60% of the total detectable vitamin D metabolites (Ouattara et al., unpublished data). Interestingly, although 24,25$(OH)_2D_3$ has long been considered as an inactive form of vitamin D, the literature from poultry (Seo et al., 1997) showed that it may have an important impact on bone metabolism, especially on cartilage formation. The implication of this peculiarity of the sow milk deserves to be further studied. In order to increase piglet survival, major focus has recently been given to the provision of milk supplements either in the form of formulas or in the form of other sources such as bovine colostrum. Use of formula supplemented with vitamin D may be a strategic tool to optimize vitamin D status in piglets, but the bioavailability of other (fat-soluble) vitamins may not be the same in milk formulas as for sow milk. For example, vitamins E and A provided as acetate-bound forms (i.e. α-tocopheryl acetate and retinyl acetate, respectively) should be hydrolysed before absorption. As previously discussed in relation to humans (Lauridsen and Jensen, 2007), premature and low body weight infants may have limited capacity for utilizing acetate-bound vitamin E due to impaired intestinal hydrolysis, and this may also be a focus point in future pig production having a considerable proportion of low birth weight piglets.

In contrast to fat-soluble vitamins, the global perinatal (*in utero* + colostrum) transfer for B vitamins and vitamin C would be favourable to foetuses and piglets (Matte et al., 2014a,b). In fact, it appears that the placental structures allow an active prenatal maternal transfer except for folates and riboflavin. In these last two cases, the maternal transfer is compensated by an important colostral transfer shortly after birth (Matte et al., 2014a,b).

2.2 The post-weaning period

Reduced capacity for absorption of fat-soluble vitamins at weaning and for two–three weeks beyond has also been a focus point in vitamin E research during the last decades. In order to overcome the challenges around weaning with impaired enzyme capacity and lipid absorption, some studies have focused on alternative strategies to vitamin E supplementation, that is, natural versus synthetic vitamin E for sows and piglets (Lauridsen et al., 2002a,b; Amazan et al., 2014). Natural forms may be RRR-α-tocopheryl acetate, or micellized vitamin E (D-α-tocopherol), which, upon supplementation, will increase the concentration of the RRR-α-tocopherol in blood, tissue and cellular membranes of the pig (Lauridsen et al., 2002a,b). Likewise, the metabolite 25(OH)D_3 has been developed and commercialized (HY-D, DSM Nutritional products) as alternative forms to vitamin D_3 (cholecalciferol) for use in swine nutrition. As shown in Fig. 1, doses greater than 200 IU of 25(OH)D_3 was more bioavailable than vitamin D_3, and as such, could be considered an equivalent of an even more advantageous source of vitamin D (Fig. 1), especially when pigs are housed indoor without exposure to sunlight. Analyses of the tissue samples (adipose tissue, white and red muscle, and liver) from the same animal experiment (Fig. 1) showed that the content of 25(OH)D_3 in the different tissues fully correlated with the serum 25(OH)D_3 level. However, the correlation between the tissue content of vitamin D_3 and serum 25(OH)D_3 was dependent on the source of ingested vitamin D_3 (Burrild et al., 2016).

In contrast to α-tocopherol, absorbed retinol is re-esterified in the enterocyte and accumulates in the liver. Because of concerns that excess vitamin A intakes in humans are associated with increased risks of hepatotoxicity, bone fracture and teratogenicity, and that elevated dietary vitamin A for pigs reduce vitamin E concentration in pork, withdrawal

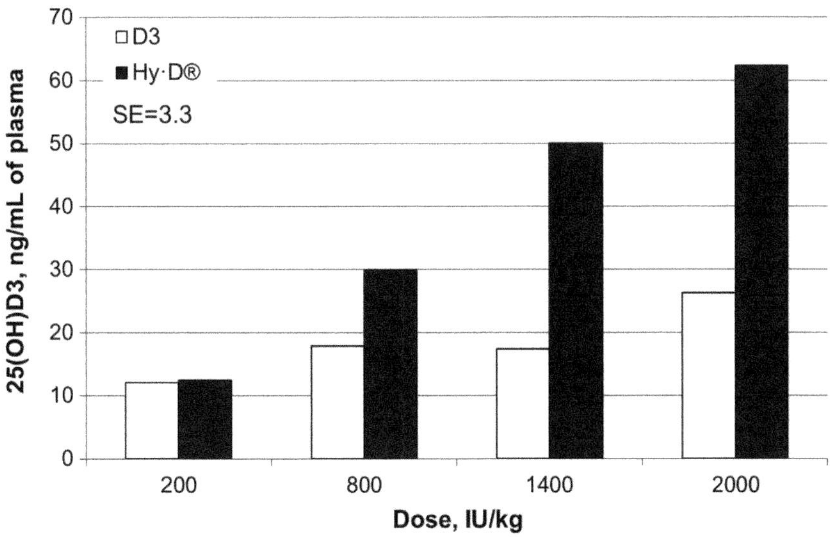

Figure 1 Concentration of 25(OH)D_3 (ng/mL) plasma of pigs after feeding an increasing dose of vitamin D in the form of cholecalciferol ('D_3') or 25$(OH)_2D_3$ (HY·D®). Modified after Lauridsen et al. (2010).

of vitamin A from pig feed has been studied, (Ayuso et al., 2015). It was concluded that the use of 7.5 times the NRC dietary vitamin A supplementation for long periods was not needed and that short- or long-term vitamin A withdrawal has the potential to reduce feed costs and increase fat and liver α-tocopherol levels without adverse effect on overall growth performance in heavy pigs (Ayuso et al., 2015).

B-vitamins are generally supplemented as synthetic forms from weaning up to slaughter age. Most of those synthetic forms do not present any particular problems in terms of bioavailability, one exception being vitamin B_{12} (Combs, 2012). Cyanocobalamin is the synthetic form of vitamin B_{12} present in most supplements, the cyanide group being used to stabilize the molecule. However, cyanocobalamin is not biologically active until the cyanide group is enzymatically removed (Herbert, 1988). Bioavailability of the synthetic form of B_{12} is inversely dependent upon the amount given, values being lower than 4% in humans and animals receiving prophylactic or therapeutic levels of supplements (LeGrusse and Watier, 1993; Scott, 1997; Matte et al., 2010). Using pigs as an animal model for humans, Matte et al. (2012) compared the net portal flux of B_{12} (an indicator of intestinal absorption) after ingestion of a natural source (cow milk) versus an equivalent amount of cyanocobalamin and observed that vitamin B_{12} in cow's milk is substantially more bioavailable (values of approximately 10% of the intake) than the synthetic form of this vitamin. Two explanations were raised for interpreting these results, one in relation to the molecular form of this vitamin in cow milk, mostly adenosylcobalamin (Farquharson and Adams, 1976; Fie et al., 1994) and another one, to the presence of specific components in milk, probably casein, facilitating the absorption of this vitamin. Such information could be a basis for designing a more efficient synthetic source of vitamin B_{12}. Opportunities for the development of new vitamin B_{12} products are conceivable because actually, in pig production, most of this vitamin (>95%) from synthetic sources is excreted in manure besides the fact that it is one the most expensive vitamin in the market.

3 Growth performance, antioxidative pressure and immunological competence

High lean meat production is associated with chronic inflammation and activation of the innate immune system vis-à-vis cellular stress. This may negatively affect adaptive immune responses. It has been furthermore shown that lean muscle pigs show abnormally high serum concentrations of reactive oxygen metabolites, as opposed to rural swine (Brambilla et al., 2002). Amadori and Zanotti (2016) have experienced that reactive oxygen metabolite levels above 20 mM H_2O_2 can easily be found in lean pigs. This level implies an oxidative stress under resting conditions similar to human beings during intense physical stress. The high release of reactive oxygen metabolites in the presence of hypoxia is related to the mitochondrial electron transport chain. Moreover, there is strong evidence that the hypoxia response is highly detrimental to T-cell homeostasis, whereby Th2 and Th17 responses are favoured at the expense of Th1 and Treg ones (Amadori and Zanotti, 2016). Thus, lean-type pigs with high growth performance under conditions of chronic oxidative stress are not likely to induce balanced and effective T-cell responses because of a defective homeostasis of the T-cell compartment.

Oxidative stress accompanies infectious diseases and plays a dual role as free radicals to protect against invading microorganism, and they can also cause tissue damage during resulting inflammation. The mechanisms by which vitamins have antioxidant activity are addressed below. However, vitamins have direct and indirect influence on the pigs' immunocompetence beyond their antioxidative activity. Nutritional immunology seeks to increase or modulate the immune response through manipulation of the level of dietary nutrients. Many reports have described how vitamins can modulate cytokine production after *in vitro* or *in vivo* supplementation (Hernandez et al., 2009); however, little research on this topic has been performed on pigs.

3.1 Fat-soluble vitamins

The fat-soluble vitamins A, D and E seem to be important for immunomodulation: Vitamin A controls immunity through retinoid acid signalling, and of special interest is the retinoic acid synthesis in the gut (for review, see Guo et al., 2015). Many studies in human and other animal species have demonstrated the interaction between vitamin A and immune competence in relation to infectious diseases. Vitamin A enhanced Th2 cytokines and improved the immune response against gastrointestinal parasites (Dawson et al., 2006; Wang et al., 2007). Cells of the immune system express the vitamin D receptor and activated macrophages to produce 1,25 dihydroxyvitamin D_3 ($1,25(OH)_2D_3$). Without sun exposure or dietary vitamin D supplementation, pigs may have a $25(OH)_2D_3$ level that is too low to support a high level of $1,25(OH)_2D_3$ synthesis by activated macrophages, which may compromise the immunoregulatory functions at the sites of inflammation. Recent research related to humans has supported the role of the active form of vitamin D ($1,25(OH)_2D_3$) in promoting normal function of innate and adaptive immune systems (Szymczak and Pawliczak, 2016), and this calls for further studies in pigs.

Vitamin E also has immunomodulatory effects, and some effects are attributed to the reversal of deleterious influence of reactive oxygen intermediates on immune functions like other antioxidants (e.g. ascorbic acid and selenium), whereas supplementation may augment antibody response. Results of Hernandez et al. (2009) revealed that vitamin E is able to suppress IL-10 production and to influence the production of IL-2, IL-4 and maybe TBX21 after *in vitro* stimulation in peripheral blood mononuclear cells isolated from healthy pigs. Table 1 gives an overview of studies performed on swine with the aim of studying vitamin E supplementation and effect on immunity of swine. Most of the studies addressed the influence on humoral immune responses. As can be deduced

Table 1 Effect of vitamin E supplementation and immunity

Animals	Vitamin E supplementation	Effect of immunity	Reference
Weaners	11, 110, 220 IU/kg	No effect on humoral immune responses	Blodgett et al., 1988
Weaners	11 vs. 220 IU/kg	Increase in antibody titre	Peplowski et al., 1981
Weaners	11, 110, 220 IU/kg	No effect on humoral immune responses	Blodgett et al., 1988
Weaners	11 vs. 220 IU/kg	No effect on antibody response to SRBC (or performance)	Bonnette et al., 1990a
Weaners	11, 110, 220, 550 IU/kg feed (at two housing temperatures/19 vs. 30°C)	No effect on humoral and cell-mediated immune response	Bonnette et al., 1990b
Gilts	13, 48 and 136 mg/kg. After weaning, pigs fed with the same diets	Immunized weaned pigs suckling sows on high vitamin E had higher AO to ovalbumin	Babinszky et al., 1991
Gilts/sows	22, 44, 88 IU/kg (gestation) and 55, 110, 220 IU/kg (lactation)	No effect on sows' immunity, but newborn pigs from high E levels (110 and 220 IU/kg) had higher lymphocyte response to PHA and Con A than pigs from sows fed 55 IU/kg	Nemec et al., 1994
Piglets	Injection (500 IU) i.m.	Higher AO to KLH in piglets injected with vitamin E from d 21 after birth	Hidiroglou et al., 1995
Sows/piglets	70, 150 or 250 mg/kg	AO response to *E. coli* in piglets affected when sows supplemented with vitamin E	Lauridsen and Jensen, 2005
Weaners	85, 150 and 300 IU/kg at varying fat composition	No effect of vitamin E on cell-mediated immune responses, but effect of fatty acid composition	Møller and Lauridsen, 2006

from Table 1, some studies reported an effect of vitamin E supplementation on humoral and/or cell-mediated immunity, while others did not observe any effect. However, the responses on the immune system in relation to vitamin supplementation may depend on the vitamin status and body reserves, and the exposure to stress (e.g. weaning and pathogen challenge) or other conditions under which the experiment is conducted. For example, we observed a dramatic reduction of the hepatic α-tocopherol content after infection challenge of pigs with *Escherichia coli* (Lauridsen et al., 2011), and using pigs that are hereditary deficient for ascorbate synthesis has demonstrated the influence of vitamin C on pigs' immune competence. The biological activity of vitamin E in relation to immunity of pigs should probably be seen and understood in light of protection of polyunsaturated fatty acids and membrane qualities that polyunsaturated fatty acids bring about. In our study (Møller and Lauridsen, 2006), dietary fatty acid composition influenced immune responses of macrophages after *E. coli* stimulation *ex vivo*, but dietary vitamin E supplementation, which influenced cellular α-tocopherol concentration, had no influence on the cytokine production of alveolar macrophages.

3.2 Water-soluble vitamins

Folates, vitamin B12 and the intermediary amino acid homocysteine. Homocysteine is a sulphur-containing amino acid derived from the hydrolysis of *S*-adenosylhomocysteine (SAH) generated from *S*-adenosylmethionine (SAM), the major cellular methyl donor (Fig. 2). Among more than hundreds of enzymatic methylation reactions mediated by SAM, there are DNA methylation which control gene transcription and genetic stability, and synthesis of phosphatidylcholine, choline, creatine as well as several neurotransmitters. However, intense methylation reactions as expected in lean pigs frequently resulted in accumulation of homocysteine (Hoffman, 2011). Elevated levels of blood plasma homocysteine are correlated with several pathologies and are recognized as an initiating factor for arteriosclerosis of coronary, cerebral and peripheral vessels (Boushey et al., 1995; Refsum et al., 1998). They also have harmful effects on embryo development (Pietrzik and Bronstrup, 1997; DiSimone et al., 2004) and cell proliferation (Chen et al., 2000).

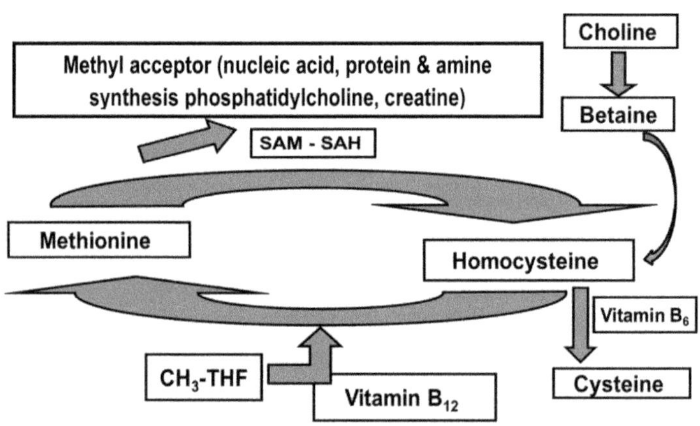

Figure 2 Transmethylation, remethylation and transsulphuration pathways of methionine. SAM, *S*-adenosylmethionine; SAH, *S*-adenosylhomocysteine; CH3-THF, methyltetrahydrofolate.

In view of the adverse effects of homocysteine on tissue integrity and its pro-oxidizing properties, organisms must rid themselves of this metabolite as quickly and as efficiently as possible. Homocysteine is metabolized via either remethylation (back to methionine) or transsulphuration (synthesis of cysteine) reactions. In remethylation, the conversion of homocysteine to methionine is catalysed by the enzyme methionine synthase, a vitamin B_{12}-dependent zinc protein (Bässler, 1997). This reaction requires the release of a methyl group from CH_3-H_4folate to H_4folate, the two main circulating forms of folic acid (vitamin B_9) in pigs (Natsuhori et al., 1996). Remethylation can also proceed through another zinc-containing enzyme, betaine-homocysteine methyltransferase. Betaine is an intermediate step in the catabolism of choline. In this last case, transmethylation capacity is limited because the tissue distribution of the enzyme is restricted to specific organs (Delgado-Reyes et al., 2001). In transsulphuration, the disposal of homocysteine is catalysed by the B_6-dependent cystathionine β-synthase, which leads to cystathionine and subsequently to cysteine and GSH biosyntheses. Therefore, removal of homocysteine is frequently impaired by a lack of folic acid, vitamin B_{12} or vitamin B_6 (Brosnan et al., 2007).

In growing and finishing pigs, even after vitamin supplementations, the plasma homocysteine concentration remained two to three times higher than the values observed in other species (<10 μM) such as dairy cows (Girard et al., 2005), rats (Sarwar et al. 2000), mice (Hofman et al., 2001), cats (Ruaux et al., 2001) and humans (Pietrzik and Brönstrup 1997). Some aspects of the remethylation pathway of methionine deserve to be further explored in pigs.

In piglets, at birth, blood plasma levels of homocysteine are very low (approximately 3 μM) and increase very rapidly at approximately 10 μM within 24 hours and at 18 μM and 30 μM at 7 and 14 days of age, respectively (Balance and House, 2005; Simard et al., 2007). During the last decade, we have showed in our laboratory that administrations of dietary supplements of folic acid and vitamin B_{12} to sows during gestation (Barkow et al., 2001; Guay et al., 2002; Simard et al., 2007) and lactation (Audet et al., 2015) can be used as efficient tools to increase maternal transfer (*in utero* or through colostrum and milk) of these vitamins to piglets which, in turn, prevents the rapid rise of homocysteine in blood plasma.

According to what was presented in the Introduction section about the evolution of hyperprolificity and the individual share of maternal vitamin transfer among piglet littermates during the perinatal period, it is likely that, during the last decades, it has included vitamins related to the chronic hyperhomocysteinaemia in piglets. Besides its multiple deleterious effects in several aspects of metabolism, homocysteine is also recognized as a powerful pro-oxidant which interferes with the immune system in stimulating, for example, the ubiquitous transcription factor NF-κB, a key regulator of pro-inflammatory cytokines such as TNF-alpha and IFN-gamma (Huang et al., 2001; Wang et al., 2000). Nevertheless, the question remained as to whether this atypical early regulation of homocysteine metabolism with high-plasma homocysteine is harmful to piglets or inversely, whether low-plasma homocysteine is beneficial. In this way, two populations of piglets (birth to 8 weeks of age) with high or low homocysteinaemia were generated by altering the transfer of folates and B_{12} from dams and by direct intramuscular (i.m.) injections of B_{12} to piglets using typical supplies of these vitamins (Audet et al., 2015). Growth performance, antibody and cellular immune responses, and some indicators of homocysteine metabolism, were then assessed according to vitamin treatments and their effects on plasma homocysteine. Indicators of cell-mediated immunity suggested a weakened immune competence with high homocysteinaemia (Audet et al., 2015). No treatment effect was reported on growth performance, but correlations were observed between plasma homocysteine and growth rate and feed conversion (feed:gain). However, these correlations were unexpectedly positive, a response which changed completely the perspective of the link between homocysteine and the immune system. Therefore, the only logical interpretation to such results would be that young 'high-performing' piglets which also generate high levels of plasma homocysteine are immunologically more fragile (Audet et al., 2015) and logically less resistant to disease challenges. The concept is important and may become a plausible explanation to the opposite relation often reported in concrete husbandry situations between superior performance and disease resistance. This side effect of high performance would be also coherent with the apparent global fragility of piglets against antigenic pressure which has developed during the last 30 years in several outbreaks of major diseases.

In order to verify if additional reductions of growth-related homocysteinaemia are possible in suckling piglets, a further study was undertaken with other nutrients related to metabolic disposal of homocysteine (Côté-Robitaille et al., 2015). The attempt was made to regulate plasma homocysteine in suckling piglets by daily oral supplementations of betaine recognized as a methyl group supplier, creatine for reducing the global demand for methyl groups coming from methionine, choline for combining both previous functions of betaine and creatine, and vitamin B_6 as an enzymatic co-factor for homocysteine catabolism. Concentrations of plasma homocysteine were reduced by 23% with the combination of all nutrients, such as betaine, pyridoxine, choline and creatine, supplied daily directly to suckling piglets during lactation. Nevertheless, the amplitude of this response on homocysteinaemia was similar to the strategy of supplementation with folic acid and vitamin B_{12} used by Audet et al. (2015). Additional studies are needed to know if a combination of approaches used by Côté-Robitaille et al. (2015) and Audet et al. (2015) could act in synergy to further enhance the clearance rate of homocysteine in suckling and post-weaned piglets and improve eventually their robustness in terms of health status.

4 Vitamins and antioxidation capacity: new perspectives

It is well known that reactive oxygen species (ROS) toxicity is controlled by complex networks of non-enzymatic (notably vitamins E and C) and enzymatic antioxidants, including the selenium (Se)-dependent glutathione peroxidase (GPX) system as well as copper (Cu), zinc (Zn)-dependent super-oxide dismutase system. In order to protect cells from oxidative reactions, vitamin E (α-tocopherol) needs to be incorporated in between the fatty acid methyl esters of the phospholipids in the cellular membranes. α-Tocopherol is mostly concentrated in membrane-rich fractions such as mitochondria and microsomes, and as reviewed by Lauridsen and Jensen (2012), it is possible by dietary means to further increase the concentration. Thus, incorporation of α-tocopherol and other antioxidants into mitochondria and other cellular compartments is important in order to maintain oxidative stability of the membrane-bound lipids and prevent damage from the ROS, and this is of major importance both for the living animal for disease prevention and for storage and processing of muscle meat.

Due to the interest in finding alternatives to the use of animal fat in the production of meat, studies on growing pigs have been conducted in order to understand the interaction between dietary fatty acid quality and antioxidants on markers of oxidative stress and antioxidant status as well as growth and organ function. More recently, the elevated cost and greater use of feed grains for biofuel production have increased the use of by-products with greater concentrations of polyunsaturated fatty acids, enhancing the potential for nutrient oxidation and oxidative stress in pig. Diets highly concentrated with PUFA as found in vegetable oils such as soya bean and linseed oils have increased potential for nutrient oxidation. Synthetic antioxidants such as butylated hydroxytoluene (BHT) and ethoxyquin are commonly used to preserve lipids and vitamins in feed and to protect the animal from oxidative stress. Lu et al. (2014a,b) reported that barrows fed with diets with an antioxidant blend (ethoxyquin and propyl gallate) or this blend with 11 IU/kg feed of vitamin E-attenuated oxidative stress in pigs provided a high-oxidant diet. However, the protective effects of supplemental vitamin E at 11 IU/kg relative to the antioxidant blend alone were minimal. In this way, Lauridsen et al. (1999) showed that a difference in 11 IU/kg (basic diet vs a basic diet supplemented with 6% rapeseed oil) actually influenced the oxidative status of muscles in pigs, as muscles of pigs on the rapeseed oil-based diets were less prone to oxidation, probably because of the content of vitamin E. However, increased susceptibility to oxidative stress because of processing (heating and mincing) of the pork enhances the requirement for antioxidant protection (Jensen et al., 1998). Thus, as also reported recently by Preveraud et al. (2014), supplementation with PUFA markedly decreases the concentration of α-tocopherol in the blood and tissues of growing pigs, and the ratio of α-tocopherol:PUFA needs to be carefully considered to meet the vitamin E requirements for pigs to ensure an optimal α-tocopherol meat enrichment. Recently, we showed (Lauridsen et al., 2013) that supplementation of PUFA (5% fish oil or sunflower oil) reduced the concentration of RRR tocopherol in plasma and muscles compared to saturated fatty acids (animal fat). In practice, it is recommended to supplement, per kg of diet, with 5 mg of vitamin E for each 10 g of dietary fat over a basic level of 30 g (Isabel et al., 2013) and that for all categories of pigs (recommended vitamin level: 100–150 IU/kg of vitamin E for piglets, sows and boars). However, this may vary depending on the proportion of PUFA in the dietary fat considering that the oxidative susceptibility increases with a factor of 2400 (from C18:0 to C18:3).

Recent focus on pig research has also been given to the interaction of antioxidative vitamins with other micronutrients, and especially with Se. In a recently published review (Surai and Fisinin, 2016), the role of selenium and its interaction with vitamins E and C as part of the antioxidant system of the body is described and explained in relation to the effect of dietary selenium (organic source vs inorganic) supplementation for protection against oxidative stress in sows and newly born piglets. The importance of storage of selenium in pig muscles and tissues to be used in stress conditions is addressed herein.

The glutathione peroxidase system (oxidized and reduced glutathione and the two enzymes, glutathione peroxidase and reductase) as mentioned above is closely related to the metabolism of the two amino acids, cysteine and methionine (Fig. 2), via the key junction metabolite, homocysteine. The *trans*sulphuration reaction provides a direct link between homocysteine and glutathione, the major redox buffer in mammalian cells. In this pathway, as mentioned previously, the disposal of homocysteine is catalysed by the B_6-dependent cystathionine β-synthase which leads to cystathionine and subsequently to cysteine and glutathione biosyntheses (Fig. 2). In fact, the two major homocysteine-utilizing enzymes, methionine synthase and cystathionine β-synthase, show reciprocal sensitivity to the oxidative conditions (Mosharov et al., 2000). In fact, reactive oxygen metabolites and peroxides upregulate the B_6-related transsulphuration pathway (homocysteine towards cysteine and eventually the glutathione peroxidase system) and downregulate the folate-B_{12}-related transmethylation (homocysteine towards methionine) (Fig. 2). Therefore, the presence (or the production) of oxidative metabolites will stimulate the pathway (transsulphuration) which will eventually neutralize them. The balance between remethylation and transsulphuration pathways during an oxidative stress regulates not only the flow of sulphur methionine and sulphur cysteine for provision of glutathione but also the flow of their seleno-methionine and seleno-cysteine for the GPX system. In fact, the animal metabolism does not distinguish between methionine and cysteine (sulphur forms) and their selenium analogues (cited by Daniels, 1996). In this case, seleno-cysteine also becomes a key factor along with seleno-homocysteine. Seleno-cysteine can be transformed in seleno-glutathione, but the size of that metabolic pool as compared to sulphur-glutathione is likely to be negligible (Fig. 2, small arrows). The importance of seleno-cysteine comes from its role for gene expression and activity of the Se-dependent enzyme, glutathione peroxidase (Flohé et al., 1997; Johansson et al., 2005). This role is due to a metabolite of seleno-cysteine called seleno-cysteinyl (a molecular form of cysteine moiety), which allows the full gene transcription of the enzyme (Johansson et al., 2005). In this process, one could think that the seleno-cysteine is cleaved and directly incorporated in the process, but this is not so (Fig. 2) as the metabolism involved is much more complicated: in fact, seleno-cysteine is mineralized through a metabolic step with the enzyme seleno-cysteine lyase (SCLY), a B_6-dependent enzyme specific to Se-cysteine which controls the flow of organic selenium towards the GPX system. The selenide produced from this reaction is then transformed into selenium phosphate before being resynthesized in the seleno-cysteinyl which is metabolically active for Se-dependent glutathione peroxidase (Yasumoto et al., 1979; Suzuki and Ogra, 2002; Johansson et al., 2005) (Fig. 2). It appears, therefore, that oxidative pressure will drive the B_6-dependent transsulphuration and 'transselenization' to produce, respectively, more GSH and further activate GPX.

Most of the reactions involved in the fate of seleno-methionine, seleno-homocysteine and seleno-cysteine for the control of the glutathione peroxidase system (Fig. 2) are B_6-dependent (Yasumoto et al., 1979). A previous report on rats has demonstrated that the response of Se-GPX to organic selenium could be dependent upon the dietary provision of vitamin B_6 (Yasumoto et al., 1979) (Fig. 3). In normal redox balance with the addition of dietary organic selenium, Se-GPX was lower at low B_6 status than after dietary supplement of vitamin B_6 (Fig. 4). The status of vitamin B_6 could play a critical role for an adequate flow of organic selenium towards the glutathione peroxidase system and might have to be taken into account for a reliable interpretation of the different studies with selenium supplementations. This last concept was addressed recently (Bueno-Dalto et al., 2015) using gilts during the peri-oestral period as a model promoting consistent and reliable oxidative stress conditions. Indeed, it is recognized that metabolic by-products such as reactive oxygen metabolites are generated during the peri-oestrus period in response to drastic enhancement of ovarian cellular energy metabolism in relation to hormone production and ovulation (Agarwal et al., 2005). Such production of reactive oxygen metabolites is particularly important for polytocous species such as pigs and probably more for modern hyperprolific lines of sows with high ovulation rates. In these animals, the oxidative pressure at the ovary level is, in fact, important enough to be detectable on enzymatic antioxidants like GPX activity at peripheric level (whole blood) during the peri-oestral period (Fig. 5) (Fortier et al., 2012). In fact, during that peri-oestral period, gene expressions of *GPX1* and *SCLY* were substantially higher in liver from gilts supplemented with organic Se and B_6 as compared to organic Se without B_6 (Bueno-Dalto et al., 2015) (Fig. 6). These last two effects confirm the crucial role of vitamin

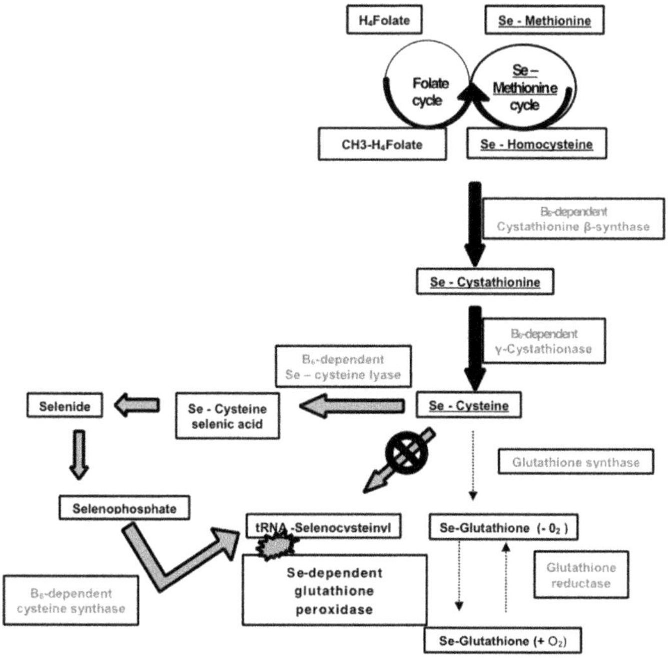

Figure 3 Se–amino acids metabolism and fate of Se–cysteine (mineralization) towards activation of Se-dependent glutathione peroxidase system.

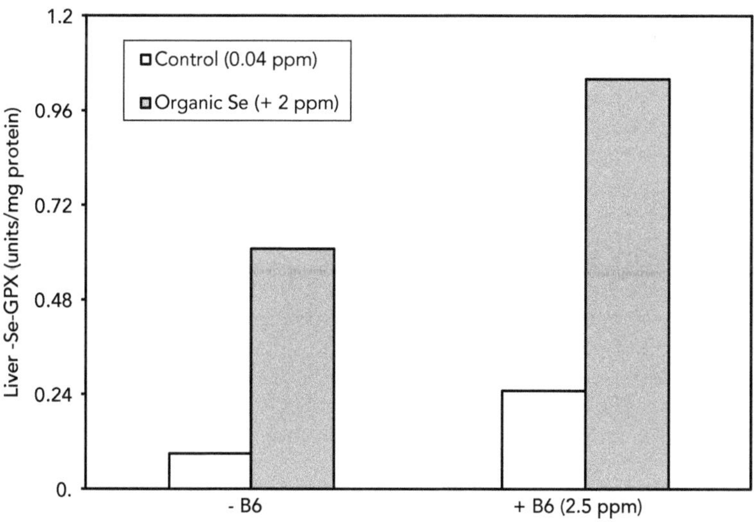

Figure 4 Effect of vitamin B_6 status on the hepatic response of Se-glutathione peroxidase (Se-GPX) to the dietary form of selenium. Adapted from Yasumoto et al. (1979).

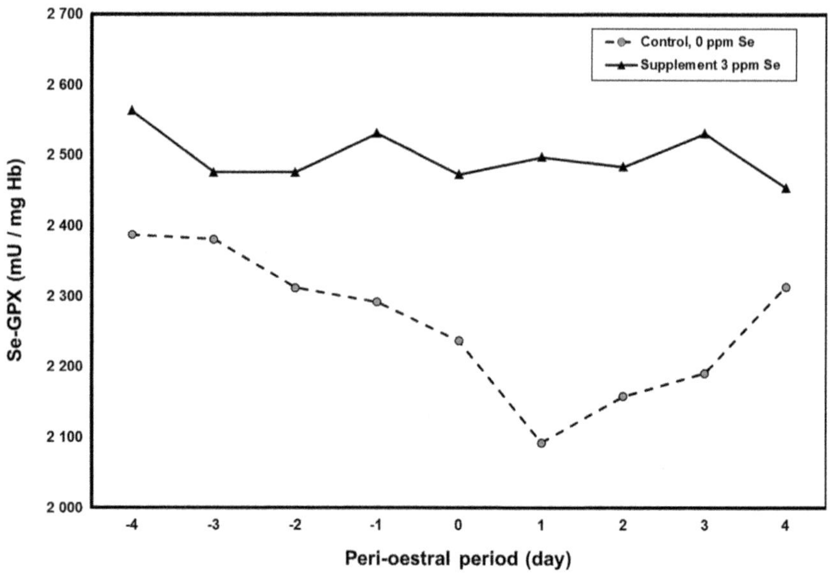

Figure 5 Blood Se-dependent glutathione peroxidase (Se-GPX) activity during the peri-oestrus period in gilts according to the level of dietary Se. The control diet contained a basal level of 0.2 mg/kg Se (unsupplemented), whereas the supplemented one represented a pooled profile from the control diet complemented with an extra 0.3 mg/kg (of inorganic selenite (MSe) or organic Se (yeast). Day 0 = third oestrus. One unit (U) of GPX activity equals 1 mmol of NADPH oxidized per minute. Adapted from Fortier et al. (2012).

B_6 in modulating the metabolic pathway of organic Se towards the GPX system and the importance of adequate levels of this vitamin for an appropriate response of GPX during episodes of oxidative pressure. This last metabolic condition appears particularly critical because B_6 and Se responses on gene expressions of these key enzymes could not be confirmed when measurements were made at a different physiological stage without any particular surges of oxidative pressure (Bueno-Dalto et al., 2016).

5 Conclusion and future trends

As mentioned in the Introduction section, the criteria used to estimate vitamin requirements for pigs have changed, during the past century, from prevention of deficiencies to optimization of performance for growth and reproduction. Such criteria can be quantitatively assessed rather easily. As presented in this chapter, the estimation of requirements for the future will have to include criteria related to other aspects potentially limiting for full expression of performances such as metabolic stress and disease resistance. This means that vitamin requirements might be 'adjustable' depending upon the risk of fragility of the animal vis-à-vis critical periods or disseminated contaminants. Besides the two levels of requirements for prevention of vitamin deficiencies and optimization of performance, there would be another one, likely at a higher level for prophylactic robustness of animals.

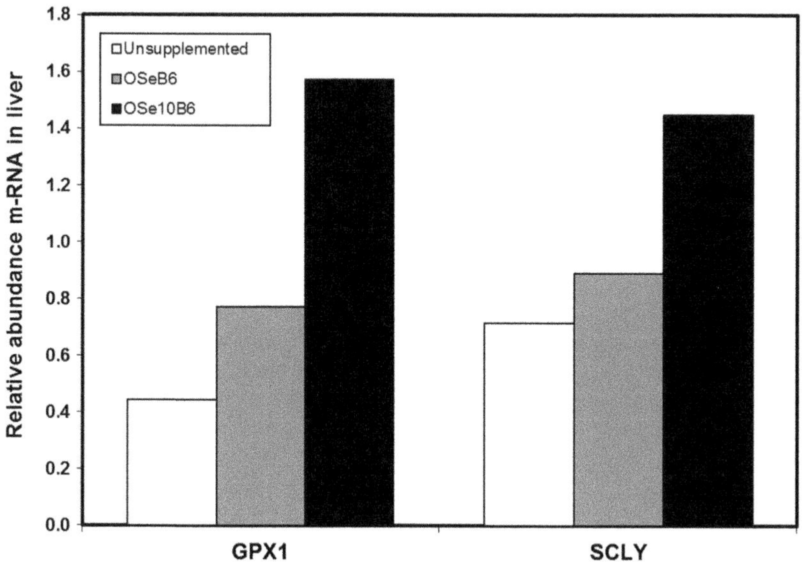

Figure 6 Gene expressions of Se-dependent glutathione peroxidase 1 (*GPX1)* and selenocysteine lyase (*SCLY)* in liver of gilts 3 days after the fourth post-pubertal oestrus. Unsupplemented = basal diet containing 0.2 mg/kg and 1.7 mg/kg of native Se and pyridoxine respectively, OSeB60 = basal diet supplemented with 0.3 mg/kg of Se-enriched yeast, OSeB610 = basal diet supplemented with 0.3 mg/kg of Se-enriched yeast and 10 mg/kg of hydro-chloride pyridoxine. Adapted from Bueno-Dalto et al. (2015).

Such a new approach to look at vitamin allowances in pigs nevertheless presents a future challenge because reliable indicators to assess the levels of metabolic stress and disease protection are still lacking.

6 Where to look for further information

The authors have previously provided an overview of vitamins and vitamin utilization in swine (see Matte and Lauridsen (2013) in the references for further details). The book chapter integrated the recent scientific advances in vitamin nutrition of pigs in an attempt to complement the basic information available in various publications on nutrient recommendation of swine (NRC, BSAS, ARC).

The authors have furthermore contributed with proceeding papers on specific vitamins in connection with seminars held by American Society of Animal Science (ASAS), Animal Nutrition Association of Canada, European Federation of Animal Science and European Society of Veterinary and Comparative Nutrition, e.g. Lauridsen (2013, 2014) and Matte (2014, 2015).

- Vitamins and vitamin utilization in swine/Matte, J. J. and Lauridsen, C. 2013. *Sustainable Swine Nutrition*. red. / Lee I Chiba. 1. udg. Wiley-Blackwell, s. 139–72.

- Update on Vitamin C, D, E and fatty acids in swine nutrition/Lauridsen, C. 2013. Proceedings of 49th Eastern Nutrition Conference. Animal Nutrition Association of Canada, Quebec, Cananda, s. 143–54.
- Triennial Growth Symposium – Establishment of the 2012 vitamin D requirements in swine with focus on dietary forms and levels of vitamin D/Lauridsen, C. 2014. Journal of Animal Science, 92 (3), s. 910–16.
- Water-soluble vitamins and reproduction in sows: beyond prolificacy/Matte, J. J. 2014. Book of Abstracts of the 65th Annual Meeting of the European Federation of Animal Science. Wageningen Academic Publishers, The Netherlands, No. 20, p. 108.
- New concepts on micronutrient requirements for sows: beyond reproductive performance/Matte, J. J. Proceeding of ESVCN 2015, Toulouse, France.
- Micronutrient requirements for sows: to be revisited for an adequate transfer to piglets? /Matte, J. J. 2015. Proceedings of Western Nutrition Conference, Animal Nutrition Association of Canada, Winnipeg, Manitoba, Pre-Conference Session.

7 References

Abedelahi, A., Salehnia, M., Allameh, A. A. and Davoodi, D. (2010), 'Sodium selenite improves the in vitro follicular development by reducing the reactive oxygen species leveland increasing the total antioxidant capacity and glutathione peroxide activity', *Hum. Reprod.*, 25, 977–85.

Agarwal, A., Gupta, S. and Sharma, R. K. (2005), 'Role of oxidative in stress in female reproduction', *Reprod. Biol. Endocrinol.*, 3, 1–28.

Al-Gubory, K. H., Fowler, P. A. and Garrel, C. (2010), 'The roles of cellular reactive oxygen species, oxidative stress and antioxidants in pregnancy outcomes', *Int. J. Biochem. Cell*, 42, 1634–50.

Amadori, M. and Zanotti, C. (2016), 'Immunoprophylaxis in intensive farming systems: the way forward', *Vet. Immunol. Immunopath.*, doi: 10.1016/j.vetimm.2016.02.011. [Epub ahead of print].

Audet, I., Girard, C. L., Lessard, M., Lo Verso, L., Beaudoin, F. and Matte, J. J. (2015), 'Homocysteine metabolism, growth performance, and immune responses in suckling and weanling piglets', *J. Anim. Sci.*, 93, 147–57.

Ayuso, M., Óvilo, C., Fernández, A., Nuñnez, Y., Isabel, B., Daza, A., López-Bote, C. J. and Rey, A. I. (2015), 'Effects of dietary vitamin A supplementation or restriction and its timing on retinol and α-tocopherol accumulation and gene expression in heavy pigs', *Anim. Feed. Sci. Tech.*, 202, 62–74.

Babinszky, L., Langhout, D. J., Verstegen, M. W., den Hartog, L. A., Joling, P. and Nieuwland, M. (1991), 'Effect of vitamin E and fat source in sows' diets on immune response of suckling and weaned piglets', *J. Anim. Sci.*, 69, 1833–42.

Ballance, D. M. and House, J. D. (2005), 'Development of the enzymes of Hcy metabolism from birth through weaning in the pig', *J. Anim. Sci.*, 83 (Suppl. 1), 160.

Barkow, B., Pietrzik, K. and Flachowsky, G. (2001), 'Effect of folic acid supplements on homocysteine in plasma of gestating sows', *Arch. Anim. Nutr.*, 54, 81–85. https://doi.org/10.1080/17450390109381967

BASF (Badische Anilin- & Soda-Fabrik) Corp. (1993), 'Vitamin Supplementation Rates for U.S. Commercial Poultry, Swine and Dairy Cattle' BASF Corp., Mount Olive, NJ.

BASF (Badische Anilin- & Soda-Fabrik) Corp. (2001), 'Dietary fortifications in vitamins for pigs and poultry in Canada, Technical seminar' BASF Corp., Toronto, ON & Richelieu, QC, Canada.

Bässler, K. H. (1997), 'Enzymatic effects of folic acid and vitamin B_{12}', *Internat. J. Vit. Nutr. Res.*, 67, 385–8.

Blodgett, D. J., Kornegay, E. T. and Schurig, G. G. (1998), 'Vitamin E-Selenium and immune response to selected antigens in swine', *Nutr. Rep. Intern.*, 38, 37–43.

Bonnette, E. D., Kornegay, E. T., Lindemann, M. D. and Notter, D. R. (1990a), 'Influence of two supplemental vitamin E levels and weaning age on performance, humoral antibody production and serum cortisol levels of pigs', *J. Anim. Sci.*, 68, 346–1353.

Bonnette, E. D., Kornegay, E. T., Lindemann, M. D. and Hammerberg, C. (1990b), 'Humoral and cell-mediated immune response and performance of weaned pigs fed four supplemental vitamin E levels and housed at two nursery temperatures', *J. Anim. Sci.*, 68, 1337–45.

Boushey, C. J., Beresford, S. A., Omenn, G. S. and Motulsky, A. G. (1995), 'A quantitative assessment of plasma homocysteine as a risk factor for vascular disease. Probable benefits of increasing folic acid intakes', *JAMA*, 274, 1049–57.

Brambilla, G., Civitareale, C., Ballerini, A., Fiori, M., Amadori, M., Archetti, L. I., Regini, M. and Betti, M. (2002), 'Response to oxidative stress as welfare parameter in swine', *Redox. Rep.*, 7, 159–63.

Brosnan, J. T., Brosnan, M. E., Bertolo, R. F. P. and Brunton, J. A. (2007), 'Methionine: A metabolically unique amino acid', *Livestock Sci.*, 112, 2–7.

Bueno-Dalto, D., Roy, M., Audet, I., Palin, M.-F., Guay, F., Lapointe, J. and Matte, J. J. (2015), 'Interaction between vitamin B6 and source of selenium on the response of the selenium-dependent glutathione peroxidase system to oxidative stress induced by oestrus in pubertal pig', *J. Trace Elem. Med. Biol.*, 32, 21–9.

Bueno-Dalto, D., Audet, I., Lapointe, J. and Matte J. J. (2016), 'The importance of pyridoxine for the impact of the dietary selenium sources on redox balance, embryo development, and reproductive performance in gilts', *J. Trace Elem. Med. Biol.*, 34, 79–89.

Burrild, A., Lauridsen, C., Faqir, N., Sommer, H. M. and Jakobsen, J. (2016), 'Vitamin D3 and 25-hydroxyvitamin D3 in pork and their relationship to vitamin D status in pigs', *J. Nutr. Sci.*, 5, e3.

CCSI (Canadian Centre for Swine Improvement) (2015), Annual Report, 36 p. https://www.ccsi.ca/main.cfm?target_page=annual.

Chen, C., Halkos, M. F., Surowiec, S. M., Conklin, B. S., Lin P. H. and Lumsden, A. B. (2000), 'Effects of homocysteine on smooth muscle cell proliferation in both cell culture and artery perfusion culture models', *J. Surg. Res.*, 88, 26–33.

Chew, B. P. (1993), 'Effects of supplemental β-carotene and vitamin A on reproduction in swine', *J. Anim. Sci.*, 71, 247–52.

Côté-Robitaille, M.-E., Girard, C. L., Guay, F. and Matte, J. J. (2015), 'Oral supplementations of betaine, choline, creatine and vitamin B6 and their influence on the development of homocysteinaemia in neonatal piglets', *J. Nutr. Sci.* 4(e31), 1–7. https://doi.org/10.1017/jns.2015.19

Daniels, L. A. (1996), 'Selenium metabolism and bioavailability', *Biol. Trace Element Res.*, 54, 185–99.

Dawson, H. D., Collins, G., Pyle, R., Key, M., Weeraratna, A., Deep-Dixit, V., Nadal, C. N. and Taup, D. D. (2006), 'Direct and indirect effects of retinoic acid on human Th2 cytokine and chemokine expression by human T lymphocytes', *BMC Immunol.*, 7, 207.

Delgado-Reyes, C. V., Wallig, M. A. and Garrow, T. A. (2001), 'Immunohistochemical detection of betaine-homocysteine S-methyltransferase in human, pig, and rat liver and kidney', *Archiv. Biochem. Biophys.*, 393, 184–6.

DiSimone, N., Riccardi, P., Maggiano, N., Piacentani, A., D'Asta, M., Capelli, A. and Caruso, A. (2004), 'Effect of folic acid on homocysteine-induced trophoblast apoptosis', *Molecul. Hum. Reprod.*, 10, 665–9.

Farquharson, J. and Adams, J. F. (1976), 'The forms of vitamin B_{12} in foods', *Br. J. Nutr.*, 36, 127–36.

Fie, M., Zee, J. A. and Amiot, J. (1994), 'Séparation et quantification des isomères de la vitamine B_{12} dans le lait et certains produits laitiers par chromatographie liquide haute performance et par radio-essai', *Sciences des aliments* 14, 763–75.

Flohr, J. R., DeRouchey, J. M., Woodworth, J. C., Tokach, M. D., Goodband, R. D. and Dritz, S. S. (2016), 'A survey of current feeding regimens for vitamins and trace minerals in the US swine industry', *J. Swine Health Prod.*, 24, 290–303.

Flohé, L., Wingender, E. and Brigelus-Flohé, R. (1997), Regulation of glutathione peroxidase. In *Oxidative Stress and Signal Transduction*. Forman, H. J. and Cadenas, H. ed., Chapman and Hall, New York, USA, pp. 415–40.

Fortier, M. E., Audet, I., Giguère, A., Laforest, J. P., Bilodeau, J. F., Quesnel, H. and Matte, J. J. (2012), 'Effect of dietary organic and inorganic selenium on antioxidant status, embryo development and reproductive performance in hyperovulatory first-parity gilts', *J. Anim. Sci.*, 90, 231–40.

Giguère, A., Girard, C. L. and Matte, J. J. (2008), 'Methionine, folic acid and vitamin B-12 in growing-finishing pigs: Impact on growth performance and meat quality', *Arch. Anim. Nutr.*, 62, 193–206. https://doi.org/10.1080/17450390802027494

Guay, F., Matte, J. J., Girard, C. L. Palin, M. F. Giguère, A. and Laforest, J. P. (2002), 'Effects of folic acid and vitamin B_{12} supplements on folate and homocysteine metabolism in pigs during early pregnancy', *Br. J. Nutr.*, 88, 253–63.

Guo, Y., Brown, C., Ortiz, C. and Noelle, R. J. (2015), 'Leukocyte homing, fate, and function are controlled by retinoic acid', *Physiol. Rev.*, 95, 125–48.

Herbert, V. (1988), 'Vitamin B-12: plant sources, requirements, and assay,' *Am. J. Clin. Nutr.*, 48, 852–8.

Hernandez, J., Soto-Canevett, E., Pinelli-Saavedra, A., Resendiz, M., Moya-Cararena, S. Y. and Klasing, K. C. (2009), 'In vitro effect of vitamin E on lectin-stimulated porcine peripheral blood mononuclear cells', *Vet. Immunol. Immunopath.*, 131, 9–16.

Hidiroglou, M., Batra, T. R., Farnworth, E. R. and Markham., F. (1995), 'Effect of vitamin E supplementation on immune status and α-tocopherol in plasma of piglets', *Reprod. Nutr. Dev.*, 35, 443–50.

Hofmann, M. A., Lalla, E., Lu, Y., Gleason, M. R., Wolf, B. M., Tanji, N., Ferran Jr., L. J., Khol, B., Rao, V., Kisiel, W., Stern, D. M. and Schmidt, A. M. (2001), 'Hyperhomocysteinemia enhances vascular inflammation and accelerated atherosclerosis in a murine model', *J. Clin. Invest.*, 107, 675–83.

Huang, R. F. S., Hsu, Y. C., Lin, H. L. and Yang, F. L. (2001), 'Folate depletion and elevated plasma homocysteine promote oxidative stress in rat livers', *J. Nutr.*, 131, 33–8.

Isabel, B., Rey, A. I. and Lopez Bote, C. (2013), *Optimum Vitamin Nutrition - In the production of quality animal foods.* 5M Publishing Benchmark House, Sheffield, United Kingdom.

Jensen, C., Lauridsen, C. and Bertelsen, G. (1998), 'Dietary vitamin E: Quality and storage stability of pork and poultry', *Trends Food Sci. Technol.*, 9, 62–72.

Johansson, L. G., Gafvelin, G. and Arner, E. S. J. (2005), 'Selenocysteine in proteins - properties and biotechnological use', *Biochim. Biophys. Acta Gen. Subj.*, 1726, 1–13.

Kovacs, C. S. (2008), 'Vitamin D in pregnancy and lactation: Maternal, fetal, and neonatal outcomes from human and animal studies', *Am. J. Clin. Nutr.*, 88(Suppl.), 520S–8S.

Lauridsen, C. and Jensen, S. K. (2005), 'Influence of supplementation of all-rac-alpha-tocopheryl acetate preweaning and vitamin C postweaning on alpha-tocopherol and immune responses of piglets', *J. Anim. Sci.*, 83, 1274–86.

Lauridsen, C. and Jensen, S. K. (2007), 'Transfer of vitamin E in milk to the newborn.' In *The Encyclopedia of Vitamin E* (eds. V. R. Preedy and R. R. Watson). Chapter 47, CAB International.

Lauridsen, C. and Jensen, S. K. (2012), 'α-Tocopherol incorporation in mitochondria and microsomes upon supranutritional vitamin E supplementation', *Genes Nutr.*, doi: 10.1007/s12263-012-0286-6.

Lauridsen, C., Engel, H., Jensen, S. K., Craig, A. M. and Traber, M. G. (2002a), 'Lactating sows and suckling piglets preferentially incorporate RRR- over all-rac-α-tocopherol into milk, plasma, and tissues', *J. Nutr.*, 132, 1258–64.

Lauridsen, C., Engel, H., Craig, A. M. and Traber, M. G. (2002b), 'Relative bioactivity of dietary RRR- and all-rac- rac-α-tocopheryl acetates in swine assessed with deuterium-labeled vitamin E', *J. Anim. Sci.*, 80, 702–7.

Lauridsen, C., Larsen, T., Halekoh and Jensen, S. K. (2010), 'Reproductive performance and bone status markers of gilts and lactating sows supplemented with two different forms of vitamin D', *J. Anim. Sci.*, 88, 202–13.

Lauridsen, C., Nielsen, J. H., Henckel, P. and Sørensen, M. T. (1999), 'Antioxidative and oxidative status in muscles of pigs fed rapeseed oil, vitamin E and copper', *J. Anim. Sci.*, 77, 105–15.

Lauridsen, C., Theil, P. K. and Jensen, S. K. (2013), 'Composition of alpha-tocopherol and fatty acids in porcine tissues after dietary supplementation with vitamin E and different fat sources', *Anim. Feed Sci. Technol.*, 179, 93–102.

Lauridsen, C., Vestergaard, E. M., Højsgaard, S., Jensen, S. K. and Sørensen, M. T. (2011), 'Inoculation of weaned pigs with E. coli reduces depots of vitamin E', *Livest. Sci.*, 137, 161–7.

Le Grusse, J. and Watier, B. (1993), *Les vitamines. Données biochimiques, nutritionnelles et cliniques. CEIV.*, Neuilly-sur-Seine, France.

Lu, T., Harper, A. F., Zhao, J., Estienne, M. J. and Dalloul, R. A. (2014a), 'Supplementing antioxidants to pigs fed diets high in oxidants: I. Effects on growth performance, liver function, and oxidative status', *J. Anim. Sci.*, 92, 5455–63.

Lu, T., Harper, A. F., Dibner, J. J. Scheffler, J. M., Corl, B. A., Estienne, M. J. Zhao, J. and Dalloul, R. A. (2014b), 'Supplementing antioxidants to pigs fed diets high in oxidants: II. Effects on carcass characteristics, meat quality, and fatty acid profile', *J. Anim. Sci.*, 92, 5464–75.

Matte, J. J., Guay, F. and Girard, C. L. (2012), 'Bioavailability of vitamin B 12 in cows' milk', *Brit. J. Nutr.*, 107, 61–6.

Matte, J. J., Guay, F., Le Floc'h, N. and Girard, C. L. (2010), 'Bioavailability of dietary cyanocobalamin (vitamin B_{12}) in growing pigs', *J. Anim. Sci.* 88, 3936–44.

Matte, J. J., Audet, I., Ouattara, B., Bissonnette, N., Talbot, G., Lapointe, J., Guay, F., Lo Verso, L. and Lessard, M. (2016), 'Sources and routes of administration of copper and vitamins A and D on metabolic status of these micronutrients in suckling piglets', *J. Anim. Sci.*, 94 (Suppl. 2), 113.

Matte J. J. and Lauridsen, C. (2013), 'Vitamins and vitamin utilization in swine.' In Chiba, Lee L. (Eds), *Sustainable Swine Nutrition*, Book chapter 6, ISUP, John Witey & Sons, Inc. Publ., 2121 State, Ames, Iowa 50014 USA.

Matte, J. J., Audet, I. and Girard, C. L. (2014a), 'Le transfert périnatal des vitamines et minéraux mineurs de la truie à ses porcelets: au-delà d'une seule insuffisance en fer?' *Journées Rech. Porcine en France*, 46, 71–6.

Matte, J. J., Audet, I. and Girard, C. L. (2014b), 'The perinatal transfer of vitamins and trace elements from sows to piglets', *J. Anim. Sci.*, 92. (Suppl 2), 153.

Matte, J. J., Audet, I., Ouattara, B., Bissonnette, N., Talbot, G., Lapointe, J., Guay, F., Lo Verso, L. and Lessard, M. (2017), 'Sources et voies d'administration du cuivre et des vitamines A et D sur le statut métabolique de ces micronutriments chez les porcelets sous la mère.' *Journées Rech. Porcine en France*, 49, (in press).

McGlone, J. J. (2000), 'Deletion of supplemental minerals and vitamins during the late finishing period does not affect pig weight gain and feed intake', *J. Anim. Sci.*, 78, 2797–2800.

Møller, S. and Lauridsen, C. (2006), 'Dietary fatty acid composition rather than vitamin E supplementation ex vivo cytokine and eicosanoid response of porcine alveolar macrophages', *Cytokine*, 35, 1–2, 6–12. https://doi.org/10.1016/j.cyto.2006.07.001

Mosharov, E., Cranford, M. R. and Banerjee, R. (2000), 'The quantitatively important relationship between homocysteine metabolism and glutathione synthesis by the transsulfuration pathway and its regulation by redox changes', *Biochemistry*, 39, 13005–11.

Natsuhori, M., Shimoda, M. and Kokue, E. (1996), 'Alteration of plasma folates in gestating sows and newborn piglets', *Am. J. Physiol.*, 270, R99–104.

Nemec, M., Butler, G., Hidiroglou, M., Farnworth, E. R. and Nielsen, K. (1994), 'Effect of supplementing gilts' diets with different levels of vitamin E and different fats on the humoral and cellular immunity of gilts and their progeny',. *J Anim. Sci.*, 72, 665–76.

Olivares, A., Daza, A., Rey, A. I. and Lopez-Bote, C. J. (2009), 'Interactions between genotype, dietary fat saturation and vitamin A concentration on intramuscular fat content and fatty acid composition in pigs', *Meat Sci.*, 82, 6–12.

Peplowski, M. A., Mahan, D. C, Murray, F. A., Moson, A. L., Canior, A. H. and Ekstrom, K. E. (1981), 'Effect of dietary and injectable vitamin E and selenium in weanling swine antigenically challenged with sheep red blood cells', *J. Anim. Sci.*, 51, 344.

Pietrzik, K. and Bronstrup, A. (1997), 'Folate in preventive medicine: A new role in cardiovascular disease, neural tube defects and cancer', *Ann. Nutr. Metab.*, 41, 331–43.

Prévéraud, D. P., Devillard, E., Rouffineau, F. and Borel, P. (2014), 'Effect of the type of dietary triacylglycerol fatty acids on α-tocopherol concentration in plasma and tissues of growing pigs', *J. Anim. Sci.*, 92, 4972–80. https://doi.org/10.2527/jas.2013-7099

Refsum, H., Ueland, P. M., Nygård, O. and Vollset, S. E. (1998), 'Homocysteine and cardiovascular disease', *Review. Annu. Rev. Med.*, 49, 31–62.

Ruaux, C. G., Steiner, J. M. and Williams, D. A. (2001), 'Metabolism of amino acids in cats with severe cobalamin deficiency', *Am. J. Vet. Res.*, 62, 1852–8.

Sahlin, A. and House, J. D. (2006), 'Enhancing the vitamin content of meat and eggs: Implications for the human diet', *Can. J. Anim. Sci.*, 86, 181–95.

Sarwar, G., Peace, R. W., Botting, H. G., L'Abbé, M. R. and Keagy P. M. (2000), 'Influence of dietary methionine with or without adequate dietary vitamins on hyperhomocysteinemia in rats', *Nutr. Res.*, 20, 1817–27.

Scott, J. M. (1997), 'Bioavailability of vitamin B12', *Eur. J. Clin. Nutr.*, 51, S49–53.

Seo, E.-G., Einhorn, T. A. and Norman, A. W. (1997), '24R,25-Dihydroxyvitamin D3: An Essential Vitamin D3 Metabolite for Both Normal Bone Integrity and Healing of Tibial Fracture in Chicks', *Endocrinol.*, 138, 3864–72. https://doi.org/10.1210/en.138.9.3864

Shaw, D. T., Rozeboom, D. W., Hill, G. M., Booren, A. M. and Link, J. E. (2002), 'Impact of vitamin and mineral supplement withdrawal and wheat middling inclusion on finishing pig growth performance, fecal mineral concentration, carcass characteristics, and the nutrient content and oxidative stability of pork', *J. Anim. Sci.*, 80, 2920–30.

Simard, F., Guay, F., Girard, C. L., Giguère, A., Laforest J. P. and Matte, J. J. (2007), 'Effects of concentrations of cyanocobalamin in the gestation diet on some criteria of vitamin B_{12} metabolism in the first-parity sows', *J. Anim. Sci.*, 85, 3294–302.

Surai, P. F. and Fisinin, V. I. (2016), 'Selenium in sow nutrition', *Anim. Feed Sci. Nutr.*, 211, 18–30.

Szymczak, I. and Pawliczak, R. (2016), 'The active Metabolite of Vitamin D3 as a Potential Immunomodulator', Scand. *J. Immunol.*, 83, 83–91. https://doi.org/10.1111/sji.12403

Theil, P. K., Lauridsen, C. and Quesnel, H. (2014), 'Neonatal piglet survival: Impact of sow nutrition around parturition on fetal glycogen deposition and production and composition of colostrum and transient milk', *Animal*, 8, 7, 1021–30.

VSP. (2016), Notat nr.1611. Danish Pig Research Centre. SEGES, Copenhagen, Denmark.

Wang, G., Siow, Y. L. and Karmin, O. (2000), 'Homocysteine stimulates nuclear factor (B activity and monocyte chemoattractant protein-1 expression in vacular smooth-muscle cells: A possible role for protein kinase C', *Biochem. J.*, 352, 817–26.

Wang, X., Allen, C. and Ballow, M. (2007), 'Retinoic acid enhances the production of IL-10 while reducing the synthesis of IL-12 and TNF-alpha from LPS-stimulated monocytes/macrophages', *J. Clin. Immunol.*, 27, 193–200.

Yasumoto, K., Iwami, K. and Yoshida, M. (1979), 'Vitamin B6 dependence of selenomethionine and selenite utilization for glutathione peroxidase in the rat', *J. Nutr.*, 109, 760–6.

Chapter 10

Modelling nutrient requirements for pigs to optimize feed efficiency

Ludovic Brossard, Jean-Yves Dourmad, Florence Garcia-Launay and Jaap van Milgen, PEGASE, INRA – Agrocampus Ouest, France

1 Introduction

Animal production is continuously facing the challenge of sustainability. In the next decades, world animal production is expected to increase by about 70% to satisfy the increased demand for animal protein, as anticipated by FAO (2011). Meeting this demand in a sustainable way requires increasing the efficiency of animal production. Additionally, livestock production systems have also to integrate the dimensions of animal health and welfare, food quality and security, environment, and consumer and citizen expectations to ensure their sustainability. In pig production, feed and feeding are major levers to control performance, with immediate and reversible effects. Feed represents a major part of the production costs (typically 60 to 70%) and thus largely affects economic results. Feed is also largely implicated in other sustainability pillars by its action on performance, animal welfare and health, product quality and environmental impact. For most nutrients, the efficiency with which animals transform dietary inputs to animal products is relatively low. For protein, this efficiency rarely exceeds 50%, while for phosphorus and energy these efficiencies are even lower. This implies that knowledge of nutritional requirements, combined with their availability in feed ingredients, is of major importance to develop feeding strategies that contribute to improving the feed efficiency.

The requirement of nutrients such as amino acids or minerals, when all other nutrients are provided at adequate levels, can be defined as the amount needed for specified production purposes such as growth, protein deposition, milk production or maintenance

http://dx.doi.org/10.19103/AS.2017.0013.10

(Fuller, 2004). Nutrient requirements are affected by factors related to the animal (e.g. genetic, age, weight, sex, social status and health), the feed (e.g. feed allowance, nutrient composition and digestibility) and housing conditions (e.g. ambient temperature and space allowance) (Noblet and Quiniou, 1999). Two methods are generally used to determine the nutrient requirements of pigs: the empirical and the factorial methods (Patience et al., 1995). A comparison of these two methods has been given by Hauschild et al. (2010b) and Pomar et al. (2013). In the empirical approach, the requirement for a nutrient is determined by feeding groups of pigs with increasing levels of this nutrient while measuring one or several performance traits during a given time interval. With this method, the nutrient requirement corresponds to the population requirement for the considered performance and time interval. The estimated requirement may depend on the measured performance trait and on the statistical model used to estimate the population requirement (Pomar et al., 2013). Differences in pig characteristics (e.g. adiposity in growing pigs and prolificacy in sows), and the effect of the environment on the requirement limits the possibility of extrapolation of results to other production situations (Pomar et al., 2003). It is why the factorial approach is generally preferred and considered as a reference method. With this method, the estimated nutrient requirement includes requirements for maintenance and production, and the efficiency of nutrient use (Fuller and Chamberlain, 1982; van Milgen and Noblet, 2003). Requirements are assumed to be the amount of the given nutrient that will allow the animal to perform normally its needed functions, without limiting growth (Pomar et al., 2013). As illustrated for lysine in Pomar et al. (2013), the requirement for a nutrient is calculated as the sum of lysine requirements for given functions (e.g. basal endogenous losses and daily gain) expressed as quantity required for the nutrient per unit of feed intake, body weight (BW) or daily gain. Whereas the empirical approach is used to obtain the requirement at a population level, the factorial approach allows estimating the requirement of an individual animal at a given stage. Applying the factorial method to determine the requirement of a population needs to account for variability among pigs in a population. Indeed, using the requirement of the average pig to feed a population of pigs implies that half of the population will get less nutrients than required and the performance of the population will be lower than expected (Pomar et al., 2003; Brossard et al., 2009; Hauschild et al., 2010b). An alternative is to use a lower nutrient efficiency for the population than for individuals (NRC, 2012) or to increase the requirement by a given percentage (van Milgen et al., 2008). However, defining the population requirement is difficult, depending on whether it is considered as the nutrient amount required to satisfy the average pig, the most demanding pigs or a given proportion of pigs.

Even if the factorial method allows for a more mechanistic determination of requirement for a given level of performance, its application is not straightforward because of the dependency of this performance level to pig characteristics, nutritional supply, housing conditions, and their interactions (Noblet et al., 2016). Modelling approaches based on the factorial approach have been developed since the 1970s to predict the response of growing pigs or sows to the nutrient supply. These models have been developed to simulate the performance of a single animal and can be used as decision support tools to assess the nutrient requirement and identify appropriate feeding strategies (e.g. van Milgen et al., 2008; NRC, 2012). However, between-animal variation has been shown to affect the population response, the efficiency of nutrient utilization and, consequently, the optimal nutrient supply for the population (Leclercq and Beaumont, 2000; Pomar et al., 2003; Brossard et al., 2009). Therefore, stochasticity has been introduced in models to deal with variability and to simulate responses of groups of pigs (e.g. Ferguson et al., 1997; Knap,

1999; Pomar et al., 2003; Vautier et al., 2013). These approaches have shown that variation among animals needs to be considered to improve nutrient efficiency on-farm. Assessing variation among animals is possible through the development of monitoring devices that allow one to characterize individual animals on-farm, and through the development of feeding devices that are controlled through real-time decision support tools.

In this chapter, we will present briefly the existing modelling approaches developed to predict nutrient requirements of a single individual (growing pig or sow) in terms of protein/amino acids, energy, and minerals. We will then present how variability among animals can be integrated in modelling nutrient requirements. We will also illustrate how these models can be used in the development of precision feeding systems that allow improvement of feed efficiency.

2 Modelling pig nutrient requirements

As described in introduction, the factorial approach allows defining requirements for a given level of performance in a given environment. However, this performance level depends on several factors such as pig characteristics (e.g. age, weight, sex, heath status and genetic potential), feeding level and feed quality, and housing conditions. Moreover, this method does not account for the possible interactions between these factors. Defining nutrient requirements more precisely implies to establish response curves between nutritional supplies and performance accordingly to the physiological status of the pig. Modelling approaches allow integrating these responses curves to predict performance (outputs from the model) from information on feed quantity and quality, animal characteristics, and housing conditions (inputs to the model) on the basis of a set of equations. Different types of models exist (France and Thornley, 1984; Sauvant, 2005). Empirical models, also called 'black box' models, relate inputs to outputs without relying on the underlying biological mechanisms. Mechanistic models include underlying biological mechanisms to calculate outputs from inputs; this implies that these models are more complex and typically use numerous equations and parameters. Models predicting outputs at a given point in time or space are called static, which contrasts with dynamic models that describe the dynamic nature of the response over a given period (e.g. the fattening period). Finally, deterministic models have a fixed set of parameters and therefore do not account for variability among animals, the same set of parameter values for inputs implying the same values for outputs. In contrast, stochastic models integrate variability in model parameters, which offers the possibility to account for variation among animals in a population.

2.1 Growing pig models

Since the 1970s, different models, most of which are dynamic, deterministic, and more or less mechanistic, have been developed to simulate pig growth and to determine nutrient requirements. They are often based on the association of empirical equations and mechanistic description of physiological or biochemical functions of the animal. The concepts used in these models have been summarized and compared by several authors (Bastianelli and Sauvant, 1997; Ferguson, 2006; Kyriazakis and Sandberg, 2006; Luiting and Knap, 2006; van Milgen et al., 2012). Whittemore and Fawcett (1974, 1976) proposed one of the first nutritional models describing pig growth, with the objective of predicting body protein and lipid deposition from the energy and protein intake. The basic concepts underlying this semi-mechanistic model can be summarized as follows (van Milgen et al., 2012):

- Growth is determined from modelling lipid and protein deposition; BW, water, and ash deposition, and fat and lean weight are calculated empirically from body protein and lipid weight.
- The model considers an upper limit to protein deposition in growing pigs, which is constant during growth (i.e. PDmax, 110 g/day from 20 to 100 kg).
- There is a minimum lipid-to-protein deposition ratio (minimum LD:PD).
- The actual protein deposition is determined by the last two factors and by the quality and quantity of ingested protein (i.e. the supply in essential amino acids).
- Feed intake is a model input; energy not used for maintenance or protein deposition is used for lipid deposition that is therefore considered as an energy sink.

An important concept used in this model was the linear–plateau relationship between ingested energy and protein deposition, with protein deposition being dependent on (linear part, minimum LD:PD) or independent of (plateau or PDmax) the energy supply.

The concepts of Whittemore and Fawcett (1974, 1976) have been used and developed further in several models such as those of Black et al. (1986), Moughan et al. (1987), Pomar et al. (1991a), Ferguson et al. (1994), de Lange (1995), Emmans (1997), Wellock et al. (2003), van Milgen et al. (2008), Yoosuk et al. (2011), NRC (2012) and Strathe et al. (2015). These models are often based on the nutrient partitioning concepts of Whittemore and Fawcett (1974, 1976), which are summarized by de Lange (1995) and van Milgen et al. (2012). Ingested nutrients are partitioned between maintenance requirements (including also endogenous losses and a minimum protein turnover rate) and requirements for protein and lipid deposition and their corresponding efficiencies. Body weight and carcass traits are determined from these depositions. Compared to the approach of Whittemore and Fawcett, various modifications have been proposed, for instance concerning the change in PDmax and the relationship between ingested energy and protein deposition during growth (van Milgen et al., 2008; NRC, 2012). In these models, feed intake can be considered as an input ('push' approach) or an output ('pull' approach). In the 'push' approach, *ad libitum* feed intake is typically modelled as a function of BW through linear, exponential, power, or gamma functions (e.g. Moughan et al., 1987; Pomar et al., 1991a; de Lange, 1995; van Milgen et al., 2008; NRC, 2012). Because lipid deposition is considered as an energy sink, energy not used for maintenance and protein deposition is deposited as lipid. Unless a specific control of energy intake is included in the model (e.g. with the Gamma function where animals eat for maintenance when they approach maturity), there will be no constraint on the way lipid deposition will evolve during growth. In the models of Black et al. (1986), Ferguson et al. (1994), Emmans (1997), Wellock et al. (2003) and Yoosuk et al. (2011), a 'pull' approach was used where feed intake is calculated to satisfy the requirements of maintenance and of the potential protein and lipid deposition. In these models, explicit response curves (e.g. Gompertz functions) are used to describe the change in the potential protein and lipid deposition during growth. Actual intake is then calculated from predicted feed intake by applying constraints such as ingestion capacity of animal, feed characteristics (e.g. nutrient composition, density and water holding capacity) or environmental conditions (e.g. temperature). In this approach, when the animal tends to maturity, protein and lipid deposition tends towards zero and feed intake corresponds to the energy expenditure for maintenance.

The different models presented here predict growth, as affected by nutrient supply, animal characteristics and, in some cases, housing conditions. These models allow one to calculate nutrient requirements to achieve this growth and identify possible limiting factors. Most of these models depend on a large number of parameters that are difficult to obtain or estimate under practical conditions. Although these models are used in a research context, practical application of models requires that these models are easily accessible and user-friendly (e.g. as a dedicated software tool) and that required model inputs can be provided by the user (e.g. TMV (Werkgroep 1991), NRC (2012) and InraPorc (Dourmad et al., 2008; van Milgen et al., 2008)). For example, the InraPorc software summarizes the phenotypic potential of a growing pig (i.e. growth potential and *ad libitum* feed intake) by five model parameters. These parameters can be estimated through a statistical routine using on-farm recorded data of body weight and feed intake. Tools such as InraPorc can be used to estimate nutrient and energy requirements and utilization, evaluate the consequence of different feeding strategies, and identify feeding strategies that allow one to improve performance traits such as feed efficiency. An example of the change in nutrient and energy utilization during growth obtained from InraPorc is given in Fig. 1 for the utilization of digestible energy (Fig. 1a) and standardized ileal digestible (SID) lysine (Fig. 1b) as a function of BW. Van Milgen et al. (2009) illustrated the use of InraPorc to adapt feeding strategies and they simulated the performance of a pig between 27 and 100 kg BW having an average daily feed intake of 2.74 kg/d and an average daily gain (ADG) of 1.10 kg/d. Diets were formulated on a least-cost basis using feed ingredient prices of May 2008 in France and a net energy content of 10 MJ/kg. A first diet (D1) was formulated to contain 9.2 g SID lysine, a level 10% higher than the requirement of the animal at the start of the studied period to account for the requirement of the population (see Section 3.1). Two other diets (D2 and D3) were formulated with lower SID lysine levels corresponding to the requirement at around 60 kg BW and at the end of the studied period. Three feeding strategies were then defined: a single-phase strategy with D1 (SP), a two-phase strategy with D1 offered until 60 kg BW and D2 for higher body weights (TP), and a multiphase strategy (MP) with a progressive replacement of D1 by D3 to follow as much as possible the changes in the SID lysine requirement. Growth performance obtained for the three strategies were identical as the diets were formulated to avoid deficiencies. However, total nitrogen (N) intake decreased from 4.79 kg for SP to 3.86 kg for MP. Consequently, N excretion decreased from 3.01 kg for SP to 2.08 for MP, and the efficiency of N utilization was increased. In addition, better adapting the nutrient supply to the requirement reduced feed costs (in comparison with the SP strategy) by 5.4% for TP and by 10.8% for MP. This example shows the interest of using modelling nutrient requirements to improve feed efficiency.

2.2 Models for reproducing sows

In contrast to models for growing pigs, only few models describing nutrient utilization in reproductive sows have been developed (Williams et al., 1985; Dourmad, 1987; Pomar et al., 1991b; Pettigrew et al., 1992; Dourmad et al., 2008; NRC, 2012; Hansen et al., 2014), and most of these are research models. Some of these are more of calculation models to determine nutrient requirements (e.g. NRC 2012). More mechanistic models also exist, such as InraPorc (Dourmad et al., 2008) and another model by Hansen et al. (2014). These two models describe energy and nutrient partitioning on a daily basis using a similar structure. The calculation of nutrient requirements during gestation and lactation,

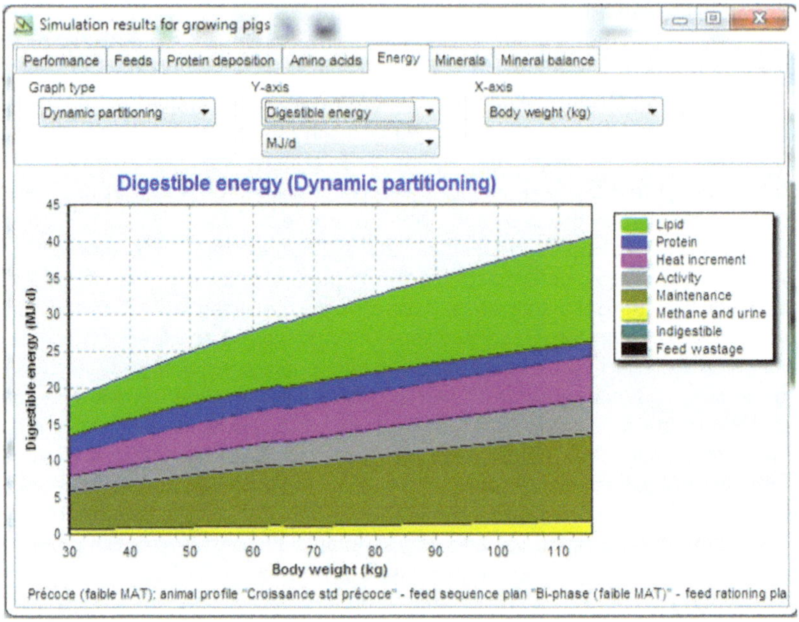

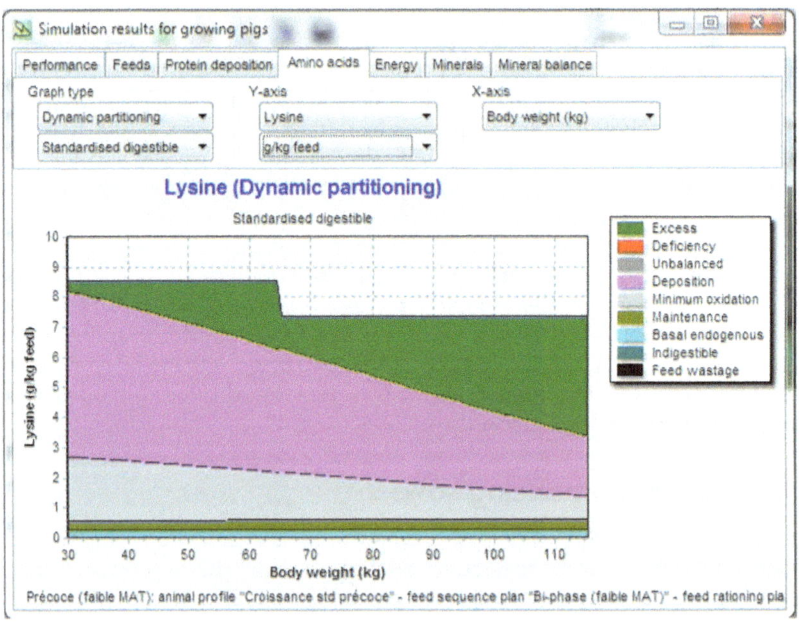

Figure 1 Simulation of changes in digestible energy (a) and lysine requirements (b) depending on body weight, for animals with an average daily gain of 763 g/d and a feed to gain ratio of 2.95 kg/kg.

that are differentiated, is based on a factorial approach. As for growing pigs, the sow models are based on body lipid and protein mass that evolve as a function of the flows of (metabolizable) energy and digestible amino acids. During gestation, requirements for maintenance (including activity and thermoregulation) and foetal/uterus and maternal growth are modelled. During lactation, nutrient and energy requirements are based on the requirements for maintenance, milk production and maternal growth. Priority of nutrient use is given to maintenance, foetal growth and milk production. Body reserves can be mobilised to contribute to these functions in case the nutrient supply is lower than the requirement during lactation. Conversely, body reserves can be (re)constituted during the next gestation when the nutrient supply exceeds the requirements for maintenance and foetal growth.

Compared to other models, the InraPorc sow model allows calibrating parameters from on-farm data of reproductive performance, feeding practices and housing conditions. These parameters can be used to calculate nutrient requirements by the factorial approach, and to simulate the dynamic response of sows to different feeding strategies. An example of the SID lysine utilization by sows over four parities is given in Fig. 2 (Dourmad et al., 2015a). Three different feeding strategies were used that differed in nutrient supplies during gestation. With the first feeding strategy (Fig. 2a) a single gestation diet was used during the entire gestation period and a lactation diet during lactation. The feeding level during gestation increased by 400 g/d during the last 3 weeks of gestation and was adjusted to body condition at mating and the target body condition at farrowing. During lactation, feed intake was assumed to be close to *ad libitum*. Diets were formulated on a least-cost basis. The Fig. 2a indicates that the SID lysine requirement increases during gestation but decreases with parity (because the sows attain mature BW). Consequently, for gestating sows receiving the same diet independently of parity, the amino acid and protein supplies exceed the requirement, especially during the beginning of gestation and more so in older sows. To test the possibility of reducing this excess, a second strategy was simulated using two different diets for gestating sows, depending on parity and gestation stage, and differing in amino acid and protein contents (Fig. 2b). The first diet contained 3.8 g SID lysine and 102 g crude protein (CP) per kg of feed and was used during the first 80 days of gestation, except for first parity sows. The second diet contained 5.5 g SID lysine and 145 g CP per kg of feed and was used in first parity sows throughout gestation, and in other sows from day 80 of gestation. Other amino acids were supplied according to the ideal protein requirement. With the two-phase strategy, total consumption of CP and SID lysine were reduced by 10 and 11%, respectively, with an associated average reduction of 15% of N excretion. These results indicate that the amino acid supplies were much better adjusted to the sow's requirements with this strategy. Further improvements have been tested using a multiphase feeding during gestation. Two gestation diets were formulated differing in amino acid and CP contents. The first and the second diets contained 3.0 g SID lysine and 99.7 g CP, and 5.5 g SID lysine and 145 g CP per kg of feed, respectively. The two diets were mixed in adequate proportions to meet, on a daily basis, the amino acid (and digestible P) requirement (Fig. 2c). Compared to the single diet feeding strategy, intake of CP and SID lysine and N excretion were reduced by 14, 17 and 2% respectively, with the multiphase strategy. Compared with the one-phase feeding strategy (and without considering the possible extra cost for applying the feed changes), the calculated feed cost was 6% lower with the two-phase strategy, and 8% lower with multiphase feeding. Modelling nutrient requirements can thus be used to develop feeding strategies to improve feed efficiency and to reduce N excretion and feed costs.

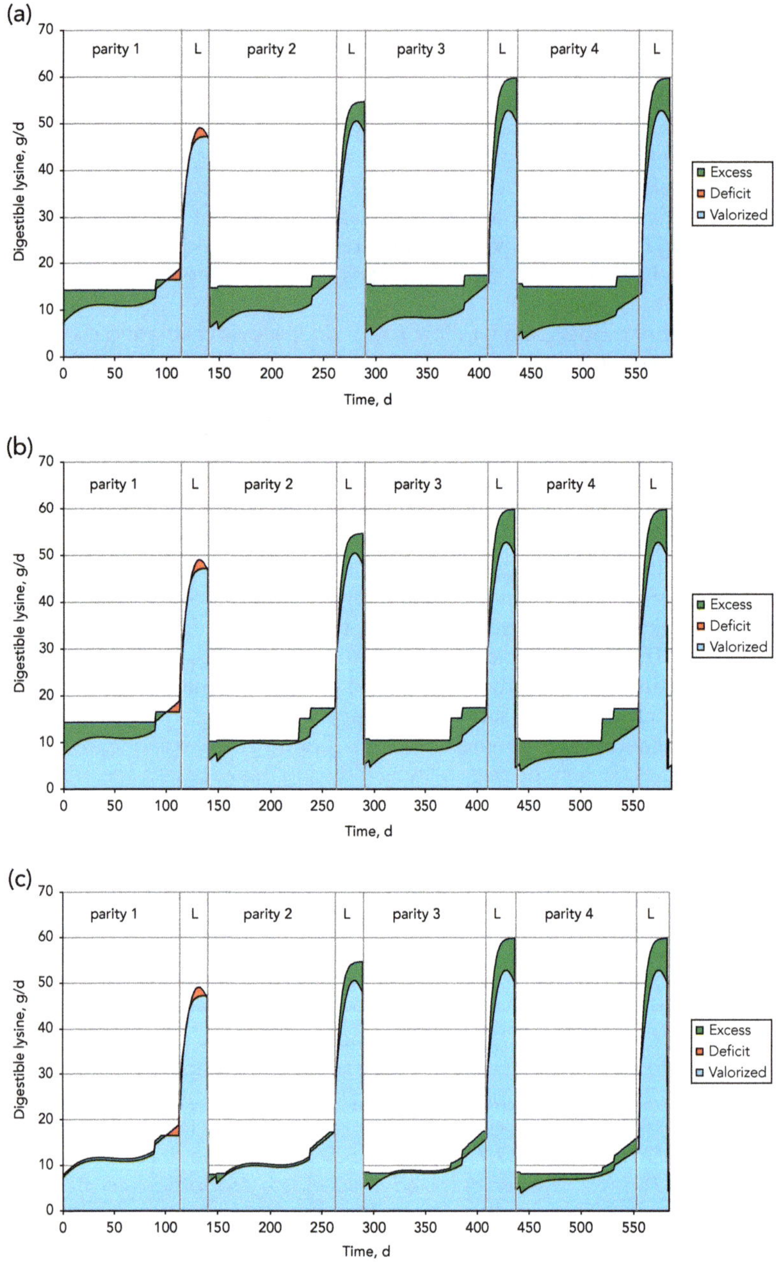

Figure 2 Simulated effect of different gestation feeding strategies on the utilization of ileal digestible lysine (a): one diet, (b): two diets in multiparous sows with a change at day 80 of gestation, (c): two diets mixed in adequate proportions to meet lysine and apparent digestible phosphorus requirements from Dourmad et al. (2015a).

2.3 Mineral requirement modelling

As described above, many current growth models for pigs consider that BW, water and ash deposition, and fat and lean weight are determined from body protein and lipid weight, as suggested by Whittemore and Fawcett (1976). Van Milgen et al. (2008) and NRC (2012) estimate empty BW and BW from protein and lipid body mass without explicitly determining water and ash mass. As indicated by Létourneau-Montminy et al. (2015), these approaches imply that body ash (i.e. all body minerals) mass and growth are driven by growth of body protein and/or lipid. Létourneau-Montminy et al. (2015) observed that pigs fed deficient diets modulated growth of soft tissues and bone mineralization independently (Pomar et al., 2006; Rousseau, 2013). Consequently, total body ash can be reduced without necessarily reducing mineral deposition and growth of soft tissues. Minerals are of great importance in pig nutrition. For instance, phosphorus (P) has an important role in bone development and metabolism of growing pigs (NRC 2012). It also has an important economic value, being the third most expensive nutrient in pig diets after energy and protein. Indeed, due to the low digestibility of dietary P of plant origin (i.e. phytate), diets are supplemented with expensive, non-renewable inorganic sources of P to meet the P requirements (Selle et al., 2011). The alternative, often used in practice, is to incorporate phytase enzymes which increase P digestibility. The high excretion of P, due to its oversupply and its low digestibility, also contributes to the environmental impact of pig production through eutrophication (Selle et al., 2011). These different elements argue to use precise modelling approaches to determine mineral requirements in order to be able to achieve an optimized supply, with minimized excretion.

The factorial approach of mineral requirements has been integrated in models for growing pigs and sows. For instance, Jondreville and Dourmad (2005) estimated P requirements for maintenance and production for different physiological stages. These principles have been integrated in models such as InraPorc for growing pigs (van Milgen et al., 2008) and sows (Dourmad et al., 2008) using the concept of apparently digestible P and considering the effect of diet form (pellet or mash) and phytase addition on digestibility. As summarized by Noblet et al. (2016), the P requirement for growth is based on the BW of animals. Export of P through milk is estimated from milk protein production while P requirement for conceptus growth is estimated from protein deposition in foetuses. The approach proposed by Jondreville and Dourmad (2005) allows one to adjust the dietary digestible P supply to pig performance and physiological status, and to evaluate the effect of performance level on apparent digestible P requirement. For instance, a decrease of 0.2 points in feed conversion ratio in growing pig requires an increase of 0.2 g/kg in dietary P concentration.

As indicated earlier, estimating mineral requirements from performance (i.e. BW or BW gain) has some limitations. Consequently, specific models have been developed, mainly for P, to allow the body mineral content to vary independently of protein and lipid mass and to allow simulating different phases of P deficiency (e.g. a phase of bone weakening while performance is maintained followed by a phase of growth reduction) and compensatory bone mineralization (Fernandez, 1995; Schulin-Zeuthen et al., 2005; Létourneau-Montminy et al., 2007, 2011, 2015; Symeou et al., 2014a,b). These models are deterministic and mechanistic as they explicitly consider the mechanisms of P intake, digestion, retention and excretion (e.g. Symeou et al., 2014a,b), and also of Ca and other minerals and natural or microbial phytase addition (e.g. Létourneau-Montminy et al., 2015). Steps of mineral digestion in the stomach, and in the small and large intestine, can be differentiated (e.g.

Symeou et al., 2014a,b) as well as the excretion in the urine and faeces. These models are mainly oriented towards research but are aimed to be included in decision support tools and can be applied to develop feeding strategies in relation to growth performance, bone mineralization and optimization of P retention while minimizing P excretion.

3 Population, variability and feed requirement modelling

3.1 Interest to include variability in pig models

As described above, several models exist to determine nutritional requirements. However, these models are mainly deterministic, representative for average animal. Furthermore, parameter values used in these models are obtained from experiments conducted with groups of pigs, for which the average animal is supposed to be representative (Pomar et al., 2013). Different arguments can be advanced in favour of considering not only the average animal but also the variability among animals. Knap (1995) pointed out some arguments for considering variability in models. For example, the profitability of production systems can be largely affected by the extent of variability of performance traits. Moreover, the change from a production system to another can have minor effects on mean production levels but important consequences on their variability. For example, offering feed *ad libitum* in the finishing period can have limited impacts on the population mean of ADG in comparison with a restricted feed allowance. However, it can induce an increase in the variability in ADG among animals, influencing the BW and lean content distribution at slaughter and thus impact carcass payment. Additionally, when using a model for replacing experimental comparisons by simulations, the inclusion of variability in models is needed as a proper significance testing of differences between production systems requires statistical tests based on knowledge of variability within and between systems. Indeed, a difference between treatment means is meaningful only if it can be compared to the variation within each treatment. A last example is the study of the relationships between performance criteria, for which covariance between these criteria and thus variability must be accounted for. Recent studies have demonstrated the importance of considering variability among animals to evaluate biological responses and in defining nutritional programmes (Pomar et al., 2003; Main et al., 2008; Brossard et al., 2009; Vautier et al., 2013). Indeed, between-animal variation determines the population response and, therefore, the overall efficiency of nutrient utilization (Pomar et al., 2003) and optimal nutrient levels (Leclercq and Beaumont, 2000; Pomar et al., 2003; Brossard et al., 2009).

Pomar et al. (2003) and Pomar (2005) illustrated this point by using a growth model based on those of Knap (1999) and Wellock et al. (2003). The inputs of the model are the diet composition and the pig genotype, which is described by three independent parameters. This model was used to simulate the growth of 2500 pigs originating from five different populations with the same mean genetic potential (i.e. the same mean values for each of the three model parameters) but with different genetic variances. Populations were generated randomly to obtain for each parameter 0, 0.5, 1, 1.5 and 2 times the estimated genetic variance from a reference population. Consequently, the null variance population corresponded to a population of fully identical animals, while other populations corresponded to more or less heterogeneous populations. The performance of pigs was simulated during one day considering that all pigs were 50 kg BW. Simulations were carried out with 11 diets, with intake of ideal protein varying from 212 to 290 g/d. For the null variance population, the response of protein deposition to increasing protein supply had a linear–plateau shape, as stated in the single animal model (Fig. 3a). With an increasing variance of parameters, the response became increasingly curvilinear–plateau and the protein level required to attain the plateau increased as the variance increased. The ideal protein requirement, defined as the nutrient level required to maximize protein deposition, increased from 235 g/d to 251 g/d when the variance increased from 0 to 1 times the reference variance. For higher variance, the requirement value was difficult

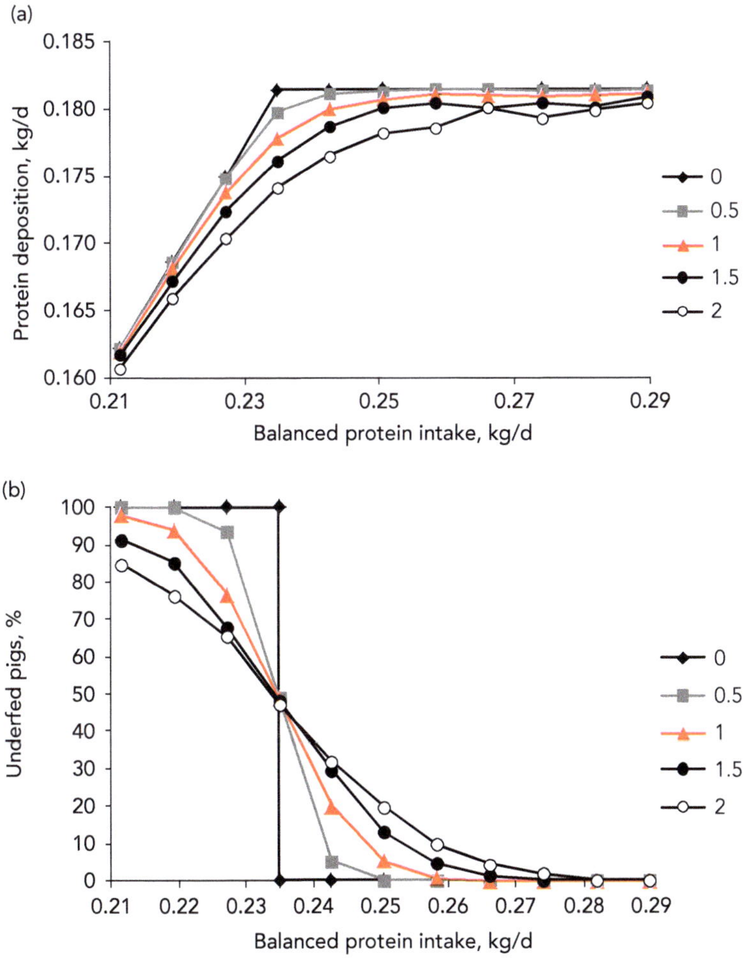

Figure 3 Effect of growth parameters variability between individual pigs (0, 0.5, 1, 1.5 and 2 times the variance of a reference population) and quantity of ingested protein on daily protein deposition (a) and percentage of under-fed pigs (b) for 50 kg BW pig population from Pomar et al. (2005).

to estimate because some animals in the population did not receive sufficient protein to express their potential (Fig. 3b). The average animal is represented by the null variance population for which the requirement is met for a supply of 235 g/d. For other populations, this supply allowed covering the requirements of 50% of the population. This percentage of pigs with their requirement met was decreased when the protein supply was reduced, the extent of this decrease being higher when the population variance increased. Similarly, the proportion of over-fed pigs increased when the protein supply increased. In the same way, Brossard et al. (2009) also showed that the SID lysine requirement varies between pigs in a population and that the percentage of pigs for which the requirement was met can vary greatly with the feeding strategy (e.g. then number of feeding phases) and the growth period.

3.2 Integrating variability in feed requirement modelling

To account for the variability among pigs in determining nutrient requirements and to study the effect of feeding strategies on performance, different models have been developed to simulate the growth of a pig population (e.g. Ferguson et al., 1997; Knap, 1999; Pomar et al., 2003; Wellock et al., 2004; Brossard et al., 2009; Symeou et al., 2016a,b). These approaches are based on the knowledge of mean values and variability (mainly through variance, rarely through covariance) of model parameters. Different methods can be used to estimate these values. Schinckel and de Lange (1996) proposed repeated measurements of body composition using ultrasounds or tomography to assess individual variability during growth. Methods of inversed modelling have also been developed to obtain these parameter values for a population or for individual animals (Knap et al., 2003; Doeschl-Wilson et al., 2006; Vautier et al., 2013). For instance, model parameters can be adjusted iteratively through optimization by comparing real data (e.g. BW, feed intake, protein or lipid mass) to model outputs to minimize the difference between predicted and real values.

Once the mean and the variance of parameter values are known, different methods can be applied to integrate variability in models. For instance, combinations of a limited number of parameters can be generated randomly from mean and variance values of parameters. Parameters are supposed to have normal and independent distributions and the coefficient of variation (CV) of parameters are fixed by authors. The lack of information constrained the authors to ignore the covariance between parameters (e.g. Ferguson et al., 1997; Pomar et al., 2003; Wellock et al., 2004; Symeou et al., 2016a,b). Indeed, Ferguson et al. (1997) considered it too costly and difficult to develop experimental procedures to obtain a correct estimation of CV and correlations between parameters. However, not considering covariance between parameters induces an overestimation of simulated variability (Pomar et al., 2003). For example, a parameter describing feed intake is undoubtedly correlated to a parameter describing growth because a pig eating more than average is also likely to grow faster than the average pig. Therefore, introducing variability in models requires knowledge not only of the variance in parameter values but also of the covariance between parameters (Ferguson et al., 1997; Kyriazakis, 1999). Consequently, other methods have been developed to integrate covariance between parameters. Morel et al. (2010, 2012) used a model in which the pig genotype is described by three parameters. Combining a correlation matrix between parameters obtained from literature and different levels of CV, they calculated a variance–covariance matrix, and thus generated different populations using this variance–covariance matrix and different mean values. Brossard et al. (2009, 2010) and Vautier et al. (2013) obtained a variance–covariance matrix for the five parameters describing the growth potential and feed intake of the InraPorc model through inversed modelling. Brossard et al. (2009) determined parameter values for 192 pigs, allowing one to study correlations between parameters and to obtain mean values and a variance–covariance matrix of the parameters. These last two elements allow the generation of virtual populations of pigs with the same characteristics as the initial population in terms of mean values and variability. For the same model parameters, Vautier et al. (2013) developed a methodology for the generation of pigs populations based on a generic variance–covariance matrix associated with distribution laws of the parameters. The generic variance–covariance matrix was calculated as a median from matrices obtained from 40 subpopulations of pigs differing in sex and cross-breeds, to ensure the genericity of the relationships between parameters and to avoid the use of a particular variance–covariance structure. This generic matrix and the distribution laws can be combined with mean parameters obtained on-farm to generate virtual populations of pigs with realistic mean performance and variability. Finally, Schinckel et al. (2003) and Strathe et al. (2009, 2010) developed stochastic growth models using nonlinear mixed equations, integrating a mean effect of the population and random effects due to individuals. Their method allows one to obtain a mean value for each parameter but also individual values and correlation and variance–covariance structures that can be used to generate populations and perform stochastic simulations and predictions of mean and standard deviation of performance.

3.3 Applying stochastic modelling to feed efficiency improvement

The stochastic growth models presented above can be used to simulate the effect of different feeding strategies not only on animal performance but also on economic results and environmental impacts. Morel et al. (2010, 2012) applied their model to investigate how pig genotype (i.e. lean, normal and fat), population size and variability, feed costs, carcass payment scheme, and feed allowance (i.e. restraint vs *ad libitum*) affect the gross margin and/or N retention that are optimized through feeding strategies (i.e. number of diets, energy and amino acid content, quantity and duration during which each diet is fed). Morel et al. (2010) randomly generated populations of 1 to 625 pigs by varying four parameters of a growth model. Variation of these parameters around the mean was modelled by a CV of 0% (single pig population), 5, 10 or 20%. To generate populations, the parameters were considered either independent or correlated with an associated covariance matrix for the parameters, as described in Section 3.2. Populations were generated for lean, normal or fat pig genotypes. Diets used changed each week; levels of digestible energy (DE) and ratio digestible lysine (Lysd)/DE used for formulation of diets were determined by nonlinear optimization with a genetic algorithm on the basis of simulated performance between 20 and 85 kg BW with an objective of maximizing gross margin (i.e. carcass value minus feed costs and piglet value). Their results showed that lean pigs allowed one to obtain the highest gross margins. The variability in gross margin between simulation runs increased with increasing population variability but decreased with increasing population size. When the covariance was introduced, difference in optimal Lysd/DE ratio varied from −4% to 50% between single pig population (average pig) and population of 125 pigs, depending on the pig type. These authors also noticed that including variance increased optimal values for Lysd/DE ratio. In conclusion, this study showed the importance of the knowledge and the inclusion of variability in the optimization of gross margin through adapted feeding strategies.

Brossard et al. (2010) used a stochastic model for an economic and environmental analysis of pig production. They simulated the effect of changing the SID Lys/net energy (NE) ratio from 85 to 115% of the average population requirement on growth performance, economic results, and N excretion in two contexts of costs of feed ingredients (high or moderate). These different Lys/NE ratios were combined with three different feeding strategies: one strategy with a single diet formulated to meet the mean requirement at the beginning of the growth period; a two-phase strategy with a grower diet and a finisher diet formulated to meet the mean requirement at the beginning and at the middle of growth period; a multiphase strategy with daily adjustments of the diet to the mean requirement of the population. Performance and N excretion were simulated from 27 to 112 kg BW with InraPorc®. Gross margin was calculated from carcass value and feed and labour costs. Maximal growth performance was observed with a SID Lys/NE ratio of 105 to 115% of the mean requirement (Table 1) illustrating the fact that the nutrient supply required to maximize growth performance is higher than mean population requirement.

Table 1 Effect of feeding strategy[1] and dietary digestible lysine to net energy ratio on feed conversion ratio, gross margin[2] and nitrogen excretion, for a virtual population of 1000 pigs from Brossard et al. (2010)

Parameter	Feeding strategy	Dietary Lys/NE content (% of reference level)						
		85	90	95	100	105	110	115
F:G (g/g)	1-phase	2.75	2.70	2.67	2.66	2.66	2.65	2.65
	2-phase	2.88	2.79	2.72	2.69	2.67	2.66	2.65
	Multiphase	3.09	2.94	2.83	2.75	2.70	2.67	2.66
Gross margin (Δ €/pig)	1-phase	−3.8	−1.8	−0.6	0.0	0.1	0.0	−0.4
	2-phase	−9.8	−5.7	−2.7	−1.1	−0.2	0.2	0.2
	Multiphase	−18.7	−12.7	−7.7	−4.0	−1.8	−0.5	0.0
Nitrogen excretion (kg/pig)	1-phase	3.14	3.27	3.44	3.61	3.84	3.88	3.88
	2-phase	3.23	3.15	3.10	3.24	3.39	3.53	3.59
	Multiphase	3.66	3.47	3.34	3.27	3.26	3.24	3.27

[1]1-phase strategy: single diet containing 9.71 MJ net energy (NE)/kg and 0.762 g SID Lys/MJ NE for the 100% reference level; 2-phase strategy: two successive diets (9.71 MJ NE), with a diet change at 112 days of age, and containing 0.762 g SID Lys/MJ NE and 0.635 g SID Lys/MJ NE, respectively, for the 100% reference level; multiphase strategy: the Lys/NE ratio was changed daily according to requirements, with an NE supply of 9.71 MJ and a SID Lys supply for the 100% reference level of 0.762 g/MJ NE and 0.531 g/MJ NE at the beginning and at the end of the growing-finishing period, respectively. [2]Difference relative to the 100% Lys/NE level of the 1-phase strategy.

Reducing the SID Lys/NE ratio below 100% of the mean requirement reduced growth performance and economic results, with the effect being more marked for the two-phase and multiphase strategy. It also increased N excretion for the multiphase strategy but decreased N excretion for the single diet strategy. Indeed, the daily adjustment of SID Lys supply below the mean requirement implied that a major part of the population encountered a SID Lys deficiency, consequently reducing the N efficiency. In contrast, with the single diet strategy, reducing SID Lys supply reduced the SID Lys oversupply and thus the N excretion. When the SID Lys/NE ratio in diets was increased above 100% of the mean requirement, economic result was improved for two-phase and multiphase strategy with an optimum with a SID Lys/NE ratio of 110 to 115% depending on cost context and strategy. For the multiphase strategy, this economic improvement was accompanied by a small reduction in N excretion, whereas the two other strategies induced higher N excretion. For growing-finishing pigs, Niemi et al. (2010) used a bio-economic stochastic dynamic programming model with a growth component and an optimization of the timing of slaughter and daily amounts of protein and energy used in the feeds. Their results suggested that a switch from the classical two-phase feeding to a multiphase feeding adjusting daily the supplies can increase the annual return by 1.35–1.88 € per pig place. For sows, as illustrated above with the results of Dourmad et al. (2015a) (Fig. 2), application of multiphase feeding could also improve feed efficiency and could reduce N excretion up to 19%. More recently, Monteiro et al. (2016) used a similar approach combined with life cycle assessment to evaluate the interest of different formulation and feeding strategies for reducing the environmental impact of pig fattening in France and Brazil. These different examples illustrate that modelling approaches accounting for individual variability are very helpful in identifying feeding strategies that allow one to improve feed efficiency and gross margin and that can contribute in reducing the environmental impact.

4 Towards precision feeding

As indicated above, modelling approaches exist that deal with variability in requirements among individual pigs. This allows one to define how to feed a population, on a basis of a reference (e.g. mean animal) and of simulations on populations generated using the knowledge of mean or individual values of parameters and/or variance–covariance relationships between parameters. Even if these approaches are useful to explore optimal feeding strategies, some difficulties can arise when considering these approaches. Depending on the approach followed, obtaining appropriate parameters or variance–covariance knowledge for a given population can be difficult. Moreover, parameters obtained for one population cannot be adapted for another population due to differences in parameter variability or housing conditions.

To account for variability among pigs without dealing with issues such as covariance structure between parameters, a new approach for nutrient requirement modelling is offered by precision feeding, concomitantly with the development of technologies in the field of precision farming. Precision feeding is based on the dynamic adjustment (if possible day by day) of dietary nutrient supplies to requirements, at a group or at an individual level. In this approach, individual pigs are treated as such and each pig/group is to be modelled individually. The purpose is to improve feed efficiency whilst reducing feed cost and environmental impact. To provide daily and individual tailored diets, this technique needs to include the following elements (Pomar et al., 2009, 2013):

- the precise evaluation of the nutritional value of feed ingredients,
- the real-time determination of nutrient requirements of individual pigs,
- the formulation of balanced diets limiting the amount of excess nutrients, and
- the concomitant adjustment of the dietary supply of nutrients that will match the estimated requirements of the individual pig.

Real-time data collection is needed in terms of BW and feed intake through automatic devices. To obtain model parameters by individual and to predict nutrient requirements and performance individually and on a daily basis using real-time data, models have to evolve to integrate 'real' growth and feed intake patterns that may differ from the expected 'theoretical' patterns. Hauschild et al. (2010a, 2012) developed a prediction model combining a statistical real-time estimation of expected BW gain and feed intake (depending on realized performance during the preceding days) with a mechanistic model predicting protein deposition and NE intake and calculating by the factorial method the amino acid and mineral requirements to sustain this performance. Knowing the requirements for a pen or an individual pig, an optimal blend of feeds can be defined and distributed to individual pigs. Pomar et al. (2010) used this model to simulate the effect of applying precision feeding at an individual level. Their results indicated that precision feeding, by daily and individual adjustment of nutrient supplies to requirements, reduced feed cost by 11% and N and P excretion by more than 38%, compared to three-phase feeding applied to the whole group. This study indicates the potential of precision feeding in improving feed efficiency. The application of precision feeding using real-time modelling of nutrient requirements was tested for growing-finishing pigs over an 84-day fattening period. Pigs received a classical three-phase feeding programme (3P) obtained by blending fixed proportions of feeds A (high nutrient density) and B (low nutrient density), or a daily-phase feeding programme in which the blended proportions of diets A and B were adjusted daily to meet the estimated nutritional requirements of each individual pig (multiphase individual feeding, MPI). The performance (ADG, average daily feed intake, gain to feed ratio and N and P retention) obtained with the MPI programme were similar to those obtained for the 3P programme. However, compared with the 3P programme, the application of MPI programme reduced the SID Lys intake by 27%, the estimated N excretion by 22%, and the estimated phosphorus excretion by 27% (Table 2; Andretta et al., 2014). In Andretta et al. (2016), the application of MPI programme reduced SID lysine intake, estimated N excretion and feeding costs by 26, 30 and 10%, respectively. Even if the reduction in excretion and feed costs is smaller than this estimated by simulation, these results confirm the possibility of improving feed efficiency (for instance in terms of feed cost and environmental impact) with the combination of real-time modelling of requirement with precision feeding. This approach has yet to be improved for a further application on-farm, to be able to integrate more information on animal (e.g. composition of BW gain, health status and behaviour) and its environment (e.g. temperature and ventilation rate) that are made available by the development of sensors, and that can modulate accuracy of predictions. Additionally, more complex objectives for feeding strategies could be integrated such as body composition, expected weight at fixed age to plan slaughter departures, and so on.

Table 2 Performance, digestible lysine intake and nitrogen and phosphorus excretion for pigs fed during 84 days with three successive diets applied for the whole group (Group) or by individual precision feeding (Precision) from Andretta et al. (2014)

	Feeding strategy	
Parameters	Group	Precision
Initial weight (kg)	40.1	42.7
Final weight (kg)	133.8	135.8
Daily feed intake (kg/d)	3.05	3.05
Ingested digestible lysine (g/d)	23.8[a]	17.4[b]
Average daily gain (kg/d)	1.11	1.10
Feed conversion ratio (kg/kg)	2.63	2.70
Nitrogen excretion (g/d)	48.1[a]	37.7[b]
Phosphorus excretion (g/d)	6.9[a]	5.1[b]

Different subscripts within a row indicate statistical difference ($P < 0.05$).

5 Case study

Using modelling approaches to determine feed requirements and to adapt feeding strategies has been shown by simulation to improve feed efficiency (see Section 3.3). Some studies have also been developed to assess the practical application of these modelling approaches. For instance, Brossard et al. (2014) used a herd modelling approach to evaluate different feeding strategies to control or reduce variability among pigs at slaughter. Indeed, the variability in BW among animals complicates the management of slaughter departures. These departures are planned to deliver each time a sufficient number of animals within BW range allowing a maximal payment of carcass. Controlling the variability of BW at slaughter is thus important to be able to deliver a maximal number of animals in the targeted BW range and to avoid too light or too heavy pigs. The applicability and accuracy of this approach was assessed in an experimental study using some of the feeding strategies evaluated by simulation. The InraPorc model was used to perform simulations on 10 batches, each of 84 cross-bred pigs (half barrows and half gilts), to characterize the effect of feeding strategies differing in the amino acid supply or feed allowance, on the mean and variation in growth rate and slaughter weight. In the simulations concerning feed allowance effect, pigs were offered feed *ad libitum* or were restricted (increase in feed allowance by 27 g/day up to a maximum of 2.4 and 2.7 kg/day for gilts and barrows, respectively). A two-phase feeding strategy was applied to all animals, with 0.9 and 0.7 g of digestible lysine per MJ of net energy (NE) in diets provided before or after 65 kg BW, respectively. Pigs were supposed to be slaughtered in two departures, with a mean BW at departure of 112 kg for the whole population. Results indicated that a feed restriction reduces the CV of BW at first departure for slaughter (BW1) and at slaughter by 34% (from 9.0 to 5.9%) and 26% (form 7.9 to 5.8%), respectively. Growth performance obtained from *in silico* simulations using *ad libitum* and restricted feeding plans was compared with results obtained in an *in vivo* experiment on a batch of 168 pigs when applying exactly the same feeding and slaughtering strategies. Actual growth was similar to that obtained by simulation. The CV of BW1 was also similar *in vivo* and *in silico* for the *ad libitum* feeding strategy but was slightly underestimated by 1 percentage point *in silico* for the restriction strategy. This study confirms the relevance of using simulations to predict the level and variability in performance of group-housed pigs, and the possibility of control of variability through feeding strategies.

6 Conclusion and future trends

A precise knowledge of nutrient requirements in growing pigs and reproductive sows is required to better adapt feeding strategies and thus increase feed efficiency and economic results, and reduce the environmental impact of pig production. Modelling approaches have been developed to integrate this knowledge and allow taking into account the possible interactions between the pig and production conditions. Recent development to integrate the effect of individual variability on requirement estimations allowed better accounting for these interactions. Even though additional research is still required, modelling approaches appear to be interesting alternatives and complements to experimentation. Modelling nutrient requirements is a powerful tool to test the effect of different feeding strategies on a set of criteria of interest (e.g. efficiency of nutrient utilization (and thus environmental impact), economic results). Testing a large set of feeding strategies by actual experiments would be too expensive and complicated in field situations. Moreover, results of virtual experiments give indications or trends for future improvements in feeding practices depending on farm situation and animal potential. Research models can be included in decision support tools for technicians and farmers and can also be directly integrated into automatized systems such as precision feeding systems. However, and even if current models already allow good predictions of animals requirements performance, further developments are still in progress for their improvement.

Current nutritional models allow defining nutrient requirements depending on pig characteristics. Some pig models for growing pigs are now including variability among animals in terms of growth and intake potentials. Also models for sows can account for variability among animals and within a given animal depending on parity or gestation/lactation stage. Future research for these models is aimed to integrate other sources of variation. For sows, the effect of ambient temperature, activity, parity or litter size on requirements are still to be refined and integrated more properly in models (Ngo et al., 2012; Dourmad et al., 2015b). In the same way, growing pig models have to be refined to better integrate the effect of ambient temperature (Wellock et al., 2003; Renaudeau et al., 2011), animal activity or capacity to deal with different stress sources linked to their social environment (e.g. group size, surface per animal, mixing period, as in the model proposed by Wellock et al., 2004) or health status. The effect of health status on resource allocation and the associated mechanisms have to be investigated to be integrated in models. Concerning minerals, comprehensive models were made available recently (e.g. Létourneau-Montminy et al., 2015) but refinements are still required to better account for changes in body mineral reserves and the mechanisms involved in the regulation of absorption. This will allow accounting for resorption–absorption phases in sows or compensation between phases in fattening pigs. Availability of sufficient and adapted data is required to support these developments and their application for improving feed/nutrient efficiency. The current development on sensors and data collection to characterize pigs at an individual level or on their environment will allow the access in real time and with a higher frequency to more precise data on classical characteristics (e.g. feed intake, BW and ambient temperature) but also to other new traits such as behaviour, body composition and health status. The prediction of nutrient requirements will then be supported by models integrating these different types and sources of data, on the basis of historical data and also in real time. It can be imagined that the requirements will be defined not only for simple production objectives (e.g. feed intake and growth) but also for multicriteria objectives such as economical return and environmental impact.

7 Where to look for further information

- A good introduction to the subject for non-specialists is provided by the article of Noblet et al. (2016, see reference above).
- Some books exist on feed efficiency and modelling: *Mechanistic modelling in pig and poultry production* (2006, Eds R. Gous, T. Morris and C. Fisher, CABI Publishing, Wallingford, UK); *Mathematical modelling in animal nutrition* (2008, Eds J. France and E. Kebreab, CABI Publishing, Wallingford, UK); *Feed Efficiency in Swine* (2012, Ed J. F. Patience, Wageningen Academic Publishers, Wageningen, The Netherlands); *Nutritional modelling for pigs and poultry* (2015, Eds N. K. Sakomura, R. M. Gous, I. Kyriazakis and L. Hauschild, Cabi Publication, Wallingford, UK).
- Every year, the European Federation of Animal Science (EAAP, www.eaap.org) organizes its annual meeting with sessions on pigs with a focus on feed use or feed efficiency, and also since 2016, a dedicated commission for PLF (precision livestock farming). Every five years, the international Workshop 'Modelling Nutrient Digestion and Utilization in Farm Animals' (Modnut) is organized (see http://www.jackiekyte.com.au/modnut2014/for the last one).
- The EU H2020 Feed-a-Gene project (www.feed-a-gene.eu/) aims to better adapt different components of monogastric livestock production systems (i.e. pigs, poultry and rabbits) to improve the overall efficiency and sustainability. This includes the modelling of biological functions with an emphasis on feed use mechanisms to better understand mechanisms of feed efficiency.
- A list of research centres readers can investigate, for example, for possible collaboration as well as to keep up with research trends:
    1. French National Institute for Agricultural Research (INRA) (www.inra.fr/)
    2. Agriculture and Agri-Food Canada (AAC) (www.agr.gc.ca/)
    3. Wageningen University and Research (WUR) (https://www.wur.nl/)
    4. Newcastle University (www.ncl.ac.uk/)

8 References

Andretta, I., Pomar, C., Rivest, J., Pomar, J., Lovatto, P. A. and Radünz, J. (2014), 'The impact of feeding growing–finishing pigs with daily tailored diets using precision feeding techniques on animal performance, nutrient utilization, and body and carcass composition', *J. Anim. Sci.*, 92(9), 3925–36.

Andretta, I., Pomar, C., Rivest, J., Pomar, J. and Radünz, J. (2016), 'Precision feeding can significantly reduce lysine intake and nitrogen excretion without compromising the performance of growing pigs', *Animal*, 10(7), 1137–47.

Bastianelli, D. and Sauvant, D. (1997), 'Modelling the mechanisms of pig growth', *Livest. Prod. Sci.*, 51(1–3), 97–107.

Black, J. L., Campbell, R. G., Williams, I. H., James, K. J. and Davies, G. T. (1986), 'Simulation of energy and amino acid utilisation in the pig', *Res. Dev. Agric.*, 3(3), 121–45.

Brossard, L., Dourmad, J.-Y., Rivest, J. and van Milgen, J. (2009), 'Modelling the variation in performance of a population of growing pig as affected by lysine supply and feeding strategy', *Animal*, 3(8), 1114–23.

Brossard, L., Quiniou, N., Dourmad, J.-Y. and van Milgen, J. (2010), 'A herd modelling approach to determine the economically and environmentally most interesting dietary amino acid level during the fattening period', In *Modelling Nutrient Digestion and Utilisation in Farm Animals* (Eds D. Sauvant, J. van Milgen, P. Faverdin and N. Friggens), pp. 335–46. Wageningen Academic Publishers, The Netherlands.

Brossard, L., Vautier, B., van Milgen, J., Salaün, Y. and Quiniou, N. (2014), 'Comparison of in vivo and in silico growth performance and variability in pigs when applying a feeding strategy designed by simulation to control the variability of weight at the end of fattening', *Anim Prod. Sci.*, 54(11–12), 1939–45.

Doeschl-Wilson, A. B., Knap, P. W. and Kinghorn, B. P. (2006), 'Evaluating animal genotypes through model inversion, In *Mechanistic Modelling in Pig and Poultry Production* (Eds R. Gous, T. Morris and C. Fisher), pp. 163–87. CABI Publishing, Wallingford, UK.

Dourmad, J. Y. (1987), 'Composition du gain de poids de la truie gestante: prévision en fonction des apports énergétiques et protéiques', *Journées Rech. Porcine*, 19, 203–14.

Dourmad, J. Y., Étienne, M., Valancogne, A., Dubois, S., van Milgen, J. and Noblet, J. (2008), 'InraPorc: A model and decision support tool for the nutrition of sows', *Anim. Feed Sci. Tech.*, 143(1–4), 372–86.

Dourmad, J. Y., van Milgen, J., Valancogne, A., Dubois, S., Brossard, L. and Noblet, J. (2015a), 'Modelling nutrient utilization in sows: a way toward the optimization of nutritional supplies', In *Nutritional Modelling for Pigs and Poultry* (Eds N. K. Sakomura, R. M. Gous, I. Kyriazakis and L. Hauschild), pp. 50–61. Cabi Publication, Wallingford, UK.

Dourmad, J.-Y., Le Velly, V., Lechartier, C., Gourdine, J.-L. and Renaudeau, D. (2015b), 'Influence de la température ambiante chez la truie allaitante, une approche par méta-analyse et modélisation', *Journées Rech. Porcine*, 47, 105–10.

Emmans, G. C. (1997), 'A method to predict the food intake of domestic animals from birth to maturity as a function of time', *J. Theor. Biol.*, 186(2), 189–99.

FAO (2011), *World Livestock 2011. Livestock in Food Security*, Food and Agriculture Organization of the United Nations, Rome.

Ferguson, N. S., Gous, R. M. and Emmans, G. C. (1994), 'Preferred components for the construction of a new simulation-model of growth, feed intake and nutrient-requirements of growing pigs', *S. Afr. J. Anim. Sci.*, 24(1), 10–17.

Ferguson, N. S., Gous, R. M. and Emmans, G. C. (1997), 'Predicting the effects of animal variation on growth and food intake in growing pigs using simulation modelling'. *Anim Sci.*, 64, 513–22.

Ferguson, N. S. (2006), 'Basic concepts describing animal growth and feed intake', in *Mechanistic Modelling in Pig and Poultry Production* (Eds R. Gous, T. Morris and C. Fisher), pp. 22–53. CAB International Publishing, Wallingford, UK.

Fernandez, J. (1995), 'Calcium and phosphorus metabolism in growing pigs. III. A model resolution', *Livest. Prod. Sci.*, 41(3), 255–61.

France, J. and Thornley,J. M. H. (1984), *Mathematical Models in Agriculture*, Butterworths, London, UK.

Fuller, M. F. and Chamberlain, A. G. (1982), 'Protein requirements of pigs', in *Recent Advances in Animal Nutrition* (Ed. W. Haresign), pp. 175–86. Butterwoths, London, U. K.

Fuller, M. F. (2004), *The Encyclopedia of Farm Animal Production*, CABI Publishing, Wallingford, Oxon, UK.

Hansen, A. V., Strathe, A. B., Theil, P. K. and Kebreab, E. (2014), 'Energy and nutrient deposition and excretion in the reproducing sow: Model development and evaluation', *J. Anim. Sci.*, 92(6), 2458–72.

Hauschild, L., Lovatto, P. A., Pomar, J. and Pomar, C. (2010a), 'Estimation journalière des besoins nutritionnels individuels des porcs en croissance en fonction de leur parcours de croissance et de consommation', *Journées Rech. Porcine*, 42, 99–105.

Hauschild, L., Pomar, C. and Lovatto, P. A. (2010b), 'Systematic comparison of the empirical and factorial methods used to estimate the nutrient requirements of growing pigs', *Animal*, 4(5), 714–23.

Hauschild, L., Lovatto, P. A., Pomar, J. and Pomar, C. (2012), 'Development of sustainable precision farming systems for swine: Estimating real-time individual amino acid requirements in growing-finishing pigs', *J. Anim. Sci.*, 90(7), 2255–63.

Jondreville, C. and Dourmad, J.-Y. (2005), 'Le phosphore dans la nutrition des porcs', *INRA Prod. Anim.*, 18(3), 183–92.

Knap, P. W. (1995), 'Aspects of stochasticity: variation between animals', in *Modelling Growth in the Pig* (Eds P. J. Moughan, M. W. A. Verstegen, M. I. Visser-Reynevelds), pp. 165–72. Wageningen Pers, The Netherlands.

Knap, P. W. (1999), 'Simulation of growth in pigs: evaluation of a model to relate thermoregulation to body protein and lipid content and deposition', *Anim. Sci.*, 68(4), 655–79.

Knap, P. W., Roehe, R., Kolstad, K., Pomar, C. and Luiting, P. (2003), 'Characterisation of pig genotypes for growth modeling', *J. Anim. Sci.*, 81(suppl. 2), E187–95.

Kyriazakis, I. (1999), 'Future directions for models in pig biology', in *A Quantitative Biology of the Pig* (Ed I. Kyriazakis), pp. 381–8. CABI Publishing, Wallingford, UK.

Kyriazakis, I. and Sandberg, F. B. (2006), 'The problem of predicting the partitioning of scarce resources during sickness and health in pigs', in *Mechanistic Modelling in Pig and Poultry Production* (Eds R. Gous, T. Morris and C. Fisher), pp. 117–42. CAB International Publishing, Wallingford, UK.

de Lange, C. F. M. (1995), 'Framework for a simplified model to demonstrate principles of nutrient partitioning for growth in the pig', In *Modelling Growth in the Pig* (Eds P. J. Moughan, M. W. A. Verstegen and M. I. Visser-Reynevelds), pp. 71–85. Wageningen Pers., Wageningen, The Netherlands.

Leclercq, B. and Beaumont, C. (2000), 'Etude par simulation de la réponse des troupeaux de volailles aux apports d'acides aminés et de protéines', *INRA Prod.Anim.*, 13(1), 47–59.

Létourneau-Montminy, M. P., Jondreville, C., Lescoat, P., Meschy, F., Pomar, C., Bernier, J. F. and Sauvant, D. (2007), 'First step of a model of calcium and phosphorus metabolism in growing pigs: Fate of ingested phosphorus in the stomach', *Livest. Sci.*, 109(1–3), 63–5.

Létourneau-Montminy, M. P., Narcy, A., Lescoat, P., Bernier, J. F., Magnin, M., Sauvant, D., Jondreville, C. and Pomar, C. (2011), 'Modeling the fate of dietary phosphorus in the digestive tract of growing pigs', *J. Anim. Sci.*, 89(11), 3596–3611.

Létourneau-Montminy, M. P., Narcy, A., Dourmad, J. Y., Crenshaw, T. D. and Pomar, C. (2015), 'Modeling the metabolic fate of dietary phosphorus and calcium and the dynamics of body ash content in growing pigs', *J. Anim. Sci.*, 93 (3), 1200–17.

Luiting, P. and Knap, P. W. (2006), 'Comparison of pig growth models – the genetic point of view', in *Mechanistic Modelling in Pig and Poultry Production* (Eds R. Gous, T. Morris and C. Fisher), pp. 260–81. CAB International Publishing, Wallingford, UK.

Main, R. G., Dritz, S. S., Tokach, M. D., Goodband, R. D. and Nelssen, J. L. (2008), 'Determining an optimum lysine:calorie ratio for barrows and gilts in a commercial finishing facility', *J. Anim. Sci.*, 86(9), 2190–2207.

Monteiro, A. N. T. R., Garcia-Launay, F., Brossard, L., Wilfart, A. and Dourmad, J.-Y. (2016), 'Effect of feeding strategy on environmental impacts of pig fattening in different contexts of production: evaluation through life cycle assessment', *J. Anim. Sci.*, 94(11), 4832–47.

Morel, P. C. H., Alexander D. L. J., Sherriff R. L., Sirisatien, D. and Wood, G. R. (2010), 'A new development in pig growth modelling', in *Modelling Nutrient Digestion and Utilisation in Farm*

Animals (Eds D. Sauvant, J. van Milgen, P. Faverdin and N. Friggens), pp. 83–90. Wageningen Academic Publishers, The Netherlands.

Morel, P. C. H., Sirisatien, D. and Wood, G. R. (2012), 'Effect of pig type, costs and prices, and dietary restraints on dietary nutrient specification for maximum profitability in grower-finisher pig herds: A theoretical approach', *Livest. Sci.*, 148(3), 255–67.

Moughan, P. J., Smith, W. C. and Pearson, G. (1987), 'Description and validation of a model simulating growth in the pig (20–90 kg liveweight)', *New Zeal. J. Agric. Res.*, 30(4), 481–9.

Ngo, T. T., Quiniou, N., Heugebaert, S., Paboeuf, F. and Dourmad, J.-Y. (2012), 'Influence du rang de portée et du nombre de porcelets allaités sur la production laitière des truies', *Journées Rech. Porcine*, 44, 195–6.

Niemi, J. K., Sevon-Aimonen, M. L., Pietola, K. and Stalder, K. J. (2010), 'The value of precision feeding technologies for grow-finish swine', *Livest. Sci.*, 129(1–3), 13–23.

Noblet, J. and Quiniou, N. (1999), 'Principaux facteurs de variation du besoin en acides aminés du porc en croissance', *Techni-Porc*, 22(4), 9–16.

Noblet, J., Dourmad, J.-Y. and van Milgen, J. (2016), 'Etat des lieux et perspectives sur la nutrition du porc', *Journées Rech. Porcine*, 48, 85–96.

NRC (2012), *Nutrient Requirements of Swine*, 11th ed., National Academy Press, Washington, DC.

Patience, J. F., Thacker, P. A. and de Lange, C. F. M. (1995), *Swine Nutrition Guide*, 2nd ed., University of Saskatchewan, Prairie Swine Center, Saskatoon, Saskatchewan Canada.

Pettigrew, J. E., Gill, M., France, J. and Close, W. H., (1992), 'A mathematical integration of energy and amino acid metabolism of lactating sows', *J. Anim. Sci.*, 70(12), 3742–61.

Pomar, C., Harris, D. L. and Minvielle, F. (1991a), 'Computer-simulation model of swine production systems. 1. Modeling the growth of young-pigs', *J. Anim. Sci.*, 69(4), 1468–88.

Pomar, C., Harris, D. L. and Minvielle, F. (1991b), 'Computer simulation model of swine production systems: II. Modeling body composition and weight of female pigs, fetal development, milk production, and growth of suckling pigs', *J. Anim. Sci.*, 69(4), 1489–1502.

Pomar, C., Kyriazakis, I., Emmans, G. C. and Knap, P. W. (2003), 'Modeling stochasticity: dealing with populations rather than individual pigs', *J. Anim. Sci.*, 81(Suppl. 2), E178–86.

Pomar, C. (2005), 'Nos modèles de simulation actuels sont-ils adéquats pour prédire la croissance corporelle et les besoins nutritionnels des porcs charcutiers?', *Journées Rech. Porcine*, 37, 307–16.

Pomar, C., Jondreville, C., Dourmad, J.-Y. and Bernier, J. (2006), 'Influence du niveau de phosphore des aliments sur les performances zootechniques et la rétention corporelle de calcium, phosphore, potassium, sodium, magnésium, fer et zinc chez le porc de 20 à 100 kg de poids vif', *Journées Rech. Porcine*, 38, 209–16.

Pomar, C., Hauschild, L., Zhang, G. H., Pomar, J. and Lovatto, P. A. (2009), 'Applying precision feeding techniques in growing-finishing pig operations. L'utilisation d'une alimentation de precision chez les porcs à l'engrais', In *Proceedings of 45th Eastern Nutrition Conference of* Canada, pp. 103–16. Animal Nutrition Association of Canada, Quebec, Canada.

Pomar, C., Hauschild, L., Zhang, G., Pomar, J. and Lovatto, P. (2010), 'Precision feeding can significantly reduce feeding cost and nutrient excretion in growing animals', in *Modelling Nutrient Digestion and Utilisation in Farm Animals* (Eds D. Sauvant, J. van Milgen, P. Faverdin and N. Friggens), pp. 327–34. Wageningen Academic Publishers, the Netherlands.

Pomar, C., Pomar, J., Rivest, J., Cloutier, L., Létourneau-Montminy, M.-P., Andretta, I. and Hauschild, L. (2013), 'Estimating real-time individual amino acid requirements in growing finishing pigs: towards a new definition of nutrient requirements in growing-finishing pigs?', in *Nutritional Modelling for Pigs and Poultry* (Eds N. K. Sakomura, R. M. Gous, I. Kyriazakis and L. Hauschild), pp. 157–74. Wallingford, GBR, CABI Publishing.

Renaudeau, D., Gourdine, J.-L. and St-Pierre, N. (2011), 'A meta-analysis of the effects of high ambient temperature on growth performance of growing-finishing pigs', *J. Anim. Sci.*, 89 (7), 2220–30.

Rousseau, X. (2013), *Optimisation de l'utilisation de phosphore alimentaire chez le porc et le poulet en croissance*, PhD Thesis. Université François-Rabelais, Tours, France.

Sauvant, D. (2005), 'La modélisation de la réponse des porcs aux pratiques alimentaires et d'élevage', *Journées Rech. Porcine*, 37, 283–90.
Schinckel, A. P. and de Lange, C. F. M. (1996), 'Characterization of growth parameters needed as inputs for pig growth models', *J. Anim. Sci.*, 74(8), 2021–36.
Schinckel, A., Li, N., Preckel, P. V., Einstein, M. and Miller, D. (2003), 'Development of a stochastic pig compositional growth model', *Prof. Anim. Scientist*, 19(3), 255–60.
Schulin-Zeuthen, M., Lopes, J. B., Kebreab, E., Vitti, D. M. S. S., Abdalla, A. L., Haddad, M. D., Crompton, L. A. and France, J. (2005), 'Effects of phosphorus intake on phosphorus flow in growing pigs: application and comparison of two models', *J. Theor. Biol.*, 236(2), 115–25.
Selle, P. H., Ravindran, V., Cowieson, A. J. and Bedford, M. R. (2011), 'Phytase and phytase', in *Enzymes in Farm Animal Nutrition* (Eds M. R. Bedford and G. G. Partidge), pp. 160–205. CABI Publishing, Wallingford, UK.
Strathe, A., Sorensen, H. and Danfaer, A. (2009), 'A new mathematical model for combining growth and energy intake in animals: the case of the growing pig', *J. Theor. Biol.*, 261(2), 165–75.
Strathe, A. B., Danfaer, A., Sorensen, H. and Kebreab, E. (2010), 'A multilevel nonlinear mixed-effects approach to model growth in pigs', *J. Anim. Sci.*, 88(2), 638–49.
Strathe, A. B., Danfaer, A., Jorgensen, H. and Kebreab, E. (2015), 'A dynamic growth model for prediction of nutrient partitioning and manure production in growing-finishing pigs: Model development and evaluation', *J. Anim. Sci.*, 93 (3), 1061–73.
Symeou, V., Leinonen, I. and Kyriazakis, I. (2014a), 'Modelling phosphorus intake, digestion, retention and excretion in growing and finishing pig: model description', *Animal*, 8(10), 1612–21.
Symeou, V., Leinonen, I. and Kyriazakis, I. (2014b), 'Modelling phosphorus intake, digestion, retention and excretion in growing and finishing pig: model evaluation', *Animal*, 8(10), 1622–31.
Symeou, V., Leinonen, I. and Kyriazakis, I. (2016a), 'Quantifying the consequences of nutritional strategies aimed at decreasing phosphorus excretion from pig populations: a modeling approach' *Animal*, 10 (4), 578–91.
Symeou, V., Leinonen, I. and Kyriazakis, I. (2016b), 'The consequences of introducing stochasticity in nutrient utilisation models: the case of phosphorus utilisation by pigs', *Brit. J. Nutr.*, 115(3), 389–98.
van Milgen, J. and Noblet, J. (2003), 'Partitioning of energy intake to heat, protein, and fat in growing pigs', *J. Anim. Sci.*, 81(E. Suppl. 2), E86–93.
van Milgen, J., Valancogne, A., Dubois, S., Dourmad, J. Y., Sève, B. and Noblet, J. (2008), 'InraPorc: A model and decision support tool for the nutrition of growing pigs', *Anim. Feed Sci. Tech.*, 143(1–4), 387–405.
van Milgen, J., Brossard, L., Valancogne, A. and Dourmad, J. Y. (2009), 'Using InraPorc to reduce nitrogen and phosphorus excretion', in *Recent Advances in Animal Nutrition* (Eds P. C. Garnsworthy and J. Wiseman), pp. 179–94. Nottingham University Press, Nottingham, UK.
van Milgen, J., Noblet, J., Dourmad, J. -Y., Labussière, E., Garcia-Launay, F. and Brossard, L. (2012), 'Precision pork production: predicting the impact of nutritional strategies on carcass quality', *Meat Sci.*, 92(3), 182–7.
Vautier, B., Quiniou, N., van Milgen, J. and Brossard, L. (2013), 'Accounting for variability among individual pigs in deterministic growth models', *Animal*, 7(8), 1265–73.
Wellock, I. J., Emmans, G. C. and Kyriazakis, I. (2003), 'Modelling the effects of thermal environment and dietary composition on pig performance: model logic and concepts', *Anim. Sci.*, 77, 255–66.
Wellock, I. J., Emmans, G. C. and Kyriazakis, I. (2004), 'Modeling the effects of stressors on the performance of populations of pigs', *J. Anim. Sci.*, 82(8), 2442–50.
Werkgroep, T. M. V. (1991), *Technisch Model Varkensvoeding (DOS ed.)*, Rosmalen (the Netherlands): Proefstation voor de varkenshouderij.
Whittemore, C. T. and Fawcett, R. H. (1974), 'Model responses of the growing pig to the dietary intake of energy and protein', *Anim. Prod.*, 19(2), 221–31.
Whittemore, C. T. and Fawcett, R. H. (1976), 'Theoretical aspects of a flexible model to simulate protein and lipid growth in pigs', *Anim. Prod.*, 22(1), 87–96.

Williams, I. H., Close, W. H. and Cole, D. J. A. (1985), 'Strategies for sow nutrition: predicting the response of pregnant animals to protein and energy intake', in *Recent Advances in Animal Nutrition* (Eds W. Haresign and D. J. A. Cole), pp. 133–47. Nottingham University Press, Nottingham, UK.

Yoosuk, S., Ong, H. B., Roan, S. W., Morgan, C. A. and Whittemore, C. T. (2011), 'A simulation model for predicting the voluntary feed intake of a growing pig', *Acta Agric. Scan. A-Anim. Sci.*, 61(4), 168–86.

Chapter 11

The use of exogenous enzymes to improve feed efficiency in pigs

M. R. Bedford and C. L. Walk, AB Vista, UK

1 Introduction

1.1 Enzymes in commercial swine diets: an overview

Exogenous feed enzymes have been in commercial use in swine diets for almost 30 years. The first enzymes used were endo-beta-glucanases in weaning piglet diets, which were employed to enable high inclusion levels of barley without incurring digestive upsets. Although the use of such enzymes is limited today in comparison with the almost universal application of phytases, their inclusion represented a significant breakthrough for the nascent enzyme industry. Although the NSP'ases dominated the market in the late 1980s and early 1990s, principally in barley- and wheat-based diets, it was the emergence of cost-effective phytases that resulted in the now almost universal deployment of exogenous enzymes in swine diets. More recently there has been interest in the application of proteases in swine diets; although this represents a very small proportion of total sales to date, the research and data presented deserve scrutiny. This review will focus on these three enzyme classes; although it has to be recognised that other enzyme classes such as amylases and alpha-galactosidases have been tested in the past, their commercial use does not warrant significant investigation and in many cases they are used in conjunction with other enzyme classes (Campbell and Bedford 1992).

http://dx.doi.org/10.19103/AS.2017.0013.11

1.2 Carbohydrases

Undigested feed is seen as being a potential waste of resources, and for monogastrics the component of significant consequential loss is that of carbohydrate, which presents itself in the faeces as non-starch polysaccharides (NSP) and undigested starch. Recovery of even a fraction of this component could significantly enhance the efficiency of feed utilisation, allow greater use of alternate but poorly digested feed ingredients and significantly reduce feed costs. As a result, a great deal of research has focused on supplementing monogastric diets with enzymes which target these undigested carbohydrates. This includes enzymes that the animal does not produce, namely NSP'ases, and to a lesser extent enzymes that the animal does produce but under certain circumstances perhaps sub-optimally, namely amylases and related dextranases. Given the majority of work has been on NSP'ases it is ironic that in humans, the majority of complex carbohydrates found in the caeca of sudden death victims is undigested starch and not NSP (Brown 1996). Although this may suggest the research focus has been misplaced, it is worthy of note that NSP play a significant role in delaying starch digestion and as a result their targeting is well founded. The following section will consider the evidence regarding the use of NSP'ases and suggest the likely mechanisms of action based on the evidence available to date.

2 NSP'ases

In advance of discussing the use of NSP'ases in swine nutrition it is worth highlighting the current state of the art in understanding of the relevance of NSP in digestion. A recent and excellent review of this topic has been published by Bach-Knudsen (2016), which highlights several salient points. The most relevant points with regard to this chapter are listed below:

1. The functional effects of fibre do not correlate well with the chemical measurements. Fibre now includes NSP, RS, lignin, oligosaccharides and waxes. Thus simply categorising the NSP content of an ingredient or diet is likely not of much value.
2. Soluble, viscous NSP may delay gastric emptying and thus reduce absorption rates of glucose and amino acids. Since the soluble NSP have also been implicated in forming a film over the glycocalyx, this would further delay absorption rates. Solubility and viscosity of NSP therefore provide information of potential value to the end user. It is important to note that the solubility of dietary NSP in the stomach of a pig may not be well reflected by results from simple water or buffer solubility tests.
3. In swine, a large proportion of the NSP is fermented, which clearly influences VFA and microbiome populations. An interesting observation is that single, purified NSP and perhaps oligosaccharides have more profound effects on the microbiome than the heterogeneous NSP coming from a single ingredient. Thus the results from models where purified NSP are investigated may not be applicable to the complex diets employed in commercial swine nutrition.
4. With specific reference to the arabinoxylans (AX), the section of the intestinal tract where they are fermented depends upon their solubility and degree of substitution. The greater the degree of substitution of the xylan backbone with arabinose and the lower the solubility of the polymer, the slower the rate of digestion of the AX. The most insoluble and highly substituted AX simply appear in the faeces. This means that the same NSP could have significant and varied effects throughout the

intestinal tract depending upon where it is fermented. The solubility and degree of substitution vary markedly between and even with a cereal species as indicated in a recent substantial review (Rodehutscord et al. 2016). For example in wheat, the soluble NSP content ranged between 10.6 and 29.7 g/kg. Thus the effects of the AX and as a result that of an NSP'ase need to be considered along the entire intestinal tract if relevant correlations with performance are being sought.

There are three key mechanisms which have been attributed to NSP'ase use (Aftab 2012). Whilst evidence exists to support each mechanism, it is clear that the relative contribution of each mechanism may well change depending upon the ingredients employed, and the age and the health status of the animal. These mechanisms are

1 The cell wall mechanism
2 Viscosity mechanism
3 Prebiotic mechanism

2.1 Cell wall mechanism

The pig does not produce any enzymes capable of degrading the NSP's which form the structure of the cell walls of all plant-based ingredients in the diet. A simplistic view is that if the sugars that make up the polysaccharides are released they become a source of energy. Glucose in the form of cellulose has little value to the pig whereas it is a rich energy source as the free sugar. However degradation of arabinoxylans to yield xylose and arabinose is clearly detrimental to both pigs and chickens (Schutte 1990; Schutte et al. 1991; Verstegen et al. 1997) and as a result is unlikely to be of benefit. Thus the focus is on disaggregation of cereal cell walls rather than their complete depolymerisation. Indeed finer grinding of the diet, which results in the destruction of a greater proportion of cell walls, resulted in better performance and nutrient digestibility, although a flaw with this work was that the more finely ground diet also varied in ingredient content (Flis and Sobotka 2005). These structures may impede digestive enzyme access to the starch and protein which form the bulk of the cell contents and which need to be degraded if digestion is to be optimised. Indeed microscopy has demonstrated quite clearly that even in milled and pelleted feed, the majority of cells from cereal sources are intact (Parkkonen et al. 1997; Tervila-Wilo et al. 1996). The average cell size is between 70 and 100 um and most particles in a pelleted diet range between 300 and 800 um. This means that feed particles, if spherical, will contain between 10 and 200 cells, the majority of them being intact, particularly in the larger particles. Microscopy has further demonstrated that the structural integrity of cell walls in the ileum is indeed reduced when NSP'ases are incorporated into poultry feed. Images of the intestinal tract show reductions in the number of intact wheat cell walls at the terminal ileum which was coupled with improvements in ileal amino acid digestibility (Torrallardona et al. 2000), although the caveat with this work was that the cereal (wheat) was fed at 94% of the diet for only five days in ileal cannulated animals. When similarly cannulated piglets were fed a 50% barley and SBM-based diet the effect of a xylanase/glucanase mixture on ileal amino acid digestibility was muted (Pujol and Torrallardona 2000), suggesting that the greater the substrate concentration, the bigger the potential benefit of the enzyme.

Thus the supposition is that NSP'ases are directly responsible for cell-wall perforation. However, a caveat regarding this mechanism has consistently been ignored. The amount of enzyme used commercially and the conditions under which it must work simply do not allow cell-wall destruction to the extent that is observed in vivo within two hours of feed ingestion. Papers supporting the cell-wall destruction hypothesis tend to use in vitro models with very high levels of enzyme and even then do not demonstrate a meaningful, greater release of starch, for example, even after four hours of incubation (Le et al. 2013). Thus some other mechanism(s) must also play a role in this regard since cell-wall breakdown has been documented in vivo, but clearly the proposed direct attack on the cell wall is unlikely to be the principal mechanism. As a consequence, in vitro work which compares NSP'ases for their ability to attack insoluble NSP (which form the structural integrity of the cell wall) may not be particularly relevant.

The recent availability of a cost-effective co-product from the ethanol industry has resulted in a significant research programme to determine the nutrient content of DDGs and to investigate the application of NSP'ases to 'upgrade' this NSP-rich product. Unlike the maize from which it is derived, the majority of starch and protein found in DDGs is embedded in poorly digested aggregates, probably formed during the drying process, rather than in complete cells (Jha et al. 2015). The starch present tends to be more amylose than amylopectin, which is more resistant to pancreatic amylase activity. Indeed it seems from the microscopy that little starch escaped the ethanol fermentation process as a result of encapsulation in complete cells as few intact cells were noted. As a result, specifically with regard to this ingredient, cell-wall degradation to the point of reducing nutrient encapsulation is unlikely to play a significant role in enhancing DDGs per se.

2.2 Viscosity

Viscosity in the intestinal tract of pigs is known to be lower than that of chickens fed similar diets (Danicke et al. 1999). Nevertheless it may still play a role in reducing the digestibility of susceptible nutrients, particularly saturated fats, if viscous grains such as rye, barley, triticale and some varieties of wheat are fed to young piglets. The rate at which substrates, enzymes and products diffuse is directly proportional to the viscosity of the solution in which they are dissolved, and as a result the rate of digestion is slowed with increased intestinal viscosity. Indeed feeding of high (Dusel et al. 1997) compared with low extract viscosity wheat led to poorer performance and the use of a xylanase restored performance, suggesting viscosity may play an anti-nutritive role.

Reduced viscosity may result in better starch and protein digestion in the anterior small intestine, leaving less for fermentation to VFA and amines in the lower SI and large intestine, resulting in less incidence of diarrhoea (Molist et al. 2014) and a tendency for greater portal flux of glucose coupled with reduced flux of VFAs (Jansman et al. 1996). Such results would also suggest that inputs into digestion would be minimised; that is, endogenous losses would be reduced. The data are equivocal, however, with some trials suggesting that this is not the case (Danicke 2001) and others that it is (Bartelt et al. 2002).

However, some work suggests that viscosity is largely irrelevant. Viscosity reduction on use of an enzyme without any concomitant improvement in nutrient digestibility or performance has been shown in barley-based diets (Jensen et al. 1998). More compelling, however, was the observation that (Bartelt et al. 2002; Jensen et al. 1998) the use of a xylanase in a wheat/rye-based diet reduced viscosity significantly in the duodenum and ileum, and this was accompanied by a significant improvement in ileal nutrient digestibility. However, in the same trial, ileal viscosity was increased to values 50–60 fold higher than that of the wheat/rye control diet through inclusion of 2% carboxymethyl-cellulose, with no apparent effects on digestibility of amino acids, dry matter or crude protein. Thus the contribution of viscosity to the anti-nutritive effect of the cereals and the role of viscosity reduction by the enzyme is clearly questionable. CMC itself is not fermented, however, unlike cereal xylans and glucans. This allows examination of the direct effects of viscosity in isolation without the confounding aspect of viscosity being a potential marker for fermentable CHO content of the diet. As such, few effects of CMC are noted, especially in relatively clean environments, where there are no pathogens present to take advantage of the reduced rate of digestion that would be expected with the viscosity increments noted (Lalles 2006; McDonald et al. 2001). Viscosity in cereal-based diets may therefore be a proxy measure of the potential prebiotic content of the diet. If the soluble, viscous arabinoxylans are depolymerised to oligosaccharides by an exogenous xylanase, then they can provide a beneficial prebiotic for the large intestinal flora. It is interesting to note that it is specifically the soluble, viscous xylans which are the potential substrate as these are not rapidly degraded in piglets as evidenced by increasing viscosity with transit through the gut. B-glucans on the other hand are processed by small intestinal flora and thus depolymerised to the extent that they may be fermented (Johansen et al. 1993) in the large intestine without the need for exogenous enzyme intervention. Indeed microbial pathways for fermentation of B, 1–4 gluco-oligosaccharides have been found in significant quantity in the caeca of chickens but there is a relative dearth of equivalent xylo-oligosaccharide pathways (Sergeant et al. 2014). Thus the contribution of viscosity to the 'antinutritive' effects of cereal grains per se in pigs may not be as great as in chickens, but it may serve as an indicator of the quantity of substrate available for enzymatic depolymerisation to fermentable oligosaccharides.

2.3 Prebiotic mechanism

Pigs, or more accurately their microbiome, can digest NSP far more efficiently than that of poultry due to the greater size and longer residence time offered in the large intestine (Choct et al. 2010). Raffinose and stacchyose, for example, are almost 90% digested in piglets three weeks post-weaning (Gdala et al. 1997). Nevertheless the ability of the pig to digest NSP is comparatively poor at weaning and increases significantly and consistently over time (Choct et al. 2010), which results in the evolution not only of the capacity but of the structure of microbiome itself (Deusch et al. 2015). The fermentative capacity of the weaning piglet may thus be more easily overwhelmed by excess soluble NSP than the older, more mature pig, an observation that fits with the association of digestive upsets in piglets with diets rich in soluble NSP (Durmic et al. 1998). The flow of 'prebiotics oligosaccharides' should thus be considered carefully and should likely be limited in early life, but should be accelerated with age.

Partial degradation of soluble and indeed insoluble cell wall NSP will result in the ultimate production of significant quantities of oligosaccharides, which otherwise would not have been generated. Adding an NSP'ase to DDGs after they had undergone an in vitro digestion process resulted in a significant increase in gas and VFA production compared with the untreated control when incubated with porcine faecal flora. This suggested a significant prebiotic release from the NSP fraction by the enzyme utilised (Jha et al. 2015). Numerous other studies have shown increased large intestinal VFA concentrations when NSP'ases have been deployed. Bedford and Schultze (1998) suggested that increased large intestinal fermentation may signal for longer gastric retention times and increased gastric motility. Such an effect may enable more complete digestion of the entire diet through more effective gastric processing. A recent publication has noted changes in PYY and incretin levels of piglets fed xylanase supplemented diets, which suggests such a mechanism may be relevant (May et al. 2015). This mechanism would also account for the fact that improvements in amino acid digestibility on use of an NSP'ase reflect the amino acid content of the whole diet and not the cereal components which form the majority of the substrate (Cowieson and Bedford 2009). Issues that need to be resolved include determination of the sections of the intestinal tract where such signalling is more relevant, which VFAs are most important, and whether such an effect is influenced by age. Interpretation of the VFA data is made more difficult by the fact that the effect of NSP'ases can be quite different in the small compared with the large intestine. Feeding NSP'ases can reduce fermentation in the small intestine by effectively removing fermentable starches (by accelerating the digestion process) whilst simultaneously increasing fermentation in the large intestine through provision of xylo-oligomers (Choct et al. 1996). Starch fermentation in the small intestine tends to result in increased lactate concentrations whereas large intestinal fermentation of fibre favours more acetate production. Clearly these two VFAs stimulate divergent reactions in the different sections of the intestinal tract and thus the responses to each with regard to entero-hormonal production will differ. There are also significant quantities of propionate and butyrate produced, and the effect that these VFAs have on both the microbiota and the different sections of the intestine also differs, adding to the complexity of the potential responses. Given the quantity, chain length and substitution of the oligomers produced will vary with the enzyme used, its dose and the main ingredients of the diet, and that these oligomers in turn will influence the total and relative concentrations of the VFA in each intestinal segment, it is not surprising that there are divergent data in the literature. Nevertheless, the prebiotic mechanism appears to fit the data in the literature reasonably well. If it is correct and dominant then it questions the need for multiple enzyme activities if provision of the optimum quantity of a given oligosaccharide from a single NSP'ase activity may be all that is needed. Performance benefits may derive from the provision of any of the manno, gluco or xylo-oligosaccharides, but the consistency of response may be more reliant on the delivery of the correct quantity of a specific oligosaccharide rather than variable delivery of multiple oligosaccharides.

3 Consistency of response and recent developments

3.1 Consistency of response

The relevance of the various modes of action may be quite different in pigs compared with poultry as suggested by the differential responses obtained with the same enzyme when fed at multiple doses to both species (Olukosi et al. 2007).

Regardless of the mechanism, however, the response to NSP'ases in swine in the literature is varied. The scale of response has, for a long time, been known to depend upon the performance of the control animals (Rosen 2000). Good performance of the control animals limits the 'headspace' for improvement although this does not preclude a favourable response (Dreschel et al. 1999; Van Lunen and Schulze 1996). The presence of antibiotics and the basal level of NSP digestibility (i.e. if NSP digestibility is high in the control animals) have been suggested as factors that limit the response to NSP'ases (Yin et al. 2001) in addition to the suggestion that pigs simply do not suffer the same scale of problems with NSP as poultry (Thacker et al. 1991) due to much lower intestinal viscosities (Jensen et al. 1998).

Age of animal, health status, cereal type and quality (Reilly et al. 2010), dietary components such as fat quantity and type, feed form (pellet vs mash) and the length of the trial all contribute to achieving not only the basal level of performance, but also the scale of response to addition of the enzyme.

The question of enzyme dose and complexity is also a point of debate. Often, papers present data which purports to show the benefits of combinations of different enzyme classes. The theory is that the cell walls that restrict digestion of their contents are complex in their structure and thus require multiple enzyme systems to enable their disassembly. Combinations of xylanases, mannanases, glucanases and pectinases have been shown to increase fibre digestibility in cannulated pigs quite markedly (Shengping et al. 2012), but in many cases it is not clear if each of the component enzymes is optimally dosed and whether each enzyme is actually pure. Addition of a pectinase to a xylanase, for example, may increase the xylanase content of the diet to optimal as a result of a contaminant xylanase activity present in the pectinase. The concept of 'multienzymes' in fact has never been conclusively proven and is challenged in a recent paper (Masey O'Neill et al. 2014). Dose optimisation, however, is clearly of relevance regardless of the mechanism considered. Too little and the effect will not be optimised, and too much and there may be excessive production of antinutritive monosaccharides such as xylose and arabinose.

3.2 Recent developments

Our work suggests that the benefit of NSP'ases in pigs takes a considerable length of time to develop and perhaps the main difference between the species is that much longer periods of feeding are required before a benefit is seen. In a semi-commercial study it was noted that no benefit of a xylanase was noted for the first 50 days of feeding a xylanase, but after this point in time the benefit began to build until the effect at the end of the experiment was significant at six points in FCR. At this point in time the pigs had reached 100 kg. More recent work (Hanczakowska et al. 2012) has shown very similar results, with performance and digestibility effects increasing with both age and dose. Such a developmental response suggests that neither the viscosity nor the cell wall response is dominant and perhaps the prebiotic effect, which would depend upon a more mature large intestine, is dominant. Reduction in diarrhoea and BUN are all indicators of improved microbial balance, and where this has been demonstrated, it seems the benefit takes some time to develop (Tapingkae et al. 2008), again suggesting a prebiotic effect dominates. Changes in fermentation patterns on use of NSP'ases has also, over time, been shown to reduce odours (Bikker et al. 2007), presumably through reduced putrefaction of proteins as the soluble oligomers produced provide CHO to compete with protein for fermentation.

Evidence accumulated to date suggests that soluble, fermentable fibres, especially if viscous, can create problems for young, weaning piglets as the disruption of digestion can result in supply of excess CHO and nitrogen to the large intestine which fuels dysbacteriosis and PWD (Molist et al. 2014). Introduction of the same fibres later on in life, however, when the microbiota has stabilised post-weaning, is beneficial to microbiome stability and drives to more SCFA production and a healthy intestine. Thus, use of exogenous xylanases may need to be delayed for several weeks post-weaning if the benefits of controlled production of fermentable, soluble NSP are to be realised.

It is clear that the response to NSP'ases is more variable in swine than it is in poultry, but the benefits are significant and offset the cost of inclusion manyfold. The use of such enzymes and their value will only increase with better understanding of how these products work and hence how best to extract value from their use.

4 Phytase

Phytase is an enzyme commonly supplemented in pig diets to degrade phytate, the storage form of phosphorus (P) in plant-based feed ingredients. The concentration of phytate in grains may vary (Fig. 1) and typical pig diets contain between 1.5 and 3.0 g/kg phytate P, depending on the ingredient composition of the diet. Through the degradation of phytate, phytase improves P digestibility (Kuhn and Manner 2012) and provides a source of organic P to pigs, while reducing P excretion (Guggenbuhl et al. 2012a) into the environment and dependence on non-renewable inorganic phosphate sources. In a review by Selle and Ravindran (2008), it was estimated that the global pig population consumes seven million tonnes of inorganic (dicalcium) phosphate annually and 64% of total dietary P consumed is lost in the faeces or urine (Poulsen et al. 1999). Extrapolating the data further, the use of phytase to replace 1.0–2.0 g/kg of inorganic P supplied in the diet could reduce inorganic phosphate supplementation by 38–76%. This is equivalent to 2.7–5.3 million tonnes of inorganic phosphate. In addition to the reduced use of inorganic rock phosphate, phytase improves P digestibility by approximately 60% as modelled by Letourneau-Montminy et al. (2011) and thereby reduces phytate P excretion into the environment by 780 000 tonnes per year, assuming pigs eat 1.3 million tonnes of phytate P per year (Selle and Ravindran 2008).

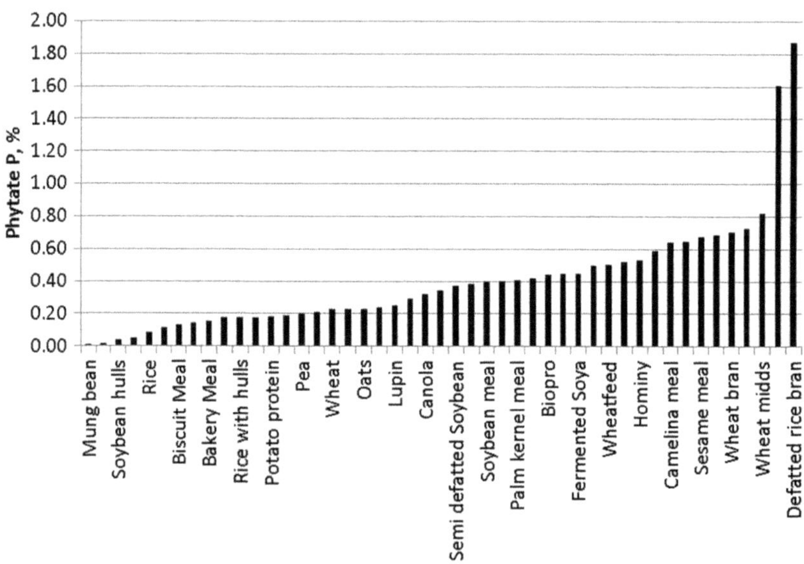

Figure 1 Phytate phosphorus content of grains as determined by NIR (n = 1250).

Benefits associated with phytate destruction and P release by phytase have been reported for over 30 years. Initial experiments with low P diets reported that cereals with high levels of endogenous phytase, such as wheat or triticale, resulted in greater phytate P availability and fewer bone disorders compared with, for example, corn-based diets which contained very little endogenous phytase (Pointillart et al. 1987). The first use of microbial phytase in pigs was reported by Simons et al. (1990), and the subsequent second- and third-generation microbial phytases have evolved to be more stable in the gastric environment and more efficacious at phytate destruction than the original microbial (Menezes-Blackburn et al. 2015) and plant-derived phytases (Zimmermann et al. 2002). Novel uses for these evolved phytases have started to emerge, with 'extra-phosphoric' effects of superdoses of phytase gaining considerable attention in the past few years (Walk et al. 2013, 2015; Santos et al. 2014). These extra-phosphoric effects of phytase are associated with the improvements in utilisation of nutrients other than P and reduced maintenance costs to the animal, which may be attributed to the mitigation of the anti-nutritive effects of dietary phytate.

In addition to being viewed as a source of P due to its six phosphate groups, phytate may also bind to other minerals such as zinc (Zn) and copper (Cu) (Persson et al. 1998) and thereby reduce the solubility and absorption of these minerals by the animal. This increases the excretion of these minerals with the manure. Phytate has also been reported to increase nitrogen (N) excretion (Veum et al. 2007), which may contribute to an increase in ammonia emissions as reviewed by Monteiro et al. (2010). The detrimental effect of phytate on N digestibility may occur because it binds and reduces the solubility of proteins and specific amino acids, increases mucin and endogenous losses in the small intestine, and interferes with endogenous protease (pepsin and pancreatic) activity in the gastrointestinal tract. The consequence of such effects is noted as reduced amino acid digestibility in pigs through the binding with these nutrients and an increase in endogenous losses (Bohlke et al. 2005; Woyengo et al. 2009), which results in reduced growth performance and feed efficiency of pigs (Woyengo et al. 2012). As a result, phytate may cost the pork industry substantially in lost performance and feed efficiency.

Phytate (myo-inositol hexakis-dihydrogen phosphate; IP6), with six phosphate groups, is not the only inositol phosphate molecule that may be considered a significant anti-nutrient and the use of standard doses of phytase will not eliminate this substance. For example, the breakdown of IP6 by standard doses of phytase results in a significant and transient increase in lower phytate esters over time (Greiner et al. 1993; Pontoppidan et al. 2012), such as myo-inositol pentakis- (IP5), tetrakis- (IP4), tris- (IP3), bis- (IP2), and monophosphate (IP1). These phytate esters can also bind to Ca, Zn, Cu and Fe and inhibit pepsin activity (Persson et al. 1998; Yu et al. 2012), suggesting that even these esters may interfere with nutrient utilisation and absorption. However, through the use of superdoses of phytase, near complete destruction of phytate and reduced concentrations of the lower phytate esters have been reported in the stomach (Fig. 2a) and small intestine (Fig. 2b) of grower-finisher pigs when compared with standard doses of phytase (Wilcock et al. 2014). As a consequence of this near complete removal of IP6 and its lower esters, the use of superdoses of phytase has resulted in significant improvements in average daily gain and feed efficiency in pigs (Walk et al. 2013, 2015; Santos et al. 2014; Wilcock et al. 2014), which was associated with improvements in nutrient utilisation and reduced maintenance costs. Minimising maintenance costs will maximise the proportion of daily digestible nutrient intake directed towards growth and subsequently improve feed efficiency (Patience et al. 2015).

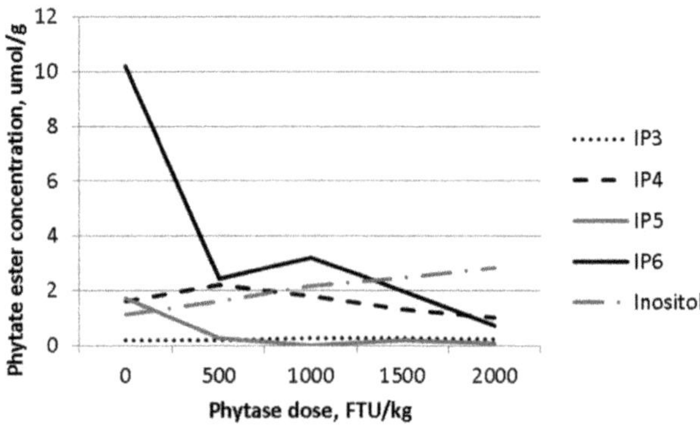

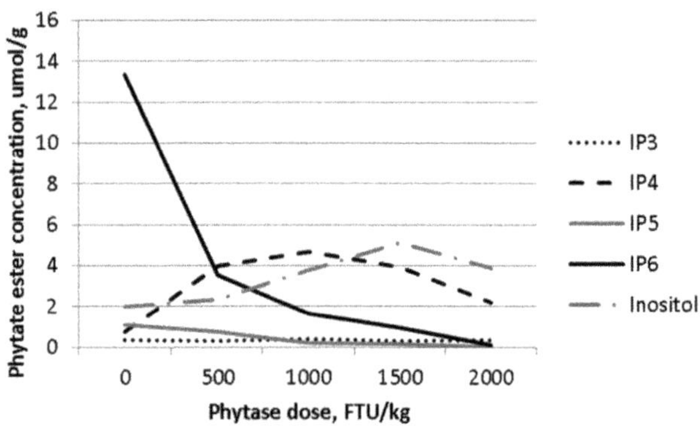

Figure 2 Influence of phytase dose on the concentration of *myo*-inositol hexakis- (IP6), pentakis- (IP5), tetrakis- (IP4), tris- (IP3), and inositol in the stomach (a) or proximal small intestine (b) of 55-kg pigs. Data is a point in time measure. In the stomach (a) there was no effect of phytase dose on IP3 ($P = 0.35$) or IP4 ($P = 0.14$), and a linear ($P < 0.0001$) effect of phytase dose on IP5 and IP6. In the small intestine (b) there was no effect of phytase dose on IP3 ($P = 0.31$), a quadratic ($P < 0.001$) effect of phytase on IP4, and a linear ($P < 0.0001$) effect of phytase dose on IP5 and IP6. There was a linear ($P < 0.01$) effect of phytase dose on inositol in both the stomach and small intestine.

Phytase use to replace inorganic phosphate has significantly improved the sustainability of pork production through the reduction of raw material use coupled with reduced P excretion in manure. This is particularly important in areas of intensive livestock production. In addition, through the use of phytase in pig diets, the excretion of other environmentally important nutrients such as Zn and N may be reduced. For example, significant reductions in ammonia (Smith et al. 2004) and tendencies for reduced total N excretion, particularly from the urine, have been reported in pigs fed phytase (Oryschak et al. 2002). Benefits are not limited to reducing mineral levels required to meet requirements. Recently, Walk et al. (2013, 2015) reported that pharmacological doses of ZnO may be reduced by 25–50% in piglet diets when fed superdoses of phytase with no detriment to performance or health status. While in growing and finishing pigs, Zn supplementation in the pre-mix may be reduced by up to 30% in the presence of phytase (EFSA FEEDAP Panel 2014).

In conclusion, a major step in sustainable pig production occurred approximately 30 years ago with the development and commercialisation of microbial phytases. This resulted in substantial reductions in rock phosphate utilisation and P excretion into the environment. It is estimated that microbial phytases are now supplied in more than 75% of the pig diets globally as a replacement for inorganic P. Further understanding of phytate destruction and subsequent developments in phytase use has led to improvements in P, Zn, Ca and amino acid utilisation, as well as substantial improvements in the feed efficiency of the animal. These benefits are considered 'extra-phosphoric' as they go beyond that of provision of P and include reductions in other environmentally sensitive nutrients such as Zn or N and improvements in the conversion of feed to lean gain. This means less feed inputs in the form of P, Zn or N and also less feed intake for the same meat output and an overall more sustainable and efficiently produced pork product.

5 Proteases

The use of exogenous protease in pig diets is not a new concept. In the early 1950s it was recognised that piglets weaned at one to five weeks of age failed to grow satisfactorily on nutrient-adequate diets due to endogenous enzyme insufficiency. Therefore in a series of experiments with piglets weaned at 6–10 days of age (2.7–3.2 kg), Lewis et al. (1955) fed nutrient-adequate diets with the addition of pancreatin, pepsin, pepsin plus pancreatin, fungal protease, papain, or a proteolytic enzyme derived from *Aspergillus oryzae*. The performance results indicated an increase in relative gain by 19, 24, 29, 24 or 29% with pancreatin, pepsin, pancreatin plus pepsin, papain, or the *Aspergillus oryzae* extract, respectively. However, supplementation of piglet diets with a fungal protease had no effect on gain (Lewis et al. 1955). The authors concluded that the results look promising for both a fundamental and an applied approach into piglet nutrition but more research is required.

Fast forward 60 years and the supplementation of exogenous protease in pig diets is still under investigation with positive, inconclusive, inconsistent or, in some cases, depressive effects on performance. The use of exogenous protease in piglet rations is still of interest, particularly with regard to improving the digestibility of plant-based protein sources. However, recent research focuses on cheaper, local, alternative sources of dietary protein and reducing the impact of protein excretion on the environment. For example, O'Doherty and Forde (1999) and Lorena-Rezende et al. (2012) evaluated the use of protease alone or with other enzymes in pea-based or cottonseed-based grow finisher (26 or 31 kg) rations, respectively. The authors reported no significant effect of protease supplementation on nutrient digestibility (O'Doherty and Forde 1999; Lorena-Rezende et al. 2012), N retention or excretion (Lorenz-Rezende et al. 2012), or growth performance (O'Doherty and Forde 1999). However, the inclusion of 40% peas or 30% cottonseed meal to grower-finisher rations significantly reduced nutrient digestibility and growth performance, indicating these ingredients contain anti-nutritive factors which were not susceptible to degradation by the added proteases.

More recently, O'Shea et al. (2014) evaluated the use of exogenous protease in grower-finisher pigs (34 or 78 kg) fed wheat- and barley-based diets with 21% rapeseed meal (RSM) and 30% wheat dried distillers grains with solubles (DDGs). Exogenous protease supplementation significantly increased apparent ileal N digestibility but significantly decreased apparent total tract N digestibility and overall feed intake by 14 kg and final body weight (BW) by 2.4 kg (O'Shea et al. 2014). There was no effect of protease on urine or faecal N. The authors hypothesised the improvement in N digestibility may be indicative of an increased amino acid uptake in the small intestine and a surplus of free amino acids which may have resulted in a reduction in feed intake via an aminostatic feedback mechanism (O'Shea et al. 2014). However, a dietary surplus of N or free amino acids is typically excreted in the urine or faeces (Monteiro et al. 2010) and there was no effect of protease supplementation on the urinary or faecal N. Therefore, the improvement in ileal N digestibility in this experiment is most likely the result of depressed intake and thus possibly reduced transit rate, rather than absorption or re-partitioning of surplus amino acids or N in the body. This resulted in a reduction in ileal digestible amino acid intake on a daily basis, despite improved amino acid digestibility, which resulted in a significant decrease in feed intake and final BW. By-product grains, such as RSM and wheat DDGs, are considered a more variable source of digestible nutrients due to differences in their production processes. Therefore, the use of exogenous enzymes, specifically proteases in diets with by-product grains or ingredients with highly variable digestible amino acid composition, requires further investigation to prevent losses in performance, increased N excretion and putrefaction by microbes which take advantage of undigested N, which can lead to dysbacteriosis.

Furthermore, in an effort to reduce the impact of pig production on the environment, Zamora et al. (2011) fed pigs (12.2 kg) a standard (20.5%) or low (14.5%) crude protein sorghum soybean meal-based diet supplemented with three doses of exogenous protease and reported no significant effects of protease supplementation on growth performance, carcass characteristics or plasma urea N. This was despite the fact that, in the standard crude protein diet, supplementation of protease resulted in a non-significant improvement in feed conversion ratio (FCR) by approximately 17% (Zamora et al. 2011) and protease supplementation resulted in a non-significant improvement of 12 FCR points in pigs fed pea-based diets (O'Doherty and Forde 1999). Other authors have evaluated the influence of exogenous protease in corn and/or wheat and soybean meal-based diets and reported no significant effects of protease supplementation on pig performance but noted improvements in crude protein and essential and branched-chain amino acid digestibility in pigs 28 days post-weaning (Guggenbuhl et al. 2012b). The lack of correlation between digestibility and performance effects is not uncommon and should not be surprising if intake is simultaneously depressed. Indeed digestibility data in the absence of intake data are of limited value.

In some cases the lack of protease effect on amino acid digestibility may relate to the age and status of the animal. For example no effect of protease supplementation was noted on amino acid digestibility in pigs 14 days post-weaning. The authors hypothesised the endogenous pancreatic protease secretions were in the exponential phase at 14 days post-weaning and the activity of the exogenous protease was not sufficient to augment the animals' own endogenous enzymes in a meaningful manner (Guggenbuhl et al. 2012b). At 28d post-weaning, the addition of the same protease increased total, indispensable and branched-chain amino acid digestibility by 1.4, 1.7 and 4.6%, respectively. Previous authors have reported linear and quadratic increases in gastric mucosa proteolytic activity and pancreatic chymotrypsin and trypsin activity as pigs' age from birth to six weeks, with a drop in enzymatic activity at weaning (Lindemann et al. 1986). Conversely, Caine et al. (1997) reported no significant effect on apparent ileal amino acid digestibility or growth performance of pigs fed synthetic diets with protease-treated soybean meal. However, the authors reported a significant effect of time on apparent ileal amino acid digestibility; digestibility was greater at 16 days post-weaning than at seven days post-weaning. This was especially apparent for the branched-chain amino acids (leucine, isoleucine and valine) in which digestibility increased by 18–20% between day 7 and day 16 post-weaning (Caine et al. 1997) and is in agreement with the increase in pancreatic enzyme activity proposed by Guggenbuhl et al. (2012b). Branched-chain amino acids are recognition sites for pancreatic proteases and thus an increase in branched-chain amino acid digestibility may correspond to an increase in the digestive and absorptive capacity of pigs' gastrointestinal tract (Caine et al. 1997). Therefore, while supplementing piglet diets with exogenous protease may be an effective means of improving apparent ileal amino acid digestibility, the timing of the response and therefore benefits to protease supplementation appear dependent on the animals' own endogenous enzyme capacity. Further research is needed to determine if there is an age-dependent effect of protease supplementation on digestibility and subsequent performance in pigs.

The results presented here suggest there may be FCR benefits associated with feeding proteases although the responses are inconsistent. This may relate to the benefits of protease supplementation being lost if there is an oversupply of dietary protein, un-balanced dietary amino acids associated with novel or by-product feed ingredients, or insufficient absorptive capacity of the animal. Thus, in diets in which exogenous proteases would be most beneficial they may sometimes fail if their effects on creating an amino acid imbalance, overloading absorptive capacity or depressing feed intake are not taken into account. The lack of data supporting the positive effect of protease supplementation on pig performance may also be attributed to a lack of understanding of how each specific protease enzyme works, how this in turn influences amino acid utilisation in the whole body, and subsequently the most appropriate ingredient and nutrient compositions to use when feeding an exogenous protease. A significant amount of work is required to identify the conditions under which specific proteases will consistently bring benefits in pig production, but with ever-increasing costs of protein sources such a breakthrough would be welcomed.

6 Lipolytic enzymes

Weaning is a stressful time for pigs with changes to their external environment, social hierarchy, and physiological and metabolic processes. Of particular interest during weaning is the effect of the diet change on digestive enzyme functions. Lindemann et al. (1986) reported a drop in pancreatic lipase, amylase, chymotrypsin and trypsin activity at weaning. Cera et al. (1990) reported significantly reduced lipase concentration in the pancreas and small intestinal lumen of pigs following weaning at 21 or 35 days of age. While lipase activity appears to be adequate to utilise milk fat in the suckling pig (Cera et al. 1990), the use of exogenous lipases may complement endogenous enzyme function immediately post-weaning, improve fat digestion and subsequently alter the gut microflora.

Lipases and esterases attack the fatty acyl linkage of water-insoluble or -soluble triacylglycerol (Dierick and Decuypere 2002; Dierick et al. 2002a) and can be found in plants, milk and fat products (Dierick and Decuypere 2002), microbial-origin products and secretions from the saliva, stomach, pancreas and small intestine (Dierick et al. 2002a,b). In most mammals, preduodenal lipase activity originates from lingual or gastric secretions and is active over a broad pH range with a preference for medium chain fatty acids (MCFAs; Dierick et al. 2002a). In newborn or suckling animals, MCFAs can provide a rapidly available energy source, stimulate intestinal hormone secretion, and are reported to act as precursors to fatty acids and derivatives with antimicrobial activity (Dierick et al. 2002a). In vitro evaluation of six different lipases from microbial or animal (calf and pig pancreas) origin indicated that microbial lipases were more stable, less selective and more efficient at releasing MCFAs than lipases of animal origin. In vitro work mimicking the poultry glandular stomach and small intestine also indicated that microbial lipases were more pH and endogenous enzyme stable than the mammalian lipases tested (Kermanshahi 1998). This may be due to the fact that lipases of animal origin are dependent on bile salts and co-lipases for activity. However, the efficacy of MCFA release by lipases appears to be dependent on the concentration of MCFAs in the fat source and can be improved with dispersion of the fat to mimic churning of the stomach (Dierick et al. 2002a). Further evaluation of the microbial lipases in in vitro simulated gastric digestion (Dierick et al. 2002a) and in vivo (Dierick et al. 2002b) gastric and duodenal digesta indicated that supplementation of exogenous lipase significantly increased MCFA concentration and reduced total bacterial counts, specifically *Lactobacillus*, *Streptococcus*, and *E. coli*. The authors reported a strong correlation ($r = 0.85$) between the amount of released MCFAs in the stomach and the inhibitory effect on the flora. However, when exogenous lipase was supplemented in piglet diets 7 days pre-weaning to 21 days post-weaning there was no significant effect on average daily gain and a numeric improvement in feed conversion from 1.79 to 1.74 (Dierick et al. 2002). Other authors have reported that lipase supplementation significantly improved total tract crude protein, gross energy and amino acid digestibility in 4.5-kg piglets (Diebold et al. 2004). More recently, Chen et al. (2014) reported supplementation of a coated or uncoated exogenous lipase significantly reduced diarrhoea rate and increased lipase activity in the pancreas and trypsin, lipase and amylase activity in the duodenal digesta content. The authors also reported significant improvements in average daily gain and gain to feed of piglets fed either the coated or the uncoated lipase compared with piglets fed a non-supplemented diet.

The use of exogenous lipase has also been investigated in older pigs to improve utilisation and digestion of rendered animal fats, which contain long-chain saturated fatty acids and are approximately 60–80% digestible depending on the unsaturated to saturated (U/S) fatty acid ratio (Dierick and Decuypere 2004). However, lipase supplementation had no effect on apparent ileal or faecal fat digestibility but significantly improved dry matter, organic matter, crude protein, energy and minor fatty acid (C6:0 and C14:0) digestibility in pigs fed 4% animal fat (Dierick and Decuypere 2004). The authors proposed the lack of an effect of the lipase on fat digestibility may be associated with the high U/S ratio and the rather high initial apparent digestibility value of the fat source used.

Based on the previous results in newly weaned and growing pigs, it could be postulated that the efficacy of exogenous lipase supplementation in pig diets may be beneficial with regard to nutrient digestion and microbial modulation. However, complete degradation of triglycerides into free fatty acids can have negative implications on growth performance and fat digestion as reported in young poultry (Kermanshahi 1998) and therefore the magnitude of the effect and benefits of feeding exogenous lipase will be dependent on a few factors:

1 the concentration of medium chain fatty acids and the U/S ratio of the fat source,
2 the source of exogenous lipase, with those of microbial origin appearing more efficacious, more gastrointestinal stable but less selective than mammalian derived lipases,

3 the pH and mean retention time in the gastrointestinal tract which will influence the solubility of the fatty acids in the fat source and the length of time the lipase has to work, and
4 the animals' own endogenous co-lipase, lipase and bile secretion and activity in the saliva, stomach, pancreas and small intestine, which may be sufficient to effectively utilise dietary fat, with an exception immediately post-weaning.

7 Assays and comparisons

The assay used for NSP'ases are not standardised, most being based on either a reducing sugar release or a colourimetric assay using a specific NSP'ase as a standard. Even with an enzyme class, for example, xylanase, there are various conditions used for not only the assay but also extraction of the enzyme from the feed. As a result there is no common basis upon which these enzymes are compared, which is in stark contrast to the phytases. Even though the conditions of the phytase assay do not allow prediction of efficacy, the industry has been able to associate a given phosphorus matrix for each product based on the FTU due to the fact that almost all products use the same assay. The plethora of assays used in the field of NSP'ase analysis has obfuscated such a view and contributed to the confusion when products are compared. Further, significant problems exist in the extraction of some but not all NSP'ases from compound animal feed, making it difficult to establish. In vivo comparison between NSP'ases are fraught with problems. Some NSP'ases favour soluble and some insoluble NSP and they often differ substantially in their ability to attack substituted NSP. As a result the relative in vivo efficacy of products can vary depending upon the main ingredients and even the samples of ingredients used in the diet. There is also a paucity of published data which clearly establishes an optimum dose of a given product and as a result comparative trials are (Ndou et al. 2015) fraught with problems. Most often they fail in establishing the ranking of such products as they rarely establish that the targeted activity was in fact delivered and do not test whether the dosage of each product employed in that particular diet was the optimum. Consequently, the commercial application of NSP'ases is not as well understood as that of phytases and there is little confidence in understanding how to compare products.

8 Conclusions

Whether the enzyme considered is an NSP'ase, phytase or protease (or indeed any other class of enzyme), the common thread of all enzyme research is that the response to, and hence the value of an enzyme, is variable. This variability is a result of the inconsistencies of animal production as a whole. Animal, ingredient and environmental factors can all conspire to enhance or degrade performance over several cycles even though the intentions may have been to maintain all such factors constant. Given the response to enzymes is clearly dependent upon how well, or poorly, the animal would have performed in the absence of the enzyme, and it is obvious that responses will vary markedly. Nevertheless, the value of feed enzymes are not only that performance on average is improved, or nutrient density can be reduced and performance maintained, but also that they reduce the variation between the best and worst animals and herds. It is evident that the full economic benefit created by exogenous enzymes is not realised in practice. With greater understanding of the factors which influence the response to exogenous enzymes, the more consistent their response in the field will be and the greater the economic return generated.

9 Where to look for further information

A comprehensive review of feed enzymes in monogastric nutrition is provided in *Enzymes in Farm Animal Nutrition, 2nd Edition*. 2010. ISBN 978-1-84593-674-7 available from CABI. This covers all aspects of enzyme production, mode of action, categories and their commercial application in monogastrics and ruminants alike.

There are also a number of review articles that may be of interest and tend to be more focussed on specific enzymes or species, a selection is given below: ADDIN EN.REFLIST

Bedford MR, Schulze H. 1998. EXOGENOUS ENZYMES FOR PIGS AND POULTRY [Review]. *Nutrition Research Reviews*, 11, 91–114

Cowieson AJ, Roos FF. 2016. Toward optimal value creation through the application of exogenous mono-component protease in diets of non-ruminants. *Animal Feed Science & Technology*, http://dx.doi.org/10.1016/j.anifeedsci.2016.04.015

Dersjant-Li Y, Awati A, Schulze H, Partridge G. 2015. Phytase in non-ruminant animal nutrition:a critical review on phytase activities in the gastrointestinal tract and influencing factors. *Journal of the Science of Food & Agriculture*, 95, 878–96

Masey O'Neill HV, Smith JA, Bedford MR. 2014. Multicarbohydrase enzymes for non-ruminants. *Asian-Australasian Journal of Animal Sciences*, 27, 290–301

Woyengo TA, Nyachoti CM. 2011. Review: Supplementation of phytase and carbohydrases to diets for poultry. *Canadian Journal of Animal Science*, 91, 177–92

10 References

Aftab, U. 2012. Exogenous carbohydrase in corn-soy diets for broilers. *World's Poultry Science Journal*, 68, 447–63.

Bach Knudsen, K. E. 2016. Carbohydrates in pig nutrition – recent advances. *Journal of Animal Science*, 94(7), 1–11. doi: 2527/jas.2015-9785.

Bartelt, J., Jadamus, A., Wiese, F., Swiech, E., Buraczewska, L. and Simon, O. 2002. Apparent precaecal digestibility of nutrients and level of endogenous nitrogen in digesta of the small intestine of growing pigs as affected by various digesta viscosities. *Archives of Animal Nutrition*, 56(2), 93–107.

Bedford, M. R. and Schulze, H. 1998. Exogenous enzymes for pigs and poultry. *Nutrition Research Reviews*, 11(1), 91–114.

Bikker, P., Dirkzwager, A., Fledderus, I. J., Trevisi, P., Le Huerou-Luron, I., Lalles, J. P. and Awati, A. 2007. Dietary protein and fermentable carbohydrates contents influence growth performance and intestinal characteristics in newly weaned pigs. *Livestock Science*, 108(1/3), 194–7.

Bohlke, R. A., Thaler, R. C. and Stein, H. H. 2005. Calcium, phosphorus, and amino acid digestibility in low-phytate corn, normal corn, and soybean meal by growing pigs. *Journal of Animal Science*, 83, 2396–403.

Brown, I. 1996. Complex carbohydrates and resistant starch. *Nutrition Reviews*, 54(11), S115–19.

Caine, W. R., Sauer, W. C., Tamminga, S., Verstegen, M. W. and Schulze, H. 1997. Apparent ileal digestibilities of amino acids in newly weaned pigs fed diets with protease-treated soybean meal. *Journal of Animal Science*, 75, 2962–9.

Campbell, G. L. and Bedford, M. R. 1992. Enzyme applications for monogastric feeds: A review. *Canadian Journal of Animal Science*, 72, 449–66.

Cera, K. R., Mahan, D. C. and Reinhart, G. A. 1990. Effect of weaning, week postweaning, and diet composition on pancreatic and small intestinal luminal lipase response in young swine. *Journal of Animal Science*, 63, 384–91.

Chen, S. Y., Liu, Z. X., He, Y. D., Chu, C. and Wang, M. Q. 2014. Effect of coated lipase supplemetation on growth, digestion and intestinal morphology in weaning piglets. *Journal of Animal and Veterinary Advances*, 13, 1093–7.

Choct, M., Dersjant-Li, Y., McLeish, J. and Peisker, M. 2010. Soy oligosaccharides and soluble non-starch polysaccharides: A review of digestion, nutritive and antinutritive effects in pigs and poultry. *Asian-Australasian Journal of Animal Sciences*, 23(10), 1386–98.

Choct, M., Hughes, R. J., Wang, J., Bedford, M. R., Morgan, A. J. and Annison, G. 1996. Increased small intestinal fermentation is partly responsible for the anti-nutritive activity of non-starch polysaccharides in chickens. *British Poultry Science*, 37, 609–21.

Cowieson, A. J. and Bedford, M. R. 2009. The effect of phytase and carbohydrase on ileal amino acid digestibility in monogastric diets: complimentary mode of action? *Worlds Poultry Science Journal*, 65, 609–24.

Danicke, S. 2001. Endogenous N-Losses in piglets estimated by a 15N isotope dilution technique:Effect of xylanase addition to a wheat and rye based diet. *Archives of Animal Nutrition*, 54, 209–23.

Danicke, S., Dusel, G., Jeroch, H. and Kluge, H. 1999. Factors affecting efficiency of NSP-degrading enzymes in rations for pigs and poultry. *Agribiological Research-Zeitschrift fur Agrarbiologie Agrikulturchemie Okologie*, 52(1), 1–24.

Deusch, S., Tilocca, B., Camarinha-Silva, A. and Seifert, J. 2015. News in livestock research – use of Omics-technologies to study the microbiota in the gastrointestinal tract of farm animals. *Computational and Structural Biotechnology Journal*, 13, 55–63.

Diebold, G., Mosenthin, R., Piepho, H.-P. and Sauer, W. C. 2004. Effect of supplementation of xylanase and phospholipase to a wheat-based diet for weanling pigs on nutrient digestibility and concentrations of microbial metabolites in ileal digesta and feces. *Journal of Animal Science*, 82, 2647–56.

Dierick, N. A. and Decuypere, J. A. 2002. Endogenous lipolysis in feedstuffs and compound feeds for pigs: effects of storage time and conditions and lipase and/or emulsifier addition. *Animal Feed Science and Technology*, 102, 53–70.

Dierick, N. A., Decuypere, J. A., Molly, K., Van Beek, E. and Vanderbeke, E. 2002a. The combined use of triacylglycerols containing medium-chain fatty acids (MCFAs) and exogenous lipolytic enzymes as an alternative for nutritional antibiotics in piglet nutrition I. In vitro screening of the release of MCFAs from selected fat sources by selected exogenous lipolytic enzymes under simulated pig gastric conditions and their effects on the gut flora of piglets. *Livestock Production Science*, 75, 129–42.

Dierick, N. A., Decuypere, J. A., Molly, K., Van Beek, E. and Vanderbeke, E. 2002b. The combined use of triacylglycerols (TAGs) containing medium-chain fatty acids (MCFAs) and exogenous lipolytic enzymes as an alternative for nutritional antibiotics in piglet nutrition II. In vivo release of MCFAs in gastric cannulated and slaughter piglets by endogenous and exogenous lipases; effects on the luminal gut flora and growth performance. *Livestock Production Science*, 76, 1–16.

Dierick, N. A. and Decuypere, J. A. 2004. Influence of lipase and/or emulsifier addition on the ileal and faecal nutrient digestibility in growing pigs fed diets containing 4% animal fat. *Journal of the Science of Food and Agriculture*, 84, 1443–50.

Dreschel, H., Hackl, W., Matthes, W., Stolken, B. and Hennig, U. 1999. Investigations on supplements of NSP-hydrolysing enzymes to diets for pigs. *Landbauforschung Volkenrode*, 193, 343–8.

Dritz, S. 2013. Economic evaluation of feed per unit of gain: Is lower always better? *The Pig Site* accessed 26 April 2016.

Durmic, Z., Pethick, D. W., Pluske, J. R. and Hampson, D. J. 1998. Changes in bacterial populations in the colon of pigs fed different sources of dietary fibre, and the development of swine dysentery after experimental infection. *Journal of Applied Microbiology*, 85, 574–82.

Dusel, G., Schulze, H., Kluge, H., Simon, O. and Jeroch, H. 1997. The effect of wheat quality measured by extract viscosity and dietary addition of feed enzymes on performance of young pigs. *Journal of Animal Science*, 75(Suppl 1), 200.

EFSA FEEDAP Panel (EFSA Panel on Additives and Products or Substances used in Animal Feed). 2014. Scientific Opinion on the potential reduction of the currently authorised maximum zinc content in complete feed. *EFSA Journal*, 12, 77.

Flis, M. and Sobotka, W. 2005. Fine particle size and enzyme supplementation as factors improving utilization of protein from diets with lowered protein contents by pigs. *Journal of Animal and Feed Sciences*, 14(Suppl 1), 341–4.

Gdala, J., Johansen, H. N., Knudsen K. E. B., Knap, I. H., Wagner, P. and Jorgensen, O. B. 1997. The digestibility of carbohydrates, protein and fat in the small and large intestine of piglets fed non-supplemented and enzyme supplemented diets. *Animal Feed Science & Technology*, 65, 15–33.

Greiner, R., Konietzny, U. and Jany, K. D. 1993. Purification and characterisation of two phytases from *Escherichia* coli. *Archives of Biochemistry and Biophysics*, 303, 107–13.

Guggenbuhl, P., Wache, Y., Simoes Nunes, C. and Fru, F. 2012a. Comparative effects of three phytases on the phosphorus and calcium use in the weaned piglet. *Journal of Animal Science*, 90, 95–7.

Guggenbuhl, P., Wache, Y. and Wilson, J. W. 2012b. Effects of dietary supplementation with a protease on the apparent ileal digestibility of the weaned piglet. *Journal of Animal Science*, 90, 152–4.

Hanczakowska, E., Swiatkiewicz, M. and Kuhn, I. 2012. Efficiency and dose response of xylanase in diets for fattening pigs. *Annals of Animal Science*, 12(4), 539–48.

Jansman, A. J. M., Schulze, H., Bosch, M. W. and Leuvenink, H. 1996. Effect of xylanase supplementation to a wheat based diet on the portal fluxes of glucose and volatile fatty acids in pigs. *Animal Science*, 74(Suppl 1), 191.

Jensen, M. S., Knudsen, K. E., Inborr, J. and Jakobsen, K. 1998. Effect of b glucanase supplementation on pancreatic enzyme activity and nutrient digestibility in piglets fed diets based on hulled and hulless barley varieties. *Animal Feed Science and Technology*, 72, 329–45.

Jha, R., Woyengo, T. A., Li, J., Bedford, M. R., Vasanthan, T. and Zijlstra, R. T. 2015. Enzymes enhance degradation of the fiber-starch-protein matrix of distillers dried grains with solubles as revealed by a porcine in vitro fermentation model and microscopy. *Journal of Animal Science*, 93, 1–13.

Johansen, H. N., Wood, P. J. and K. E. Bach Knudsen. 1993. Molecular weight changes in the (1->3)(1->4) beta-D-Glucan of Oats incurred by the digestive processes in the upper gastrointestinal tract of pigs. *Journal of Agricultural and Food Chemistry*, 41, 2347–52.

Kühn, I. and Manner, K. 2012. Performance and apparent total tract phosphorus and calcium digestibility in grower-finisher pigs fed diets with and without phytase. *Journal of Animal Science*, 90, 143–5.

Kermanshahi, H. 1998. The potential of dietary lipases to improve fat utilization in young birds. PhD thesis Department of Animal and Poultry Science, University of Saskatchewan.

Lalles, J. P. 2006. High-viscosity carboxymethylcellulose reduces carbachol-stimulated intestinal chloride secretion in weaned piglets fed a diet based on skimmed milk powder and maltodextrin. *British Journal of Nutrition*, 95(3), 488–95.

Le, D. M., Fojan, P., Azem, E., Pettersson, D. and Pedersen, N. R. 2013. Visualization of the anticaging effect of ronozyme WX xylanase on wheat substrates. *Cereal Chemistry*, 90(5), 439–44.

Letourneau-Montminy, M. P., Narcy, A., Lescoat, P., Magnin, M., Bernier, J. F., Sauvant, D., Jondreville, C. and Pomar, C. 2011. Modeling the fate of dietary phosphorus in the digestive tract of growing pigs. *Journal of Animal Science*, 89, 3596–611.

Lewis, C. J., Catron, D. V., Liu, C. H., Speer, V. C. and Ashton, G. C. 1955. Enzyme supplementation of baby pig diets. *Journal of the Science of Food and Agriculture*, 3, 1047–50.

Lindemann, M. D., Cornelius, S. G., El Kandelgy, S. M., Moser, R. L. and Pettigrew, J. E. 1986. Effect of age, weaning and diet on digestive enzymes levels in the piglet. *Journal of Animal Science*, 62, 1298–307.

Lorena-Rezende, I. M. B., Dutra Jr., W. M., Rezende, F. M., Palhares, L. O., Ludke, M. C. M. M. and Rabello, C. B. V. 2012. Digestibility of cottonseed meal with or without addition of protease and phytase enzymes in swine diet. *Acta Scientiarum*, 34, 259–65 (Portuguese).

Masey O'Neill, H. V., Smith, J. A. and Bedford, M. R. 2014. Multicarbohydrase enzymes for non-ruminants. *Asian-Australasian Journal of Animal Sciences*, 27(2), 290–301.

May, K., O'Sullivan, S. E., Brameld, J. M., Masey O'Neill, H. V., Parr, T. and Wiseman, J. 2015. Xylanase supplementation in feed reduces incretin and PYY levels in piglets. *Proceedings of the Nutrition Society*, 74, E294.

McDonald, D. E., Pethick, D. W., Mullan, B. P. and D. J. Hampson. 2001. Increasing viscosity of the intestinal contents alters small intestinal structure and intestinal growth, and stimulates proliferation of enterotoxigenic Escherichia coli in newly-weaned pigs. *British Journal of Nutrition*, 86(4), 487–98.

Menezes-Blackburn, D., Gabler, S. and Greiner, R. 2015. Performance of seven commercial phytases in an in vitro simulation of poultry digestive tract. *Journal of the Science of Food and Agriculture*, 63, 6142–9.

Molist, F., van Oostrum, M., Perez, J. F., Mateos, G. G., Nyachoti, C. M. and van der Aar, P. J. 2014. Relevance of functional properties of dietary fibre in diets for weanling pigs. *Animal Feed Science and Technology*, 189, 1–10. doi: 10.1016/j.anifeedsci.2013.12.013.

Monteiro, D. O., Pinheiro, V. M. C., Mourao, J. L. M. and Rodrigues, M. A. M. 2010. Strategies for mitigation of nitrogen environmental impact from swine production. *Revista Brasileira de Zootecnia*, 39, 317–25.

Ndou, S. P., Kiarie, E., Agyekum, A. K., Heo, J. M., Romero, L. F., Arent, S., Lorentsen, R. and C. M. Nyachoti. 2015. Comparative efficacy of xylanases on growth performance and digestibility in growing pigs fed wheat and wheat bran- or corn and corn DDGS-based diets supplemented with phytase. *Animal Feed Science and Technology*, 209, 230–9.

O'Doherty, J. V. and Forde, S. 1999. The effect of protease and alpha-galactosidase supplementation on the nutritive value of peas for growing and finishing pigs. *Irish Journal of Agricultural Food Research*, 38, 217–26.

Olukosi, O. A., Bedford, M. R. and Adeola, O. 2007. Xylanase in diets for growing pigs and broiler chicks. *Canadian Journal of Animal Science*, 87(2), 227–35.

Oryschak, M. A., Simmins, P. H. and Zilstra, R. T. 2002. Effect of dietary particle size and carbohydrase and/or phytase supplementation on nitrogen and phosphorus excretion of grower pigs. *Canadian Journal of Animal Science*, 82, 533–40.

O'Shea, C. J., Mc Alpine, P. O., Solan, P., Curran, T., Varley, P. F., Walsh, A. M. and O'Doherty, J. V. 2014. The effect of protease and xylanase enzymes on growth performance, nutrient digestibility, and manure odour in grower-finisher pigs. *Animal Feed Science and Technology*, 189, 88–97.

Parkkonen, T., Tervila-Wilo, A., Hopeakoski-Nurminen, M., Morgan, A. J., Poutanen, K. and Autio, K. 1997. Changes in wheat microstructure following in vitro digestion. *Acta Agriculturae Scandinavica*, 47, 43–7

Patience, J. F., Rossoni-Serao, M. C. and Gutierrez, N. A. 2015. A review of feed efficiency in swine: biology and application. *Journal of Animal Science and Biotechnology* 6:33. doi: 10.1186/s40104-015-0031-2.

Persson, H., Turk, M., Nyman, M. and Sandberg, A. S. 1998. Binding of Cu2+, Zn2+, and Cd2+ to inositol tri-, tetra-, penta-, and hexaphosphates. *Journal of Agricultural and Food Chemistry*, 46, 3194–200.

Pointillart, A., Fourdin, A. and Fontaine, N. 1987. Importance of cereal phytase activity for phytate phosphorus utilisation by growing pigs fed diets containing triticale or corn. *Journal of Nutrition*, 117, 907–13.

Pontoppidan, K., Glitsoe, V., Guggenhuhl, P., Quintana, A. P., Nunes, C. S., Pettersson, D. and Sandberg, A. S. 2012. In vitro and in vivo degradation of myo-inositol hexakisphosphate by a phytase from Citrobacter braakii. *Archives of Animal Nutrition*. 66, 431–44.

Poulsen, H. D., Jongbloed, A. W., Latimer, P. and Fernandez, J. A. 1999. Phosphorus consumption, utilisation and losses in pig production in France, The Netherlands, and Denmark. *Livestock Production Science*, 58, 251–9.

Pujol, S. and Torrallardona, D. 2000. Ileal and Faecal digestibility of nutrients in piglets fed non-supplemented and enzyme-supplemented barley diets. In J. E. Lindberg and B. Ogle (Eds),

Digestive Physiology of Pigs, Proceedings of the 8th Symposium edn, Swedish University of Agricultural Sciences, Uppsala, Sweden, June 20–22, CABI, pp. 181–3.
Reilly, P., Sweeney, T., O'Shea, C., Pierce, K. M., Figat, S., Smith, A. G., Gahan, D. A. and O'Doherty, J. V. 2010. The effect of cereal-derived beta-glucans and exogenous enzyme supplementation on intestinal microflora, nutrient digestibility, mineral metabolism and volatile fatty acid concentrations in finisher pigs. *Animal Feed Science & Technology*, 158, 165–76.
Rodehutscord, M., Ruckert, C., Maurer, H. P., Schenkel, H., Schipprack, W., Bach Knudsen, K. E., Schollenberger, M., Laux, M., Eklund, M., Siegert, W. and R. Mosenthin. 2016. Variation in chemical composition and physical characteristics of ceral grains from different genotypes. *Archives of Animal Nutrition*, 70(2), 87–107.
Santos, T. T., Walk, C. L., Wilcock, P., Cordero, G. and Chewning, J. 2014. Performance and bone characteristics of growing pigs fed diets marginally deficient in available phosphorus and a novel microbial phytase. *Canadian Journal of Animal Science*, 94, 493–7.
Schutte, J. B. 1990. Nutritional implications and metabolizable energy value of D- Xylose and L-Arabinose in chicks. *Poultry Science*, 69, 1724–30.
Schutte, J. B., Beelan, G. M., Derkson, G. B. and Wiebenga, J. 1991. 'Digestive & Utilization of D-xylose in pigs as affected by age, frequency of feeding and dietary level. In M. W. A. Verstegen and J. Huisman (Eds), *Digestive Physiology in Pigs*, Centre for Agri. publishing & Documentation, Wageningen, N. L., pp. 411–21.
Selle, P. H. and Ravindran, V. 2008. Phytate-degrading enzymes in pig nutrition. *Livestock Science*, 113, 99–122.
Sergeant, M. J., Constantinidou, C., Cogan, T. A., Bedford, M. R., Penn, C. W. and Pallen, M. J. 2014. Extensive microbial and functional diversity within the chicken cecal microbiome. *PLoS One*, 9(3), e91941–13.
Shengping, W., Fang, J., Xiangshan, T., Jie, W., Zhan, Z., Zhou, Y., Wu, X., Su, D., Liu, G., Ding, J. and Cai, L. 2012. Non-starch polysaccharides (NSP) Enzyme improves the nutrient digestibility of weaned piglet fed a cottonseed meal based diets. *Journal of Animal and Veterinary Advances*, 11(16), 2976–9.
Simons, P. C. M., Versteegh, H. A. J., Jongbloed, A. W., Kemme, P. A., Slump, P., Bos, K. D., Wolters, M. G. E., Beudeker, R. F. and Verschoor, G. J. 1990. Improvement of phosphorus availability by microbial phytase in broilers and pigs. *British Journal of Nutrition*, 64, 525–40.
Smith, D. R., Moore Jr., P. A., Haggard, B. E., Maxwell, C. V., Daniels, T. C., VanDevander, K. and Davis, M. E. 2004. Effect of aluminium chloride and dietary phytase on relative ammonia losses from swine manure. *Journal of Animal Science*, 82, 605–11.
Tapingkae, W., Yachai, M., Visessanguan, W., Pongtanya, P. and Pongpiachan, P. 2008. Influence of crude xylanase from *Aspergillus niger* FAS128 on the *in vitro* digestibility and production performance of piglets. *Animal Feed Science & Technology*, 140, 141–54.
Tervila-Wilo, A., Parkkonen, T., Morgan, A. J., Hopeakoski-Nurminen, M., Poutanen, K., Heikkinen, P. and Autio, K. 1996. In vitro digestion of wheat microstructure with xylanase and cellulase from Trichoderma reesei. *Journal of Cereal Science*, 24, 215–25.
Thacker, P. A., Campbell, G. L. and Grootwassink, J. 1991. The effect of enzyme supplementation on the nutritive value of rye-based diets for swine. *Canadian Journal of Animal Science*, 71, 489–96.
Torrallardona, D., Nielsen, J. E. and Brufau, J. 2000. Apparent ileal digestibility of protein and amino acids in wheat supplemented with enzymes for growing pigs. In J. E. Lindberg & B. Ogle (Eds), *Digestive Physiology of Pigs*, Proceedings of the 8th Symposium edn, Swedish University of Agricultural Sciences, Uppsala, Sweden, June 20–22, CABI, pp. 184–6.
Veum, T. L., Ledoux, D. R. and Raboy, V. 2007. Low-phytate barley cultivars improve the utilisation of phosphorus, calcium, nitrogen, energy and dry matter in diets fed to young swine. *Journal of Animal Science*, 85, 961–71.
Van Lunen, T. A. and Schulze, H. 1996. Influence of Trichoderma longibrachiatum xylanase supplementation of wheat and corn based diets on growth performance of pigs. *Canadian Journal of Animal Science*, 76, 271–3.

Verstegen, M. W. A., Schutte, J. B., Hel, W. V. D., Polziehn, R., Schrama, J. W. and Sutton, A. L. 1997. Dietary xylose as an energy source for young pigs. *Journal of Animal Physiology and Animal Nutrition*, 77, 180–8

Walk, C. L., Srinongkote, S. and Wilcock, P. 2013. Influence of a microbial phytase and zinc oxide on young pig growth performance and serum minerals. *Journal of Animal Science*, 91, 286–91.

Walk, C. L., Wilcock, P. and Magowan, E. 2015. Evaluation of the effects of pharmacological zinc oxide and phosphorus source on weaned piglet growth performance, plasma minerals, and mineral digestibility. *Animal* 9, 1145–52.

Wilcock, P., Bradley, C. L., Chewning, J. J. and Walk, C. L. 2014. The effect of superdosing phytase on inositol and phytate concentration in the gastrointestinal tract and its effect on pig performance. *Journal of Animal Science*, suppl (abstract), 92, 383.

Woyengo, T. A., Cowieson, A. J., Adeola, O. and Nyachoti, C. M. 2009. Ileal digestibility and endogenous flow of minerals and amino acids: responses to dietary phytic acid in piglets. *British Journal of Nutrition*, 102, 428–33.

Woyengo, T. A., Weihrauch, D. and Nyachoti, C. M. 2012. Effect of dietary phytic acid on performance and nutrient uptake in the small intestine of piglets. *Journal of Animal Science*, 90, 543–9.

Yin, Y.-L., McEvoy, J. D., Schulze, H. and K. J. McCracken. 2001. Effects of xylanase and antibiotic addition on ileal and faecal apparent digestibilities of dietary nutrients and evaluating HCl-insoluble ash as a dietary marker in growing pigs. *Animal Science*, 72, 95–103.

Yu, S., Cowieson, A., Gilbert, C., Plumstead, P. and Dalsgaard, S. 2012. Interactions of phytate and myo-inositol phosphate esters (IP1-5) including IP5 isomers with dietary protein and iron and inhibition of pepsin. *Journal of Animal Science*, 90, 1824–32.

Zamora, V., Figueroa, J. L., Reyna, L., Cordero, J. L., Sanchez-Torres, M. T. and Martinez, M. 2011. Growth performance, carcass characteristics and plasma urea nitrogen concentration of nursery pigs fed low-protein diets supplemented with glucomannans or protease. *Journal of Applied Animal Rersearch*, 39, 53–6.

Zimmermann, B., Lantzsch, H. J., Mosenthin, R., Sconer, F. J., Biesalski, H. K. and Drochner, W. 2002. Comparative evaluation of the efficacy of cereal and microbial phytases in growing pigs fed diets with marginal phosphorus supply. *Journal of the Science of Food and Agriculture*, 82, 1298–304.

Chapter 12

The use of growth promoters in pig nutrition

John M. Brameld, David M. Brown and Tim Parr, University of Nottingham, UK

1 Introduction

1.1 Growth promoters and their commercial use

The term 'growth promoter' is used to describe an agent that induces growth/anabolism when given to animals and includes beta-adrenergic agonists (BAs), anabolic steroids and growth hormone (GH). All of them tend to increase lean deposition and decrease fat deposition, which is why they are also sometimes referred to as metabolic modifiers. Despite the term 'growth promoter', they do not necessarily always result in increased growth rates, but they do tend to alter body composition and improve feed efficiency, as determined by the gain-to-feed ratio.

In relation to pig production, only BA and GH (also called somatotropin or ST) are used commercially, because anabolic steroids appear to have reduced efficacy in pigs compared to ruminants (Heitzman 1979). Use of all growth promoters in commercial livestock production was banned across the European Union in 1988. However, some growth promoters are licenced (i.e. legal) for use in livestock across other parts of the world, including North and South Americas, Southern Africa, India, Australia, New Zealand and some parts of Asia. For example, the United States allows the use of the BA, ractopamine, for pig production and China allows the use of GH (recombinant porcine GH, rpGH), but not BA, while Australia and New Zealand allow the use of both. There is a lack of information on the commercial use of growth promoters in some parts of the world though.

http://dx.doi.org/10.19103/AS.2017.0013.12

1.2 History and safety issues

Synthetic BAs were originally developed as bronchodilators for the treatment of asthma (Solis-Cohen 1900). Over 80 years later, it was shown that the above therapeutic doses of BA also stimulate a potent nutrient-repartitioning effect, increasing the deposition of lean mass, while limiting the accretion of fat mass (Baker et al. 1984; Emery et al. 1984). Importantly, the growth advantages of BA could be achieved by oral administration of the drug, making it ideal for use as a feed additive to improve livestock growth performance.

In relation to the discovery of GH, studies in the 1930s showed that pituitary gland extracts could increase the growth rates in animals. The structure of human GH was identified in 1966 by the Chinese-born biochemist, Choh Hao Li, working in the United States. Much of the work demonstrating the anabolic effects of GH in pigs was carried out in the 1970s and 1980s, with the first pGH transgenic pig being described in 1988 (Vize et al. 1988).

In 1988, the complete ban on the use of growth promoters across Europe related to concerns about the safety of the meat for human consumption, since residues were sometimes found in the meat from animals following recent administration. However, subsequent studies with ractopamine have shown that meat contains very low levels of residues, with only liver and kidney containing levels above the recommended threshold. For this reason, there is no requirement for a withdrawal period prior to slaughter. In 1999, ractopamine became the first FDA-approved BA for use in finishing pigs in the United States.

In some countries, growth promoters have now been used for over 20 years, with no apparent increase in reports of human safety issues, suggesting that they are relatively safe when used according to the recommended guidelines. In the case of GH, the treatment is a protein hormone administered by injection, meaning the risks following human consumption of the meat are thought to be low, since any remaining residual GH would be digested by the proteolytic enzymes in the gut, as it happens to all dietary proteins.

2 Beta-adrenergic agonists

2.1 Introduction to beta-adrenergic receptors and different types of agonists

Synthetic BAs are analogues of the endogenous catecholamines, adrenaline and noradrenaline, and, therefore, bind to the beta-adrenergic receptors. There are three subtypes of beta-adrenergic receptors (beta-1, beta-2 and beta-3), which display tissue-specific enrichment.

It is commonly reported that skeletal muscle contains predominantly beta-2 receptors, with only a small proportion of beta-1 receptors (Kim et al. 1991; Lynch and Ryall 2008). However, in pig skeletal muscle, beta-1 and beta-2 receptors display much closer relative proportions at an approximately 60:40 ratio, respectively (Liang and Mills 2002). The anabolic effect of BAs on skeletal muscle has mostly been attributed to signalling via the beta-2 receptor subtype (Choo et al. 1992; Hagg et al. 2016). As such, beta-2 selective agonists such as clenbuterol and formoterol (Baker 2010) have been extensively examined and observed to display excellent anabolic effects *in vivo*, even at low doses (Ryall et al.

2006; Burniston et al. 2006). To date, ractopamine is the only BA approved for use in pig production. Ractopamine is a mixed beta-1 and beta-2 receptor agonist (Mills et al. 2003) that improves pig carcass composition by increasing lean mass and reducing fat mass (Apple et al. 2007).

For a detailed account of downstream beta-adrenergic receptor signalling, we refer the reader to an excellent review by Lynch and Ryall (2008). Briefly, ligand binding at the beta-receptor is followed by activation of the adenylate cyclase pathway, which generates intracellular cyclic AMP (cAMP) and subsequent activation of protein kinase A (PKA). The cAMP response element-binding protein is then activated by PKA, causing transcription of genes containing cAMP response elements. The PKA also phosphorylates numerous proteins involved in calcium handling, glycogen mobilization and proteolytic enzyme activity. The cAMP–PKA pathway remains the most well-documented beta-adrenergic receptor signalling pathway and is considered at least partially responsible for the muscle hypertrophic effects.

2.2 Effects of BA on growth and feed efficiency

The majority of studies investigating the effects of administering the commercially available BA, ractopamine, to growing pigs were performed or published between 1991 and 2014 (see Table 1). Many, but not all, show an increase in growth rates (ADG) as well as an increase in feed efficiency (gain-to-feed ratio), with some studies also showing a decrease in feed intake. The majority of studies are in barrows (i.e. castrated males), with fewer studies in gilts (young females) and very few in intact boars. It was observed in a recent meta-analysis on ractopamine studies (Pompeu et al. 2017) that the barrows showed a greater response than the gilts. One study (Dunshea et al. 1993) directly compared the responses in barrows, gilts and boars and showed improvements in growth and feed efficiency in barrows and gilts, but no effects in boars. Ractopamine is normally mixed in with the feed at doses ranging from 5 to 20 ppm (mg/kg feed), with the normal commercial use being 10 ppm. The majority of studies administer ractopamine for four weeks towards the end of the finishing period (60–100 kg). It should be noted that some studies included a control group where ractopamine was not included in the diet, but the diet contained a reduced protein content (either expressed as % crude protein or % lysine). Hence, the magnitude of the growth and feed efficiency responses are often greater in those studies compared with studies where the control diet was the same as the ractopamine group.

2.3 Mechanisms of action in skeletal muscle

Regulation of protein synthesis and degradation

Post-natal skeletal muscle growth is largely accounted for by hypertrophy (increased muscle fibre size), rather than hyperplasia (increased muscle fibre number). Accordingly, BA perturbs proteostasis in favour of net protein accretion via increased protein synthesis (Emery et al. 1984; Bardocz et al. 1992) and/or decreased protein degradation (Reeds et al. 1986; Bohorov et al. 1987; Higgins et al. 1988; Parr et al. 1992). The BA-mediated alterations to proteostasis likely occur in a time-dependent manner (Koopman et al. 2010).

Activation of v-akt murine thymoma viral oncogene homologue (Akt) – mechanistic target of rapamycin (mTOR) signalling (a potent regulator of both protein synthesis and

Table 1 Summary of studies investigating effects of ractopamine on average daily gain (ADG), average daily feed intake (ADFI) and feed efficiency (gain-to-feed ratio) in growing pigs

Reference	Sex	Breed	Housing	Length of study	CP in diet	DE or ME of diet	How fed?	Ractopamine	ADG	ADFI	Feed efficiency
Gu et al. 1991	Barrow	Various cross-breeds	Individual	59–100, 73–114 or 86–114 kg	18.50%	3.59 Mcal/kg	ad lib	20 ppm	NS	NS	NS
Mitchell et al. 1991	Barrow	Yorkshire × Duroc	Individual	60–103 kg	17.60%	3.52 Mcal/kg	ad lib	20 ppm	NS	Decreased	NS
Mitchell et al. 1991	Barrow	Yorkshire × Duroc	Individual	60–103 kg	11.61%	3.68 Mcal/kg	ad lib	20 ppm	NS	Decreased	NS
Mitchell et al. 1991	Barrow	Yorkshire × Duroc	Individual	60–103 kg	17.60%	3.52 Mcal/kg	Rest (67%) 1x/d	30 ppm	NS	NS	NS
Bark et al. 1992	Barrow	High lean growth	Individual	63–104 kg	16.70%	3.20 Mcal/kg	ad lib	20 ppm	Increased ($P = 0.01$, $C = 0.89$ kg/d)	NS	Increased ($P = 0.01$)
Bark et al. 1992	Barrow	Low lean growth	Individual	63–104 kg	16.70%	3.20 Mcal/kg	ad lib	20 ppm	Increased ($P = 0.01$, $C = 0.63$ kg/d)	NS	Increased ($P = 0.01$)
Dunshea et al. 1993	Barrow	LW × Landrace	Individual	60–90 kg	18.3% (1% Lys)	3.47 Mcal/kg	ad lib	20 ppm	Increased ($P < 0.01$, $C = 1.16$ kg/d)	NS	Increased ($P < 0.01$)
Dunshea et al. 1993	Gilt	LW × Landrace	Individual	60–90 kg	18.3% (1% Lys)	3.47 Mcal/kg	ad lib	20 ppm	Increased ($P < 0.01$, $C = 1.03$ kg/d)	NS	Increased ($P < 0.01$)
Dunshea et al. 1993	Boar	LW × Landrace	Individual	60–90 kg	18.3% (1% Lys)	3.47 Mcal/kg	ad lib	20 ppm	NS	NS	NS
Williams et al. 1994	Barrow	Cross-breed	Individual	67.8–104 kg	16.3–21.9%	3.29–3.68 Mcal/kg	Once/d (rest)	44.7 mg/d	Increased ($P < 0.01$, $C = 0.76–0.90$ kg/d)	NS	Increased ($P < 0.01$)
Williams et al. 1994	Gilt	Cross-breed	Individual	67.8–104 kg	16.3–21.9%	3.29–3.68 Mcal/kg	Once/d (rest)	44.7 mg/d	Increased ($P < 0.01$, $C = 0.74–0.88$ kg/d)	NS	Increased ($P < 0.01$)
Crome et al. 1996	Barrow & Gilt	Eli-Lilly cross-breed	Groups of 4 (2 x B + 2 x G)	68–107 kg	16.30%	3.56 Mcal/kg	ad lib	10 or 20 ppm	Increased in linear manner (C = 0.81 kg/d)	Decreased (linear)	Increased (linear)

Crome et al. 1996	Barrow & Gilt	Eli-Lilly cross-breed	Groups of 4 (2 x B + 2 x G)	85–125 kg	16.30%	3.56 Mcal/kg	ad lib	10 or 20 ppm	Increased in linear manner (*C* = 0.84kg/d)	Decreased (linear)	Increased (linear)
Lutz and Stahly 1997	Gilt	High lean growth	Individual	70–114 kg	?	?	ad lib	20 ppm	Increased ($P = 0.01$)	NS	Increased ($P = 0.01$)
Armstrong et al. 2004	Barrow	Dekalb cross-breed	Individual	6–34d (all 110 kg at end)	18.50%	3.22 Mcal/kg	ad lib	5 ppm	Increased (all except 27d time point, *C* = 0.85–0.95 kg/d)	Increased d34 only	Increased (all time points)
Armstrong et al. 2004	Barrow	Dekalb cross-breed	Individual	6–34d (all 110 kg at end)	18.50%	3.22 Mcal/kg	ad lib	10 ppm	Increased (all except 34d time point, *C* = 0.85–0.95 kg/d)	NS	Increased (all time points)
Armstrong et al. 2004	Barrow	Dekalb cross-breed	Individual	6–34d (all 110 kg at end)	18.50%	3.22 Mcal/kg	ad lib	20 ppm	Increased (all except 6d time point, *C* = 0.85–0.95 kg/d)	Decreased d6 only	Increased (all time points)
See et al. 2004	Barrow & Gilt	Newshams hybrid	Groups of 5 (same sex)	71–109 kg (6 wks)	20.65%	3.77Mcal/kg	ad lib	11.7 ppm	Increased ($P < 0.05$, *C* = 0.93kg/d)	NS	Increased (wks 0–4, not 5–6)
Barker et al. 2005	Barrow	PIC cross-breed	Groups of 12	21d (starting at 100kg)	1% Lys	3.28 Mcal/kg	ad lib (self-feeder)	10 ppm	Increased ($P = 0.01$, *C* = 0.80 kg/d)	NS	Increased ($P = 0.01$)
Barker et al. 2005	Gilt	PIC cross-breed	Groups of 12	21d (starting at 100 kg)	1% Lys	3.28 Mcal/kg	ad lib (self-feeder)	10 ppm	Increased ($P = 0.01$, *C* = 0.80 kg/d)	NS	Increased ($P = 0.01$)
Carr et al. 2005b	Barrow & Gilt	PIC cross-breed	Groups of 4 (same sex)	80–109 kg	16.5–18.4%	2.96–3.36 Mcal/kg	ad lib	10 ppm	NS	Decreased	Increased ($P < 0.05$)
Mimbs et al. 2005	Barrow	Cross-breed	Groups of 5	28d (starting at 80 kg)	19.00%	3.35 Mcal/kg	ad lib	10 ppm	NS	Decreased ($P < 0.01$)	Increased ($P < 0.01$)
Patience et al. 2009	Barrow & Gilt	PIC 337 × C22	Group housed	28d (starting at 86 kg)	(0.63–0.83% Lys)	3.3 Mcal/kg	ad lib	5 ppm	Increased ($P < 0.001$)	NS	Increased ($P < 0.05$)
Rikard-Bell et al. 2009a	Boar & Gilt	Myora LW and Myora Landrace	groups of 11 or 12	31d (starting at 17wks of age)	(0.56 g lys)	3.15 Mcal/kg	ad lib	5 and 10 ppm	Increased ($P = 0.002$)	NS	Increased ($P < 0.001$)
Bundy et al. 2011	Barrow	Cross-breed	Individual	4wks (176lb at start)	13.33%	1.52 Mcal/lb	ad lib	10 ppm	NS	NS	NS

(*Continued*)

Table 1 (*Continued*)

Reference	Sex	Breed	Housing	Length of study	CP in diet	DE or ME of diet	How fed?	Ractopamine	ADG	ADFI	Feed efficiency
Hinson et al. 2011	Barrow	PIC C22 × 337	Individual	21d (starting at 100 kg)	13.14–19.07%	3.32–3.54 Mcal/kg	ad lib	7.4 ppm	Increased (P < 0.001, 1.2 vs. 1.0 kg/d)	Decreased (P = 0.069, 3.07 v 3.29 kg/d)	Increased (P < 0.001)
Bohrer et al. 2013	Barrow & Gilt	Berkshire	Groups of ten (mixed sex)	28d (starting 80kg)	13.1–17.1%	3.49–3.54 Mcal/kg	ad lib	7.4 and 10 ppm	Increased (P < 0.01)	NS	Increased (P < 0.01)
Brana et al. 2013	Gilt & Barrow	PIC 337 × C23	Groups of 22 (same sex)	28d (starting at 126d of age)	17%	3.22 Mcal/kg	ad lib	5.5 ppm	Increased (P < 0.05)	Increased in gilts; NS in castrates	Increased (P < 0.05)
Garbossa et al. 2013	Barrow & Gilt	Cross-breed	Groups of 2 (mixed sex)	28d (starting at 77 kg)	16%	3.3 Mcal/kg	ad lib	5, 10, 15, 20 ppm	Increase in linear manner (P < 0.01)	NS	Increased (linear response plateau)
Gerlemann et al. 2014	Barrow and Gilt	PIC C22 × 337	Groups of 22 (same sex)	7, 21 and 35d (starting at 100 kg)	13.22–16.98%	3.42–3.3 Mcal/kg	ad lib	7.4 ppm	Increased (P < 0.01, d0–7 and 8–21)	Decreased (P < 0.01, d8–21 and 22–35)	Increased (P < 0.0001, particularly d8–21)
Graham et al. 2014	Barrow & Gilt	PIC 327 × 1,050	Groups of 8 (2 x B + 2 x G)	d49–73 (Phase 3)	13.5–20.4%	3.27–3.36 Mcal/kg	ad lib	10 ppm	Increased (P < 0.001, 1.1 v 0.9 kg/d)	NS	Increased (P < 0.001)
Puls et al. 2014	Barrow	Genetiporc G Performer × Fertilis 25	Groups of 4	23d (67–132 kg)	14.3–17.3%	3.31 Mcal/kg	ad lib	9.92 ppm	Increased (P = 0.03, 1.37 v 1.23 kg/d)	NS	Increased (P < 0.001)
Paulk et al. 2015	Barrow & Gilt	TR4 × PIC 1050	Groups of 5	33d (starting at 90 kg)	16%	?	ad lib	10 ppm	Increased (P<0.05, 1.24 v 1.15 kg/d)	Decreased (P = 0.001, 3.88 v 3.5 kg/d)	Increased (P < 0.05)
Brown et al. 2016b	Gilt	PIC C23 × 337	Individual	1, 3, 7, 13 or 27d (77–103 kg)	16.70%	3.35 Mcal/kg	45 minutes ad lib (2x/d)	20 ppm	NS	NS	NS
Needham et al. 2017	Barrow	PIC cross-breed	Individual	4wks (from 16 to 20wks; 58 to 89 kg)	16.1–25.6%	13.3–13.8 MJDE/kg	ad lib	10 ppm	Increased (P < 0.001, 1.50 v 1.36 kg/d)	NS	Increased (P = 0.007 interaction with dietary protein)

References are in year order. Abbreviations: CP, Crude Protein; DE, Digestible Energy; ME, Metabolisable Energy.

protein degradation; Laplante and Sabatini 2013) has been implicated in BA-mediated muscle growth because the mTOR inhibitor, rapamycin, blocks clenbuterol-induced muscle growth (Kline et al. 2007). However, others have shown that rapamycin did not inhibit muscle growth induced by overexpression of the beta-2 adrenergic receptor in mouse muscle (Hagg et al. 2016). Whether mTOR is involved in BA-mediated muscle growth, therefore, remains unclear. The BAs increase phosphorylation of Akt (which is upstream of mTOR; Koopman et al. 2010) and, therefore, compensatory Akt-dependent, but mTOR-independent mechanisms of BA-induced muscle growth may exist.

Reductions in myofibrillar protein degradation by BA are likely to be caused by inhibition of the proteolytic systems. For example, BA increased expression of the specific endogenous calpain inhibitor, calpastatin (Higgins et al. 1988; Bardsley et al. 1992; Parr et al. 1992, 2001; Sensky et al. 1996), and reduced calpain activity, particularly calpain-1 (Higgins et al. 1988; Koopman et al. 2010). However, other proteolytic systems have been described as being affected, such as the proteasome system. This system utilizes the multi-catalytic protease complex, but is critically dependent on the proteins selected for degradation being ubiquitinated through the action of the ubiquitin system ligases, muscle RING finger 1 (MuRF1) and muscle atrophy F-box (Atrogin-1) (Bodine and Baehr 2014). In BA-treated animals, reduced expression of Atrogin-1 has been described (Koopman et al. 2010). We have previously described changes in the activity of some of the multi-catalytic protease complex subunits in response to BA (Mareko et al. 2013); with caspase-like activity being significantly reduced, the chymotrypsin-like activity increased with proteasome activity, and an increase in total proteasome activity was observed. There is currently a lack of consistency in the observations of the effects of BA on individual proteolytic systems, apart from the calpain proteolytic system, where a consistent increase in calpastatin and, thereby, an associated decrease in the calpain system's activity are observed (Bardsley et al. 1992; Parr et al. 2016).

Novel mechanisms

The use of modern transcriptomic technologies is providing novel insights into the molecular mechanisms underpinning the BA response in skeletal muscle (Spurlock et al. 2006; Pearen et al. 2009; Brown et al. 2016b). We examined changes to the skeletal muscle transcriptome of pigs treated with the BA, ractopamine, for 1, 3, 7, 13 and 27 days to identify temporal alterations in gene expression (Brown et al. 2016b). A key finding of the study was that BA treatment for seven days caused a 2-fold increase in protein abundance of phosphoglycerate dehydrogenase (Phgdh), the initiating enzyme of the serine/one-carbon/glycine (SOG) biosynthesis pathway. Gene expressions of all three enzymes in the SOG biosynthesis pathway, Phgdh, phosphoserine aminotransferase (Psat1) and phosphoserine phosphatase (Psph), were increased by BA treatment, peaking at d 3 and declining thereafter. Hyperactivity of the SOG biosynthesis pathway has been shown to generate anabolic intermediates during accelerated cell growth (Possemato et al. 2011; Locasale 2013). The role of Phgdh and other SOG biosynthesis pathway enzymes in the context of regulating post-natal muscle growth is currently being explored.

Our analysis of the pig skeletal muscle transcriptome during prolonged BA administration also revealed a distinct 'stress' response, with a coordinated increase in expression of more than 20 activating transcription factor 4 (ATF4) target genes, such as mitochondrial phosphoenolpyruvate carboxykinase (Parr et al. 2015; Brown et al. 2016a). The ATF4-inducible genes are typically increased to promote cellular survival and adaptation to cellular stress.

The majority of ATF4 target genes induced by BA were involved in amino acid metabolism/ biosynthesis, suggesting that nutrient limitation or altered metabolic flux may occur during BA-mediated muscle growth. However, we have been unable to detect an increase in ATF4 protein expression in skeletal muscle of BA-treated pigs (unpublished data), but the very short half-life of ATF4 makes measuring this protein very challenging.

2.4 Effects of BA on fibre type and meat quality

Muscle fibres can be histochemically classified, according to their actomyosin ATPase activity, into three types, I, IIA and IIB, which have contractile and associated metabolic characteristics of slow-twitch oxidative, fast-twitch oxidative and fast-twitch glycolytic, respectively (Brooke and Kaiser 1970). When considered at the molecular level, the differences in fibre type are, in part, determined by the myosin heavy chain (MyHC) isoform that they express, as this protein is a determinant of muscle contractile speed and, thereby, the associated metabolism required to support this. There are four MyHC isoforms in adult pig muscle – types I, IIa, IIx and IIb – which has led to a revision in the classification of fibre types as types I, IIA, IIX and IIB, with a rank order of contraction speed of I<IIA<IIX<IIB (Schiaffino and Reggiani 1996). In addition to their growth-promoting effects in pigs, BAs also stimulate a transition in muscle fibre-type composition, shifting towards a faster, more glycolytic fibre type (Depreux et al. 2002; Baker et al. 2006; Gunawan et al. 2007; Brown et al. 2016b). The BA-mediated muscle fibre-type transitions are coordinated by altering mRNA expression of the MyHC isoforms. We have observed dynamic transitions in MyHC isoform gene expression with BA treatment in sheep and pigs using cimaterol and ractopamine, respectively (Hemmings et al. 2015; Brown et al. 2016b). The BA treatment in pigs increased mRNA abundance of MyHC IIB and decreased expression of MyHC IIA and IIX (Gunawan et al. 2007; Brown et al. 2016b). In sheep, six days of BA treatment was sufficient to induce *de novo* mRNA expression of MyHC IIB, an isoform not normally expressed in sheep skeletal muscle (Hemmings et al. 2015). Skeletal muscle of BA-treated animals also displays a profound metabolic switch from an oxidative to a more glycolytic phenotype (Vestergaard et al. 1994; Kitaura et al. 2001; Hemmings et al. 2015; Costa-Lima et al. 2015; Brown et al. 2016b). Whether the transitions in contractile and metabolic parameters of the muscle fibre are influential in the growth-modifying effects of BA, or merely an epiphenomenon, remains unclear.

Intramuscular fat (IMF) influences a variety of meat-quality traits, including flavour, juiciness, tenderness and/or firmness, and the overall acceptability of meat (for review see Hocquette et al. 2010). As there is an association between the metabolic characteristics of muscle and fibre type, it would be expected that IMF content would be lower in muscles that contain a high proportion of fast fibre types, particularly those containing a high proportion of type IIB fibres, for example in pigs. However, there does not appear to be an established relationship between the quantity of intramuscular adipocytes and fibre type (Lefaucheur 2009; Hocquette et al. 2010). More recent work has reported that there was a positive relationship between IMF content and size of type IIB fibres (Kim et al. 2013), which has also been reported previously (Henckel et al. 1997). Type IIB fibres have a higher dependence on glycolytic metabolism; therefore, they have high glycogen content. As a result, muscles that have a high proportion of these fibre types tend to have an increased rate of post-mortem pH decline, along with a lower ultimate pH, as well as decreased water-holding capacity (higher drip loss) and a paler colour (Choe et al. 2008; Larzul et al. 1997; Ryu and Kim 2005). The fast fibre types tend to have greater cross-sectional area (CSA) than the slow fibres, with the type IIB fibre having the greatest CSA. This feature is related to their high power output, which is supported by glycolytic metabolism. The greater the muscle fibre CSA, the greater their power output (Schiaffino and Reggiani 1996). Muscles containing fibres with a high CSA have been reported as having a lower meat quality in pigs (Karlsson et al. 1993; Kim et al. 2013), but the relationship between meat quality and CSA per se is not absolutely clear. Studies that have reported this relationship has often done so in animals that have variant genes or are subject to growth-promoter treatments, which show a change in CSA, but this is also associated with alterations in metabolic activity (Lefaucheur 2009). As indicated above, there are well-documented effects of BA on fibre-type distributions and these were some of the earliest observations of the effects of these growth-promoting agents (Beermann et al. 1987). In pigs treated over a four-week period with ractopamine, there was a sustained induction of MyHCIIb in longissimus muscle (Gunawan et al. 2007). This change in gene expression takes place early in the treatment, within a day for MyHCIIb. Likewise, in pigs following seven days of treatment with ractopamine, we demonstrated an increase in MyHCIIb mRNA in longissimus muscle, while there was a trend for MyHCIIa mRNA to decrease (Brown et al. 2012) – a transition that is likely not to be conducive to good meat quality. In addition, there are more immediate direct effects of BA on muscle, with the BA-induced glycogenolytic effect causing less glycogen to be available at slaughter for glycolysis and lactic acid production, which results in the meat tending to become dark, firm and dry (DFD; Warriss et al. 1990), due to the elevated ultimate pH (pH measured 24 h after death).

The final tenderness of meat depends on the degree of alteration of the muscle structural and associated proteins post-mortem (Hopkins and Thompson 2002). There are several endogenous proteolytic systems present in muscle, which could participate in post-mortem proteolysis, but it is the calpain system for which there is the strongest evidence demonstrating a role in post-mortem proteolysis (Kemp et al. 2010; Koohmaraie and Geesink 2006; Sentandreu et al. 2002). Experimental evidence suggests calpain-1 has the most significant role in carrying out post-mortem proteolysis and meat tenderization (Koohmaraie and Geesink 2006). A consistent observation of the calpain system's involvement in tenderness is that high levels of calpastatin are associated with poorer quality meat; the model being that high levels of calpastatin reduce the activity of calpain (predominantly calpain-1), thereby reducing the proteolysis required for tender meat (Kemp et al. 2010). Our studies on a random selection of commercially slaughtered pigs have shown that a high level of calpastatin (both activity and protein levels) in the first few hours after slaughter are associated with an increased incidence of toughness at eight days post-mortem (Sensky et al. 1998; Parr et al. 1999). As described above, BA treatment results in reduced protein degradation and, importantly, in the context of factors affecting meat quality, BA increases calpastatin activity. The BAs, therefore, reduce myofibrillar protein degradation resulting in an associated increase in shear force (Kretchmar et al. 1990; Koohmaraie et al. 1991; Pringle et al. 1993).

Hence, several of the metabolic changes stimulated by BA are not conducive to producing carcasses with high meat quality. After reviewing and carrying out a meta-analysis on a large number of studies of BA-treated pigs, Dunshea et al. (2005) concluded that the use of BA had a clear negative impact on pork quality. However, the range and magnitude of the effects were dependent on the type of BA used. For example, the potent BA, cimaterol, caused a decrease in IMF (−17.2%), which was associated with an increase in drip loss (+97.1%) and increased toughness (+25%), the latter being recorded as an increased shear force. Ractopamine, the commonly used BA in pigs had no effect on both IMF and drip loss, but had increased shear force (+11.6%). For salbuterol, there was no effect on IMF, but drip loss was decreased (−21.7%) and shear force increased (+6.5%). The consistent observations that BA-treated pigs produce meat with elevated shear force, indicating tough meat, were also supported by consumer-based studies, which reported a significant decrease in meat tenderness. Although some of the earlier studies described an increase in ultimate pH and associated DFD meat (Warriss et al. 1990), the meta-analysis of Dunshea et al. (2005) found there was a very small elevation in pH (+0.02 units). As indicated by Dunshea et al. (2005), it must be remembered that most of the studies describing the effects of BA on meat quality are done at relatively high doses, for example 20 ppm ractopamine, while commercial use tends to be lower, with ractopamine being used commercially at a maximum of 10 ppm. The consequence of this is that effects of commercial doses are likely to be less severe than those used for research.

3 Growth hormone

3.1 Introduction to the growth hormone-insulin-like growth factor axis

Growth hormone is a protein hormone produced by the somatotroph cells in the anterior pituitary of the brains of all mammals (see Brameld 1997). In most mammals (except rodents), GH is released into the blood in pulses, with the frequency and amplitude of the pulses increasing at night when the animal is partially fasted. As the name suggests, its main function is in the regulation of growth, although it does so in an indirect way, by stimulating the synthesis of another protein hormone, insulin-like growth factor (IGF)-1, which was originally called somatomedin C. The GH binds to specific GH receptors (GHR) on the membranes of various cell types, but particularly liver cells (hepatocytes), leading to increased mRNA expression and synthesis of IGF-1, which is quickly released into the blood. The IGF-1 is the main functional component of the GH–IGF axis, binding to a specific receptor (called the IGF type 1 receptor or IGF1R) on cell membranes, resulting in the stimulation of processes involved in growth, including cell proliferation and protein synthesis. This receptor is particularly seen in tissues like cartilage and skeletal muscle, although IGF1Rs are found on the membranes of most cell types in the body. As well as the IGF-dependent effects, GH also has direct (IGF-independent) effects on adipose tissue. Binding to the GHR on adipocytes tends to stimulate lipolysis (breakdown of the storage form of fat, triacylglycerol or TAG) and inhibit lipogenesis (synthesis of TAG and fatty acids), resulting in a net mobilization of fat and release of free fatty acids into the blood circulation. Hence, the overall effect of GH administration to an animal tends to be an increase in lean tissues and decrease in fat.

3.2 Effects of GH on growth and feed efficiency

The majority of studies investigating the effects of GH administration to pigs were performed or published between 1985 and 1992 (see Table 2). Many but not all show an increase in growth rates as well as an increase in feed efficiency (gain-to-feed ratio), with some studies also showing a decrease in feed intake. The majority of studies were in barrows, with fewer studies in gilts and very few in intact boars. Even so, there do not appear to be any differences in responsiveness to GH between the sexes (Table 2).

Table 2 Summary of studies investigating effects of porcine growth hormone (pGH) on average daily gain (ADG), average daily feed intake (ADFI) and feed efficiency (gain-to-feed ratio) in growing pigs

Reference	Sex	Breed	Housing	Length of study	CP in diet	DE of diet	How fed?	Dose of pGH	ADG	ADFI	Feed efficiency
Machlin 1972	?	Yorkshire	Individual	46–95 kg	?	?	ad lib	0.22 mg/kg/d	NS (*C* = 0.70 kg/d, GH = 0.68kg/d)	–	NS
Machlin 1972	?	Yorkshire × Hampshire	Groups of 9	45–94 kg	16%	?	ad lib	0.13 mg/kg/d	Increased ($P < 0.01$, C = 0.74 kg/d)	–	Increased
Chung et al. 1985	Barrow	Yorkshire	Individual	30d (starting 32 kg)	16%	?	ad lib	0.022 mg/kg/d	Increased ($P < 0.01$, *C* = 0.91kg/d)	NS	Increased
Etherton et al. 1987	Barrow	Yorkshire × Duroc	Groups of 2	35d (starting 40 kg)	18%	?	ad lib	0.01 mg/kg/d	NS (*C* = 0.90kg/d, GH = 0.98kg/d)	–	NS
Etherton et al. 1987	Barrow	Yorkshire × Duroc	Groups of 2	35d (starting 40 kg)	18%	?	ad lib	0.03 mg/kg/d	NS (*C* = 0.90 kg/d, GH = 0.95 kg/d)	–	NS
Etherton et al. 1987	Barrow	Yorkshire × Duroc	Groups of 2	35d (starting 40 kg)	18%	?	ad lib	0.07 mg/kg/d	Increased ($P < 0.05$, *C* = 0.90, GH = 1.03 kg/d)	–	Increased
Bryan et al. 1989	Gilt	Yorkshire	Groups of 4	30d (starting 72 kg)	?	?	ad lib	0.07 mg/kg/d	Increased ($P = 0.03$, *C* = 0.77 kg/d)	–	Increased
Bryan et al. 1989	Gilt	Duroc	Groups of 4	30d (starting 72 kg)	?	?	ad lib	0.07 mg/kg/d	NS (*C* = 0.78 kg/d, GH = 0.88 kg/d)	–	Increased
Campbell et al. 1990a	Boar	LW × Landrace	Individual	60–90 kg	20%	3.5 Mcal/kg	ad lib	0.1 mg/kg/d	Increased ($P < 0.01$, *C* = 1.17 kg/d)	Decreased	Increased
Campbell et al. 1990a	Gilt	LW × Landrace	Individual	60–90 kg	20%	3.5 Mcal/kg	ad lib	0.1 mg/kg/d	Increased ($p < 0.01$, C = 0.99kg/d)	Decreased	Increased
Campbell et al. 1990b	Boar	LW × Landrace	Individual	30–60 kg	8.3–23.8%	3.44–3.48 Mcal/kg	14% rest (2x/d)	0.09 mg/kg/d	Increased when CP > 14% ($P < 0.01$ GH × CP, C14%CP = 0.78 kg/d)	NS	Increased when CP>15%
Caperna et al. 1990	Barrow	Yorkshire × Duroc × Landrace	Individual	42d (starting 30 kg)	11%	3.85 Mcal/kg	20% rest (1x/d)	0.1 mg/kg/d	Increased ($P < 0.01$, C = 0.45 kg/d)	–	Increased
Caperna et al. 1990	Barrow	Yorkshire × Duroc × Landrace	Individual	42d (starting 30 kg)	19%	3.83 Mcal/kg	20% rest (1x/d)	0.1 mg/kg/d	Increased ($P < 0.01$, *C* = 0.53 kg/d)	–	Increased

(*Continued*)

Table 2 *(Continued)*

Reference	Sex	Breed	Housing	Length of study	CP in diet	DE of diet	How fed?	Dose of pGH	ADG	ADFI	Feed efficiency
Caperna et al. 1990	Barrow	Yorkshire × Duroc × Landrace	Individual	42d (starting 30 kg)	27%	3.81 Mcal/kg	20% rest (1x/d)	0.1 mg/kg/d	Increased ($P < 0.01$, $C = 0.51$ kg/d)	–	Increased
Grant et al. 1991	Barrow	Cross-breed	Groups of 4	24d (starting 90 kg)	14%	13.3 MJ/kg	ad lib	0.05 mg/kg/d	NS ($C = 0.89$ kg/d, GH = 0.82 kg/d)	–	–
Grant et al. 1991	Barrow	Cross-breed	Groups of 4	25d (starting 90 kg)	20%	13.2 MJ/kg	ad lib	0.05 mg/kg/d	NS ($C = 0.81$ kg/d, GH = 0.91 kg/d)	–	–
Azain et al. 1992	Barrow & Gilt	Yorkshire × Landrace or Yorkshire × Duroc dams and Hampshire × Duroc sires	Groups of 4 (same sex)	38d (starting 72–87 kg)	16.00%	3.21 Mcal/kg	ad lib	2 mg/d	NS ($C = 0.85$kg/d, GH = 0.93kg/d)	Decreased	Increased
Becker et al. 1992	Barrow & Gilt	Landrace × Yorkshire × Duroc	Individual	28–35d (starting 77 kg)	18%	3.66 Mcal/kg	ad lib	200 mg implant	NS ($C = 1.0$kg/d, GH = 1.03kg/d)	Decreased	Increased
Becker et al. 1992	Barrow & Gilt	Landrace × Yorkshire × Duroc	Individual	28–35d (starting 77 kg)	18%	3.66 Mcal/kg	ad lib	200 mg implant	NS ($C = 0.79$kg/d, GH = 0.79kg/d)	Decreased	Increased
Bryan et al. 1992	Gilt	Yorkshire	Groups of 4	30d (starting 72 kg)	?	?	ad lib	5 mg/d	Increased ($P < 0.05$, $C = 0.83$ kg/d)	NS	Increased
Evock-Clover et al. 1992	Barrow	Cross-breed	Individual??	7–8wks (starting 35 kg)	16%	3.5 Mcal/kg	15% rest (1x/d)	0.06 mg/kg/d	Increased ($P < 0.05$, $C = 0.70$ kg/d)	NS	Increased
Evock-Clover et al. 1992	Barrow	Cross-breed	Individual??	7–8wks (starting 35 kg)	16%	3.5 Mcal/kg	15% rest (1x/d)	0.12 mg/kg/2d	Increased ($P < 0.05$, $C = 0.70$ kg/d)	NS	Increased
Krick et al. 1992	Barrow	PIC cross	Individual	30–90 kg	22.50%	3.50 Mcal/kg	ad lib	0.05–0.2 mg/kg/d	Increased ($P < 0.01$, $C = 0.89$ kg/d)	Decreased	Increased
Krick et al. 1992	Boar	PIC cross	Individual	30–90 kg	22.50%	3.50 Mcal/kg	ad lib	0.05–0.2 mg/kg/d	Increased ($P < 0.01$, $C = 0.92$ kg/d)	Decreased	Increased
Brameld et al. 1996	Boar	LW × Landrace	Groups of 8/ individually fed	7d	9.9–19.4%	13.88–13.90 MJ/kg	10% rest (2x/d)	0.05 mg/kg/d	NS (slightly higher when CP > 13%)	NS	–

Reference	Sex	Breed	Housing	Length of study	CP in diet	DE of diet	How fed?	Dose of pGH	ADG	ADFI	Feed efficiency
Klindt et al. 1998	Barrow	Meishan	Individual	28d (starting at 34 kg)	22%	3.76 Mcal/kg	ad lib	0.033 mg/kg/d	Increased (*P* < 0.01, *C* = 0.53 kg/d)	NS	Increased
Dunshea et al. 1999	Boar & Gilt	LW × Landrace	Groups of 8	4wks (from 80 kg)	20%	14.6 MJDE/kg	ad lib	5 mg/d	Increased particularly in Gilts (*P* = 0.015 for interaction)	Decreased	Increased, particularly in Gilts
McCauley et al. 2003	Boar, Barrow & Gilt	Cross-breed	Individual	4wks (from 18wks/80 kg)	20%	14.6 MJDE/kg	ad lib	5 mg/d	Increased (*P* < 0.001, particularly 14–28d)	–	Increased, particularly in Gilts & Barrows
Suster et al. 2004	Boar	LW × Landrace	Individual	35d (starting at 64 kg)	20.6%	3.3 Mcal/kg	ad lib	5 mg/kg	Increased (*P* < 0.001)	Decreased	Increased (*P* < 0.001)
Dunshea 2007	Gilt	LW × Landrace		5wks (starting at 59 kg)	19.7%	3.3 Mcal/kg	ad lib	5 mg/d	Increased (*P* < 0.001)	Increased (*P* = 0.004)	Increased (*P* < 0.001)
Moore et al. 2012	Barrow	LW × Landrace × Duroc	Groups of 6	28d (starting at 68 kg)	21–24%	3.2–3.4 Mcal/kg	ad lib	5 mg/d	Increased (*P* = 0.066)	Decreased (*P* = 0.002)	Increased (*P* = 0.002)
Brown et al. 2016b	Gilt	PIC C23 × 337	Individual	1, 3, 7, 13 or 27d (77–103 kg)	16.70%	3.35 Mcal/kg	45 minutes ad lib x2/d	10 mg/2d	NS	NS	NS

References are in year order. Abbreviations: CP, Crude Protein; DE, Digestible Energy; LW, Large White.

The GH and its vehicle control (often saline) are normally administered as a small (<1 mL) daily intramuscular or subcutaneous injection, with doses ranging from 0.01 to 0.2 mg/kg/day. Most studies administered 0.05 or 0.1 mg/kg/day, although some studies tended to simply give a set daily dose (e.g. 2 mg/d). Interestingly, larger effects were observed when the GH was administered via daily injection rather than as a continual slow release from an osmotic mini-pump. This is thought to relate to the daily injection more closely mimicking the normal pulsatile release of GH from the anterior pituitary. Reporcin is a commercially available recombinant porcine GH (rpGH) preparation used across Australasia (Australia, New Zealand, Malaysia and Indonesia), and there are three suggested administration regimens – a daily injection, an injection every other day or injections three days per week (e.g. Monday, Wednesday and Friday). The latter is to make it more practical for large commercial producers, but this is at the expense of a reduced response (Dunshea 2012; Hellams et al. 2000), although the company do also supply a gas-powered injection system that speeds up the process (~600 injections per hour). As for ractopamine, it should be noted that some pGH studies compare to a control group where the protein content of the diet is reduced (either expressed as % crude protein or % lysine). Hence, the magnitude of the growth and feed efficiency responses are often greater in those studies compared with studies where the control diet was the same as for the GH group.

3.3 Influence of genetic selection on GH responsiveness

As indicated in the previous section, there have been few research trials published since the early 1990s. We performed a study in 2011–12 (Brown et al. 2016) using a modern genotype of pig (C23 × 337 progeny purchased from PIC) and appeared to get reduced responses to GH in terms of feed efficiency and growth rates. However, we did administer GH every two days rather than every day and did still see the significant increase in liver weight that is a characteristic response in pigs given GH (Brameld et al. 1996). Tremendous improvements in growth rates and efficiencies have been achieved between the 1980s and now, relating to genetic selection and improvements in feed formulation. Many of the traits selected for (e.g. growth rate, lean:fat ratio and feed efficiency) are regulated by the GH–IGF axis, which might suggest that the modern pig may indeed be less responsive to GH administration. In addition, we know that the effects of GH are dependent upon the diet, particularly energy and protein/amino acid balance. Hence, the majority of research studies tend to provide high-energy, high-protein diets, meaning that the control group also grows well and efficiently (if fed the same diet). The real test of the responsiveness of modern pigs might best be achieved by feeding suboptimal diets instead and this is likely to be informative for the future if, as predicted, the availability of high-quality, human edible feed ingredients become problematic or overly expensive.

3.4 Mechanisms of action

Indirect effects via IGF-1 (cell proliferation, protein synthesis, protein degradation)

Although the effect of IGF-1 can be mediated through a range of receptors, it is the IGF-1R which has the highest affinity (Denley et al. 2005) and subsequently activates multiple signalling pathways. The net deposition of protein often results from an increase in protein synthesis; however, as described for BA, alterations in protein degradation are also observed at times. The GH-stimulated increases in IGF-1 potentially mediates its effects through the mitogen-activated protein kinase kinase/extracellular signal-regulated kinase (MAPK/ERK) pathway, which tends to be associated with proliferative growth, as well as the Akt–mTOR–ribosomal protein S6 kinase (S6K) pathway. Although IGF-1-stimulated protein synthesis can be influenced by both pathways, the effect of IGF-1 on protein degradation is mediated by the latter pathway (Clemmons 2009). The Akt–mTOR–S6K pathway controls protein degradation in skeletal muscle by phosphorylating Akt in the forkhead box protein (FOXO) transcription factors (Glass 2005), thereby reducing the expression of the ubiquitin system ligases, muscle RING finger 1 (MuRF1) and muscle atrophy F-box (Atrogin-1). These two ubiquitin ligases are important in muscle protein turnover as they target myofibrillar proteins for degradation through the ubiquitin–proteasome system (Bodine and Baehr 2014).

Direct effects of GH on adipose tissue

As indicated above, GH has direct (IGF-independent) effects on adipose tissue leading to the release of free fatty acids into the blood circulation and a net reduction in body fat. However, the effects of GH on lipolysis and lipogenesis are thought to involve interactions with other regulators. Hence, GH is thought to enhance catecholamine (i.e. noradrenaline)-stimulated lipolysis and to inhibit insulin-stimulated lipogenesis. The latter may relate to GH administration inducing a degree of insulin resistance, since one of the effects of giving GH to pigs is an increase in plasma insulin concentrations, as well as plasma IGF-1 concentrations (Brameld et al. 1996). We are currently analysing blood metabolomics data from pigs treated with GH or BA over a 27-day time course, and both BA and GH increased plasma-free fatty acid and glycerol concentrations, but with slightly different timescales (unpublished data), while only GH caused a sustained increase in plasma glucose concentrations (unpublished data).

Other effects and novel mechanisms

As mentioned previously, a very consistent phenotypic response to GH administration in mammals is a rapid increase in liver weights. We recently showed that this was apparent after just one day of treatment in pigs (Brown et al. 2016). In that study, we also showed that both BA and GH increased the expression of genes in the SOG biosynthesis pathway in skeletal muscle. While a clear time profile for muscle gene expression was observed in BA-treated pigs, increases were also observed in GH-treated pigs. In fact, there was a greater response to GH than BA at d 1, which was still present at d 3 (when the peak was observed in BA-treated pigs), but subsequently lost by d 7 (Brown et al. 2016). As indicated in the section on novel mechanisms for BA, these genes may represent novel genes involved in muscle hypertrophy and/or feed efficiency and are currently under further investigation.

3.5 Lack of effects of GH on fibre type and meat quality

The GH has a weaker effect than BA on muscle growth, with GH only having limited stimulation of hypertrophy (Solomon et al. 1990). In the majority of studies examining the effects of GH treatment, there appears to be no effect on fibre-type distribution, particularly in pigs (Aalhus et al. 1997; Oksbjerg et al. 1995). Some studies reported that GH caused an increase in fast oxidative fibres and a decrease in fast glycolytic fibres in pigs (Solomon et al. 1991), but other studies have reported the opposite observations (Whipple et al. 1992). In one of our studies, pigs treated with GH for seven days showed a trend ($P < 0.1$) for a decrease in MyHC IIb gene expression, but no changes in muscle weights (Brown et al. 2012). However, in another experiment where we treated pigs with GH for up to 27 days, there was no effect on the expression of any the MyHC isoforms (Brown et al. 2016). Overall, there appears to be no consistent effect of GH on muscle fibre type, unlike BA.

Administration of GH is known to have effects on fat synthesis with reports that subcutaneous fat depth can decrease by up to 70% (Krick et al. 1992). In addition, IMF is also decreased in animals treated with GH, with reductions reported in pigs being down to 55% (Goodband et al. 1993). Although all muscles so far examined decrease their IMF, the effect appears to be greater in some muscles than others. For example, Lefaucheur et al. (1992) examined the longissimus dorsi (mixed fibre type) and the semispinalis (relatively slow-twitch fibre type) and found there was a 27 or 50% reduction in IMF respectively with the same dose of GH. However, there has not been a comprehensive analysis of changes in IMF across different muscle types, as the majority of studies examine the longissimus dorsi and the values for this muscle vary dependent on the dose of GH used (see Dunshea et al. 2005, for a review).

Unlike BA, GH has not been reported as consistently influencing protein degradation systems. This is not surprising, as the main signalling pathway activated by GH–IGF-1 axis stimulates protein synthesis rather than causing a decrease in protein degradation. However, the reports in the literature are again inconsistent. Some studies report similar increases in both protein synthesis and breakdown, but an overall net increase in nitrogen balance (Tomas et al. 1992), while others (Bush et al. 2003) report that protein synthesis is increased, but protein degradation is unaffected, and some (Vann et al. 2000) describe a decrease in protein degradation. As described above, the calpain proteolytic system plays a major role in the development of meat tenderness, but there are very few studies examining the effects of GH on the calpain proteolytic system. Ji et al. (1998) examined the effect of GH on pigs and reported that the treatment resulted in positive growth effects associated with a decrease (rather than an increase) in calpastatin gene expression. Although meat quality was not measured, a decrease in calpastatin would be expected to have a positive or no effect on meat tenderness.

Generally, GH appears to have only slight or no effects on meat tenderness (Aalhus et al. 1997). Dunshea et al. (2005) carried out a comprehensive review and meta-analysis on the effects of pGH on pig meat quality, in which they indicated that GH decreased IMF by 12% and increased shear force by 9%, along with a 6% reduction in drip loss. Their conclusion was that GH causes a small increase in shear force and the sensory perception of meat from treated pigs is tough, but it was unlikely that this would be detected by general consumers (Dunshea et al. 2005). It appears that the sensory perceptions of toughness caused by GH treatment are unlikely to be due to negative effects on the calpain proteolytic system. However, a potential explanation for the increased sensory perception of toughness could be due to the effect of GH that increases the rates of deposition of collagen and other connective tissue proteins within the muscle (Harper 1999), as well as a reduction in IMF, which together could enhance the perception of toughness.

4 Effects of combining ractopamine and reporcin

Very few studies have actually combined the administration of GH and BA to pigs, but what has been done does suggest it might further improve the responses. As indicated above, they do work in slightly different ways, so are likely to complement each other and may even help in overcoming some of the negative effects of BA on meat quality. A study in Australia (Rikard-Bell et al. 2009b) combined the inclusion of ractopamine in the feed (5 ppm) for four weeks with daily reporcin (5 mg) injections for the last two weeks and observed additional improvements in feed efficiency over that obtained with ractopamine alone. The effects on gilts were greater than on intact boars, mainly because boars are already very efficient and growing rapidly. It would be interesting to see whether this combination might completely counter the negative effects of castration (surgical or chemical/immunological) on growth and feed efficiency, since ractopamine has been shown to have greater effects in barrows than gilts.

5 Global food security and the future use of growth promoters

Since the publication of the UK government's Foresight report in 2011 (Foresight 2011), global food security has been in the news regularly. That report and numerous other publications have made predictions as to what will happen to the world human population and, therefore, the global food requirements through to the year 2050. The predictions suggest that by 2050, the world population will increase to nearly 10 billion, an increase of 25% from the present (2017) population of approximately 7.5 billion. At the time of the Foresight report, malnutrition was responsible for about 35% of all deaths among children under five, while in many countries around 25–30% of the adult population were obese. Clearly, this means food distributions around the world are unequal, but we will need to provide sufficient, healthy food for those extra 2.5 billion people. While reductions in food waste will help achieve some of this, we will still need to simply produce more food, but without using more land or water. As many countries develop, the normal diet of the population changes, with consumption of meat and animal products often increasing. Currently, there are suggestions that people in some parts of the world (e.g. UK and USA) should reduce their consumption of meat and animal products, in order to counter the increasing prevalence of obesity. Even if this happens, the likely increasing consumption in developing countries means that global production of meat and other animal products will still need to increase.

Tremendous improvements in livestock productivity have already been achieved since the 1950s, particularly in relation to eggs, milk, pork and chicken. The improvements relating to pigs and poultry have been achieved by a combination of genetic selection and optimization of nutrition, but the latter relies heavily on using high-quality ingredients (e.g. wheat, maize and soya) that can also be consumed by humans. Hence, the competition for these human edible ingredients is likely to increase, potentially meaning that they will become more expensive. We, therefore, need to find ways of improving the efficiency of the various livestock, to increase the amount of product relative to the amount of feed consumed. We also need to use more by-products and 'poorer quality' (i.e. human inedible) feed ingredients, while at the same time maintaining current levels of growth and feed efficiency, as well as the quality and nutritional value of the meat. This is where the use of growth promoters and other technologies (e.g. use of feed enzymes and feed processing) are likely to become important tools for sustainable pig meat production in the future.

6 Where to look for further information

Further information on the commercially available BA, ractopamine (paylean), and GH, reporcin, is available from the manufacturer's websites (https://www.elanco.us/products-services/swine/feed-efficiency-finishing-swine.aspx; http://zamira.com.sg/wp-content/uploads/2014/09/Reporcin-brochure.pdf).

7 References

Aalhus, J. L., Best, D. R., Costello, F. and Schaefer, A. L. (1997) The effects of porcine somatotropin on muscle fibre morphology and meat quality of pigs of known stress susceptibility. *Meat Sci.* 45:283–95.

Apple, J. K., Rincker, P. J., McKeith. F. K. and et al. (2007) Review: Meta-analysis of the ractopamine response in finishing swine. *Prof. Anim. Sci.* 23:179–96.

Armstrong, T. A., Ivers, D. J., Wagner, J. R., Anderson, D. B. and et al. (2004) The effect of dietary ractopamine concentration and duration of feeding on growthperformance, carcass characteristics, and meat quality of finishing pigs. *J. Anim. Sci.* 82:3245–53.

Azain, M. J., Bullock, K. D., Kasser, T. R. and Veenhuizen, J. J. (1992) Relationship of mode of porcine somatotropin administration and dietary fat to the growth performance and carcass characteristics of finishing pigs. *J. Anim. Sci.* 70:3086–95.

Baker, D. J., Constantin-Teodosiu, D., Jones, S. W. and et al (2006) Chronic treatment with the beta(2)-adrenoceptor agonist prodrug BRL-47672 impairs rat skeletal muscle function by inducing a comprehensive shift to a faster muscle phenotype. *J. Pharmacol. Exp. Ther.* 319:439–46. doi:10.1124/jpet.106.107045.Rajab.

Baker, J. G. (2010) The selectivity of beta-adrenoceptor agonists at human beta1-, beta2- and beta3-adrenoceptors. *Br. J. Pharmacol.* 160:1048–61. doi:10.1111/j.1476-5381.2010.00754.x.

Baker, P., Dalrymple, R. and Ingle, D. (1984) Use of beta-adrenergic agonist to alter muscle and fat deposition in lambs. *J. Anim. Sci.* 59:1256–61.

Bardocz, S., Brown, D. S., Grant, G. and et al. (1992) Effect of the beta-adrenoceptor agonist clenbuterol and phytohaemagglutinin on growth, protein synthesis and polyamine metabolism of tissues of the rat. *Br. J. Pharmacol.* 106:476–82.

Bardsley, R. G., Allcock, S. M., Dawson, J. M. and et al. (1992) Effect of beta-agonists on expression of calpain and calpastatin activity in skeletal muscle. *Biochimie* 74:267–73.

Bark, L. J., Stahly, T. S., Cromwell, G. L. and Miyat, J. (1992) Influence of genetic capacity for lean tissue growth on rate and efficiency of tissue accretion in pigs fed ractopamine. *J. Anim. Sci.* 70:3391–400.

Barker, M. R., Dritz, S. S., Tokach, M. D., Goodband, R.D. and Grosebeck, C. N. (2005) Effects of ractopamine HCl on growth performance and within-pen weight variation in finishing pigs. *J. Swine Health Prod.* 13:318–21.

Becker, B. A., Knight, C. D., Buonomo, F. C., Jesse, G. W., Hedrick, H. B. and Baile, C. A. (1992) Effect of a hot environment on performance, carcass characteristics, and blood hormones and metabolites of pigs treated with porcine somatotropin. *J. Anim. Sci.* 70:2732–40.

Beermann, D. H., Butler, W. R., Hogue, D. E., Fishell, V. K., Dalrymple, R. H., Ricks, C. A. and Scanes, C. G. (1987). Cimaterol-induced muscle hypertrophy and altered endocrine status in lambs. *J. Anim. Sci.* 65:1514–24.

Bodine S. C. and Baehr, L. M. (2014) Skeletal muscle atrophy and the E3 ubiquitin ligases MuRF1 and MAFbx/atrogin-1. *Am. J. Physiol. Endocrinol. Metab.* 307:E469–E484.

Bohorov, O., Buttery, P. J., Correia, J. H. R. D. and et al. (1987) The effect of the B-2-adrenergicagonist clenbuterol or implantation with oestradiol plus trenbolone acetate on protein metabolism in wether lambs. *Br. J. Nutr.* 57:99–107.

Bohrer, B. M., Kyle, J. M., Little, K. L. and et al. (2013) The effects of a step-up ractopamine feeding program on growth performance and low-sodium ham characteristics of purebred Berkshire pigs. *J. Anim. Sci.* 91:5535–43. doi:10.2527/jas.2013-6854.

Brameld, J. M. (1997) Molecular mechanisms involved in the nutritional and hormonal regulation of growth in pigs. *Proc. Nut. Soc.* 56:607–19.

Brameld, J. M., Atkinson, J. L., Saunders, J. C., Pell, J. M., Buttery, P. J. and Gilmour, R. S. (1996) Effects of Growth Hormone administration and dietary protein intake on Insulin-like growth factor-I and Growth Hormone Receptor mRNA expression in porcine liver, skeletal muscle and adipose tissue. *J. Anim. Sci.* 74:1832–41.

Brana, D. V., Rojo-Gomez, G. A., Ellis, M. and Cuaron, J. A. (2013) Effect of gender (gilt and surgically and immunocastrated male) and ractopamine hydrochloride supplementation on growth performance, carcass, and pork quality characteristics of finishing pigs under commercial conditions. *J. Anim. Sci.* 91:5894–904. doi:10.2527/jas.2013-6545.

Brooke, M. H. and Kaiser, K. K. (1970). Muscle fiber types – How many and what kind. *Arch. Neurol.* 23:369–79.

Brown, D., Ryan, K., Daniel, Z., Mareko, M. H. D., Parr, T. and Brameld, J. (2012). Beta-adrenergic agonist and growth hormone have differential effects on muscle fibre composition in growing gilts. *Proc. Br. Soc. Anim. Sci.* 3(1): 66.

Brown, D. M., Ryan, K. J., Daniel, Z. C. and et al. (2016a) Abstract: 5-03 (Identification of an ER-stress-like response to the anabolic agent, Ractopamine) of the 9th International Conference on Cachexia, Sarcopenia and Muscle Wasting, Berlin, Germany, 10–11 December 2016 (part 1). *J. Cachexia Sarcopenia Muscle* 7:626–62. doi:10.1002/jcsm.12164.

Brown, D. M., Williams, H., Ryan, K. J. P. and et al. (2016b) Mitochondrial phosphoenolpyruvate carboxykinase (PEPCK-M) and serine biosynthetic pathway genes are co-ordinately increased during anabolic agent-induced skeletal muscle growth. *Sci. Rep.* 6:28693. doi:10.1038/srep28693.

Bryan, K. A., Hammond, J. M., Canning, S., Mondschein, J. and et al. (1989) Reproductive and growth responses of gilts to exogenous porcine pituitary growth hormone. *J. Anim. Sci.* 67:196–205.

Bryan, K. A., Hagen, D. R. and Hammond, J. M. (1992) Effect of frequency of administration of exogenous porcine growth hormone on growth and carcass traits and ovarian function of prepubertal gilts. *J. Anim. Sci.* 70:1454–63.

Bundy, J. W., Miller, P. S., Moreno, R., Burkey, T. E., Hinkle, E. and Tran, H. (2011) Lysine Requirement for Barrows Fed Ractopamine. Nebraska Swine Reports, pp. 27–9.

Burniston, J. G., Clark, W. A., Tan, L.-B. and et al. (2006) Dose-dependent separation of the hypertrophic and myotoxic effects of the beta(2)-adrenergic receptor agonist clenbuterol in rat striated muscles. *Muscle Nerve* 33:655–63. doi:10.1002/mus.20504.

Bush, J. A., Burrin, D. G., Suryawan, A., O'Connor, P. M., Nguyen, H. V, Reeds, P. J. and et al. (2003). Somatotropin-induced protein anabolism in hindquarters and portal-drained viscera of growing pigs. *Am. J. Physiol. Endocrinol. Metabol.* 284:E302–12.

Campbell, R. G., Johnson, R. J., King, R. H. and Taverner, M. R. (1990a) Effects of gender and genotype on the response of growing pigs to exogenous administration of porcine growth hormone. *J. Anim. Sci.* 68:2674–81.

Campbell, R. G., Johnson, R. J., King, R. H., Taverner, M. R. and Meisinger, D. J. (1990b) Interaction of dietary protein content and exogenous porcine growth hormone administration on protein and lipid accretion rates in growing pigs. *J. Anim. Sci.* 68:3217–25.

Caperna, T. J., Steele, N. C., Komarek, D. R., McMurtry, J. P. and et al. (1990) Influence of dietary protein and recombinant porcine somatotropin administration in young pigs: Growth, body composition and hormone status. *J. Anim. Sci.* 68:4243–52.

Carr, S. N., Rincker, P. J., Killefer, J., Baker, D. H. and et al. (2005) Effects of different cereal grains and ractopamine hydrochloride on performance, carcass characteristics, and fat quality in late-finishing pigs. *J. Anim. Sci.* 83:223–30.
Choe, J. H., Choi, Y. M., Lee, S. H., Shin, H. G., Ryu, Y. C., Hong, K. C. and et al. (2008). The relation between glycogen, lactate content and muscle fiber type composition, and their influence on postmortem glycolytic rate and pork quality. *Meat Sci.* 80:355–62.
Choo, J. J., Horan, M. A., Little, R. A. and et al. (1992) Anabolic effects of clenbuterol on skeletal muscle are mediated by beta 2-adrenoceptor activation. *Am. J. Physiol.* 263:E50–6.
Chung, C. S., Etherton, T. D. and Wiggins, J. P. (1985) Stimulation of swine growth by porcine growth hormone. J Anim Sci 60:118–30.
Clemmons, D. R. (2009). Role of IGF-I in skeletal muscle mass maintenance. *Trends Endocrinol. Metabol.* 20:349–56.
Costa-Lima, B. R. C., Suman, S. P., Li, S., Beach, C. M. and et al. (2015) Dietary ractopamine influences sarcoplasmic proteome profile of pork Longissimus thoracis. *Meat Sci.* 103:7–12
Crome, P. K., McKeith, F. K., Carr, T. R., Jones, D. J. and et al. (1996) Effect of ractopamine on growth performance, carcass composition, and cutting yields of pigs slaughtered at 107 and 125 kilograms. *J. Anim. Sci.* 74:709–16.
Denley, A., Cosgrove, L. J., Booker, G. W., Wallace, J. C. and Forbes, B. E. (2005). Molecular interactions of the IGF system. *Cytokine Growth Factor Rev.* 16:421–39.
Depreux, F. F. S., Grant, A. L., Anderson, D. B. and et al. (2002) Paylean alters myosin heavy chain isoform content in pig muscle. *J. Anim. Sci.* 80:1888–94.
Dunshea, F. R. (2007) Porcine somatotropin and cysteamine hydrochloride improve growth performance and reduce back fat in finisher gilts. *Aust. J. Exp. Agric.* 47:796–800. doi:10.1071/EA06243.
Dunshea, F. R. (2012) Emerging technologies with the potential to improve feed efficiency in swine. In J. F. Patience (Ed.) *Feed Efficiency in Swine.* Wageningen Academic Publishers, The Netherlands, pp. 259–75.
Dunshea, F. R., Cox, M. L., Borg, M. R. and Harris, D. R. (1999) Porcine somatotropin (Reporcin®) improves growth performance and decreases backfat in pigs under commercial conditions. *Manipulating Pig Production VII* 123.
Dunshea, F. R., King, R. H., Campbell, R. G., Sainz, R. D. and Kim, Y. S. (1993) Interrelationships between sex and ractopamine on protein and lipid deposition in rapidly growing pigs. *J. Anim. Sci.* 71:2919–30.
Dunshea, R. R., D'Souza, D. N., Pethic, D. W., Harper, G. S. and Warner, R. D. (2005) Effects of dietary factors and other metabolic modifiers on quality and nutritional value of meat. *Meat Sci.* 71:8–38.
Emery, P. W., Rothwell, N. J., Stock, M. J. and et al. (1984) Chronic effects of beta 2-adrenergic agonists on body composition and protein synthesis in the rat. *Biosci. Rep.* 4:83–91.
Etherton, T. D., Wiggins, J. P., Evock, C. M., Chung, C. S. and et al. (1987) Stimulation of pig growth performance by porcine growth hormone: Determination of the dose-response relationship. *J. Anim. Sci.* 64:433–43.
Etherton, T. D. and Bauman, D. E. (1998). Biology of somatotropin in growth and lactation of domestic animals. *Physiol. Rev.* 78:745–61.
Evock-Clover, C. M., Steele, N. C., Caperna, T. J. and Solomon, M. B. (1992) Effects of frequency of recombinant porcine somatotropin administration on growth performance, tissue accretion rates, and hormone and metabolite concentrations in pigs. *J. Anim. Sci.* 70:3709–20.
Foresight (2011) The Future of Food and Farming: Challenges and choices for global sustainability. Final Project Report. The Government Office for Science, London. https://www.gov.uk/government/uploads/system/uploads/attachment_data/file/288329/11-546-future-of-food-and-farming-report.pdf.
Garbossa, C. A. P., Sousa, R. V. de, Cantarelli, V. de S. and et al. (2013) Ractopamine levels on performance, carcass characteristics and quality of pig meat. *Rev. Bras. Zootec.* 42:325–33. doi:10.1590/S1516-35982013000500004.

Gerlemann, G. D., Allee, G. L., Rincker, P. J., Ritter, M. J., Boler, D. D. and Carr, S. N. (2014) Impact of ractopamine hydrochloride on growth, efficiency, and carcass traits of finishing pigs in a three-phase marketing strategy. *J. Anim. Sci.* 92:1200–7.

Glass, D. J. (2005). Skeletal muscle hypertrophy and atrophy signaling pathways. *Int. J. Biochem. Cell Biol.* 37(10):1974–84.

Goodband, R. D., Nelssen, J. L., Hines, R. H., Kropf, D. H., Stoner, G. R., Thaler, R. C. and et al. (1993). Interrelationships between porcine somatotropin and dietary lysine on growth performance and carcass characteristics of finishing swine. *J. Anim. Sci.* 71:663–72.

Graham, A. B., Goodband, R. D., Tokach, M. D., Dritz, S. S. and et al. (2014) The interactive effects of high-fat, high-fiber diets and ractopamine HCl on finishing pig growth performance, carcass characteristics, and carcass fat quality. *J. Anim. Sci.* 92:4585–97.

Grant, A. L., Helferich, W. G., Kramer, S. A., Merkel, R. A. and Bergen, W. G. (1991) Administration of growth hormone to pigs alters the relative amount of insulin-like growth factor-I mRNA in liver and skeletal muscle. *J. Endocrinol.* 130:331–8.

Gu, Y., Schinckel, A. P., Forrest, J. C., Kuei, C. H. and Watkins, L. E. (1991) Effects of ractopamine, genotype, and growth phase on finishing performance and carcass value in swine: I. Growth performance and carcass merit. *J. Anim. Sci.* 69:2685–93.

Gunawan, A. M., Richert, B. T., Schinckel, A. P. et al (2007) Ractopamine induces differential gene expression in porcine skeletal muscles. *J. Anim. Sci.* 85:2115–24. doi:10.2527/jas.2006-540.

Hagg, A., Colgan, T. D., Thomson, R. E. and et al. (2016) Using AAV vectors expressing the β2-adrenoceptor or associated Gα proteins to modulate skeletal muscle mass and muscle fibre size. *Sci. Rep.* 6:23042. doi:10.1038/srep23042.

Harper, G. S. (1999). Trends in skeletal muscle biology and the understanding of toughness in beef. *Aust. J. Agric. Res.* 50:1105–29.

Heitzman, R. J. (1979) The efficacy and mechanism of action of anabolic agents as growth promoters in farm animals. *J. Steroid Biochem.* 11(1):927–30.

Hellams, S. J., Bryden, W. L., Harris, D. and et al. (2000) Strategic use of growth hormone and pig performance. *Asian Austr. J. Anim. Sci.* 13D:10.

Hemmings, K. M., Daniel, Z. C. T. R., Buttery, P. J. and et al. (2015) Differential effects of short-term β agonist and growth hormone treatments on expression of myosin heavy chain IIB and associated metabolic genes in sheep muscle. *Animal* 9:285–94. doi:10.1017/S175173111400233X.

Henckel, P., Oksbjerg, N., Erlandsen, E., Barton-Gade, P. and Bejerholm, C. (1997) Histo- and biochemical characteristics of the longissimus dorsi muscle in pigs and their relationships to performance and meat quality. *Meat Sci.* 47:311–21.

Higgins, J. A., Lasslett, Y. V., Bardsley, R. G. and et al. (1988) The relation between dietary restriction or clenbuterol (a selective beta 2 agonist) treatment on muscle growth and calpain proteinase (EC 3.4.22.17) and calpastatin activities in lambs. *Br. J. Nutr.* 60:645–52.

Hinson, R. B., Wiegand, B. R., Ritter, M. J., Allee, G. L. and Carr, S. N. (2011) Impact of dietary energy level and ractopamine on growth performance, carcass characteristics, and meat quality of finishing pigs. *J. Anim. Sci.* 89:3572–9.

Hocquette, J. F., Gondret, F., Baéza, E. and et al. (2010) Intramuscular fat content in meat-producing animals: Development, genetic and nutritional control, and identification of putative markers. *Animal* 4:303–19.

Ji, S. Q., Frank, G. R., Cornelius, S. G., Willis, G. M. and Spurlock, M. E. (1998) Porcine somatotropin improves growth in finishing pigs without altering calpain 3 (p94) or alpha-actin mRNA abundance and has a differential effect on calpastatin transcription products. *J. Anim Sci.* 76:1389–95.

Karlsson, A., Enfalt, A. C., Essen-Gustavsson, B., Lundstrom, K., Rydhmer, L. and Stern, S. (1993). Muscle histochemical and biochemical properties in relation to meat qual- ity during selection for increased lean tissue growth rate in pigs. *J. Anim. Sci.* 71:930–8.

Kemp, C. M., Sensky, P. L., Bardsley, R. G., Buttery, P. J. and Parr, T. (2010). Tenderness–an enzymatic view. *Meat Sci.* 84:248–56.

Kim, G. D., Jeong, J. Y., Jung, E. Y., Yang, H. S., Lim, H. T. and Joo, S. T. (2013). The influence of fiber size distribution of type IIB on carcass traits and meat quality in pigs. *Meat Sci.* 94:267–73.

Kim, Y. S., Sainz, R. D., Molenaar, P. and Summers, R. J. (1991) Characterization of beta 1- and beta 2-adrenoceptors in rat skeletal muscles. *Biochem. Pharmacol.* 42:1783–9.

Kitaura, T., Tsunekawa, N. and Hatta, H. (2001) Decreased monocarboxylate transporter 1 in rat soleus and EDL muscles exposed to clenbuterol. *J. Appl. Physiol.* 91:85–90.

Klindt, J., Yen, J. T., Buonomo, F. C., Roberts, A. J. and Wise, T. (1998) Growth, body composition, and endocrine responses to chronic administration of insulin-like growth factor I and(or) porcine growth hormone in pigs. *J. Anim. Sci.* 76:2368–81.

Kline, W. O., Panaro, F. J., Yang, H. and et al. (2007) Rapamycin inhibits the growth and muscle-sparing effects of clenbuterol. *J. Appl. Physiol.* 102:740–7. doi:10.1152/japplphysiol.00873.2006.

Koohmaraie, M. and Geesink, G. H. (2006). Contribution of postmortem muscle biochemistry to the delivery of consistent meat quality with particular focus on the calpain system. *Meat Sci.* 74:34–43.

Koohmaraie, M., Shackleford, S. D., Muggli-Crockett, N. E. and Stone, R. T. (1991) Effect of the beta-adrenergic agonist L644,969 on muscle growth, endogenous proteinase activities, and postmortem proteolysis in wether lambs. *J. Anim. Sci.* 69:4823–35.

Koopman, R., Gehrig, S. M., Léger, B. and et al. (2010) Cellular mechanisms underlying temporal changes in skeletal muscle protein synthesis and breakdown during chronic {beta}-adrenoceptor stimulation in mice. *J. Physiol.* 588:4811–23. doi:10.1113/jphysiol.2010.196725.

Kretchmar, D. H., Hathaway, M. R., Epley, R. J. and Dayton, W. R. (1990). Alterations in postmortem degradation of myofibrillar proteins in muscle of lambs fed a beta-adrenergic agonist. *J. Anim. Sci.* 68: 1760–72.

Krick, B. J., Roneker, K. R., Boyd, R. D., Beermann, D. H., David, P. J. and Meisinger, D. J. (1992) Influence of genotype and sex on the response of growing pigs to recombinant porcine somatotropin. *J. Anim. Sci.* 70:3024–34.

Laplante, M. and Sabatini, D. M. (2013) mTOR signaling in growth control and disease. *Cell* 149:274–93. doi:10.1016/j.cell.2012.03.017.mTOR.

Larzul, C., Lefaucheur, L., Ecolan, P., Gogué, J., Talmant, A., Sellier, P. and et al. (1997). Phenotypic and genetic parameters for longissimus muscle fiber characteristics in relation to growth, carcass, and meat quality traits in Large White pigs. *J. Anim. Sci* 75:3126–37.

Lefaucheur, L., Ecolan, P. and Bonneau, M. (1992). Contractile and metabolic characteristics of skeletal muscle in pigs administered exogenous somatotropin. In *43rd Annual Meeting of the European Association for Animal Production*, pp. 1–4.

Lefaucheur, L. (2010). A second look into fibre typing – Relation to meat quality. *Meat Sci.* 84:257–70.

Liang, W. and Mills, S.E. (2002) Quantitative analysis of β-adrenergic receptor subtypes in pig tissues 1. *J. Anim. Sci.* 80:963–70.

Locasale, J. W. (2013) Serine, glycine and one-carbon units: Cancer metabolism in full circle. *Nat. Rev. Cancer* 13:572–83. doi:10.1038/nrc3557.

Lutz, T. R. and Stahly, T. S. (1997) *Effect of Ractopamine on the Optimum Dietary Phosphorus Regimen for Pigs.* Swine Research Reports, Iowa State University, University Extension.

Lynch, G. S. and Ryall, J. G. (2008) Role of beta-adrenoceptor signaling in skeletal muscle: Implications for muscle wasting and disease. *Physiol. Rev.* 88:729–67. doi:10.1152/physrev.00028.2007.

Machlin, L. J. (1972) Effect of porcine growth hormone on growth and carcass composition of the pig. *J. Anim. Sci.* 35:794–800.

Mareko, M. H. D., Ryan, K., Brown, D., Brameld, J. and Parr T. (2013) Effects of 27 days of treatment with growth promoting agents on adult pig longissimus muscle protease systems and post-mortem shear force after conditioning. *Proc. Br. Soc. Anim. Sci.* 4(1): 151.

McCauley, I., Watt, M., Suster, D., Kerton, D. J. and et al. (2003) A GnRF vaccine (Improvac®) and porcine somatotropin (Reporcin®) have synergistic effects upon growth performance in both boars and gilts. *Aus. J. Agr. Res.* 54:11–20.

Mills, S. E., Kissel, J., Bidwell, C. A. and et al. (2003) Stereoselectivity of porcine beta-adrenergic receptors for ractopamine stereoisomers. *J. Anim. Sci.* 81:122–9.

Mimbs, K. J., Pringle, T. D., Azain, M. J., Meers, S. A. and Armstrong, T. A. (2005) Effects of ractopamine on performance and composition of pigs phenotypically sorted into fat and lean groups. *J. Anim. Sci.* 83:1361–9.
Mitchell, A. D., Solomon, M. B. and Steele, N. C. (1991) Influence of level of dietary protein or energy on effects of ractopamine in finishing swine. *J. Anim. Sci.* 69:4487–95.
Moore, K. L., Mullan, B. P. and D'Souza, D. N. (2012) The interaction between ractopamine supplementation, porcine somatotropin and moisture infusion on pork quality. *Meat. Sci.* 92:125–31. doi:10.1016/j.meatsci.2012.04.022.
Needham, T., Hoffman, L. C. and Gous, R. M. (2017) Growth responses of entire and immunocastrated male pigs to dietary protein with and without ractopamine hydrochloride. *Animal* doi:10.1017/S1751731117000271.
Oksbjerg, N., Petersen, J. S., Sorensen, M. T., Henckel, P., Agergaard, N., Bejerholm, C. and et al. (1995). The influence of porcine growth hormone on muscle fibre characteristics, metabolic potential and meat quality. *Meat Sci.* 39:375–85. https://doi.org/10.1016/0309-1740(94)00012-v
Parr, T., Bardsley, R. G., Gilmour, R. S. and et al. (1992) Changes in calpain and calpastatin mRNA induced by beta-adrenergic stimulation of bovine skeletal muscle. *Eur. J. Biochem.* 208:333–9.
Parr, T., Mareko, M. H. D., Ryan, K., Hemmings, K. M., Brown, D. and Brameld, J. (2016) The impact of growth promoters on muscle growth and the potential consequences for meat quality. *Meat Sci.* 120:93–9
Parr, T., Ryan, K., Williams, H., Harris, D., Brown, D., Mareko, M., Emes, R., Giles, T., Wattis, J., Dryden, I., Lu, C., Hodgman, C. and Brameld, J. (2015) Skeletal muscle hypertrophy in pigs is associated with an increased expression of serine biosynthetic pathway genes along with genes associated with an endoplasmic reticulum stress response. *Proc. Nutr. Soc.* 74(OCE5):E278. doi:10.1017/S0029665115003250.
Parr, T., Sensky, P. L., Bardsley, R. G. and et al (2001) Calpastatin expression in cardiac and skeletal muscle and partial gene structure. *Arch. Biochem. Biophys.* 395:1–13.
Parr, T., Sensky, P. L., Scothern, G., Bardsley, R. G., Buttery, P. J., Wood, J. D. and Warkup, C. C. (1999). Immunochemical study of the calpain system in porcine longissimus muscle with high and low shear force values. *J. Anim. Sci.* 77(suppl 1):164.
Patience, J. F., Shand, P., Pietrasik, Z. and et al. (2009) The effect of ractopamine supplementation at 5 ppm of swine finishing diets on growth performance, carcass composition and ultimate pork quality. *Can. J. Anim. Sci.* 89:53–66. doi:10.4141/CJAS07152.
Paulk, C. B., McKinney, L. J., Hancock, J. D. and Williams et al. (2015) Effects of diet mix time and ractopamine hydrochloride on finishing pig growth and carcass performance. *J. Anim. Sci.* 93:1689–94.
Pearen, M. A., Ryall, J. G., Lynch, G. S. and et al. (2009) Expression profiling of skeletal muscle following acute and chronic beta2-adrenergic stimulation: Implications for hypertrophy, metabolism and circadian rhythm. *BMC Genomics* 10:448. doi:10.1186/1471-2164-10-448.
Pompeu, M. A., Rodrigues, L. A., Cavalcanti, L. F. L., Fontes, D. O. and Toral, F. L. B. (2017) A multivariate approach to determine the factors affecting response level of growth, carcass and meat quality traits in finishing pigs fed ractopamine. *J. Anim. Sci.* doi: e10n.2c52e7/jas2016.1181.
Possemato, R., Marks, K. M., Shaul, Y. D. and et al. (2011) Functional genomics reveal that the serine synthesis pathway is essential in breast cancer. *Nature* 476:346–50. doi:10.1038/nature10350.
Pringle, T. D., Calkins, C. R., Koohmaraie, M. and Jones, S. J. (1993). Effects over time of feeding a b-adrenergic agonist to wether lambs on animal performance, muscle growth, endogenous muscle proteinase activities, and meat tenderness. *J. Anim. Sci.* 71:636–44.
Puls, C. L., Rojo, A., Ellis, M., Boler, D. D., McKeith, F. K. and et al. (2014) Growth performance of immunologically castrated (with Improvest) barrows (with or without ractopamine) compared to gilt, physically castrated barrow, and intact male pigs. *J. Anim. Sci.* 92:2289–95.
Reeds, P. J., Hay, S. M., Dorwood, P. M. and et al. (1986) Stimulation of muscle growth by clenbuterol: Lack of effect on muscle protein biosynthesis. *Br. J. Nutr.* 56:249–58.
Rikard-Bell, C., Curtis, M. A., van Barneveld, R. J. and et al. (2009a) Ractopamine hydrochloride improves growth performance and carcass composition in immunocastrated boars, intact boars, and gilts. *J. Anim. Sci.* 87:3536–43. doi:10.2527/jas.2009-2002.

Rikard-Bell, C. V., Pluske, J. R., van Barneveld, R. J., Mullan, B. P. and et al. (2009b) Combining a ractopamine feeding regime and porcine somatotropin has additive effects on finisher pig performance. *Manipulating Pig Production XII* 70.

Ryall, J. G., Sillence, M. N. and Lynch, G. S. (2006) Systemic administration of beta2-adrenoceptor agonists, formoterol and salmeterol, elicit skeletal muscle hypertrophy in rats at micromolar doses. *Br. J. Pharmacol.* 147:587–95. doi:10.1038/sj.bjp.0706669.

Ryu, Y. C. and Kim, B. C. (2005). The relationship between muscle fiber characteristics, postmortem metabolic rate, and meat quality of pig longissimus dorsi muscle. *Meat Sci.* 71:351–7.

Schiaffino, S. and Reggiani, C. (1996). Molecular diversity of myofibrillar proteins: Gene regulation and functional significance. *Physiol. Rev.* 76:371–423.

See, M. T., Armstrong, T. A. and Weldon, W. C. (2004) Effect of a ractopamine feeding program on growth performance and carcass composition in finishing pigs. *J. Anim. Sci.* 82:2474–80.

Sensky, P. L., Parr, T., Bardsley, R. G. and et al. (1996) The relationship between plasma epinephrine concentration and the activity of the calpain enzyme system in porcine longissimus muscle. *J. Anim. Sci.* 74:380–7.

Sensky, P. L., Parr, T., Scothern, G. P., Perry, A., Bardsley, R. G., Buttery, P. J., Wood, J. D. and Warkup, C. (1998). Differences in the calpain enzyme system in tough and tender samples of porcine longissimus dorsi. In *Proceedings of the British Society of Animal Science*, p. 14.

Sentandreu, M. A., Coulis, G. and Ouali, A. (2002). Role of muscle endopeptidases and their inhibitors in meat tenderness. *Trends Food Sci. Tech.* 13:400–21.

Solis-Cohen, S. (1900) The use of adrenal substance in the treatment of asthma. *J. Am. Med. Assoc.* 34:1164–6.

Solomon, M. B., Campbell, R. G., Steele, N. C. and Caperna, T. J. (1991). Effects of exogenous porcine somatotropin administration between 30 and 60 kilograms on longissimus muscle fiber morphology and meat tenderness of pigs grown to 90 kilograms. *J. Anim Sci.* 69:641–5.

Spurlock, D. M., McDaneld, T. G. and McIntyre, L. M. (2006) Changes in skeletal muscle gene expression following clenbuterol administration. *BMC Genomics* 7:320. doi:10.1186/1471-2164-7-320.

Suster, D., Leury, B. J., King, R. H. and et al. (2004) Interrelationships between porcine somatotropin (pST), betaine, and energy level on body composition and tissue distribution of finisher boars. *Aust. J. Agric. Res.* 55:983–90. doi:10.1071/AR04029.

Tomas, F. M., Campbell, R. G., King, R. H., Johnson, R. J., Chandler, C. S. and Taverner, M. R. (1992). Growth hormone increases whole-body protein turnover in growing pigs. *J. Anim. Sci.* 70: 3138–43.

Vann, R. C., Nguyen, H. V., Reeds, P. J., Burrin, D. G., Fiorotto, M. L., Steele, N. C. and et al. (2000). Somatotropin increases protein balance by lowering body protein degradation in fed, growing pigs. *Am. J. Physiol. Endocrinol. Metab.* 278:E477–83.

Vestergaard, M., Henckel, P., Oksbjerg, N. and et al. (1994) The effect of cimaterol on muscle fiber characteristics, capillary supply, and metabolic potentials of longissimus and semitendinosus muscles from young Friesian bulls. *J. Anim. Sci.* 72:2298–306.

Vize, P. D., Michalska, A. E., Ashman, R. and et al. (1988) Introduction of a porcine growth hormone fusion gene into transgenic pigs promotes growth. *J. Cell Sci.* 90:295–300.

Warriss, P. D., Kestin, S. C., Rolph, T. P. et al. (1990b). The effects of the beta-adrenergic agonist salbutamol on meat quality in pigs. *J. Anim. Sci.* 68:128–36.

Whipple, G., Koohmaraie, M., Dikeman, M. E., Crouse, J. D., Hunt, M. C. and Klemm, R. D. (1990) Evaluation of attributes that affect longissimus muscle tenderness in Bos taurus and Bos indicus cattle. *J. Anim. Sci.* 68:2716–28.

Whipple, G., Hunt, M. C., Klemm, R. D., Kropf, D. H., Goodband, R. D. and Schricker, B. R. (1992) Effects of porcine somatotropin and supplemental lysine on porcine muscle histochemistry. *J. Muscle Foods* 3:217–27.

Williams, N. H., Cline, T. R., Schinckel, A. P. and Jones, D. J. (1994) The impact of ractopamine, energy intake, and dietary fat on finisher pig growth performance and carcass merit. *J. Anim. Sci.* 72:3152–62.

Chapter 13

Use of probiotics and prebiotics in pig nutrition in the post-weaning period

Ingunn Stensland, Linley Valley Pork, Australia and John R. Pluske, Murdoch University, Australia

1 Introduction

It is evident that the microbiota, that is, the ecological community of commensal, symbiotic and pathogenic microorganisms, is immensely important to the host, particularly in the gastrointestinal tract (GIT). The microbiota, particularly bacteria, plays a part in key physiological processes and can affect or influence the host in many ways including prevention of establishment of pathogenic bacteria, modulation of the immune system (both systemically and at the level of the GIT), detoxification, production of vitamins and short-chain fatty acids and facilitation of digestion and absorption processes. The succession of the microbiota starts in the young pig during the birthing process and also immediately after birth, but changes rapidly in the period after weaning that is coincidental with the myriad of changes that occurs during this process (Frese et al. 2015). Antibiotics, especially the so-called antibiotic growth promotants (AGP) and/or minerals such as zinc oxide (ZnO) and/or copper sulphate ($CuSO_4$), have been traditionally included in pig diets after weaning to help with the prevention and/or control of diseases such as post-weaning colibacillosis. However, due to concerns in relation to the development of antimicrobial resistance, as well as accumulation of minerals such as zinc in the environment, other strategies have been researched as either replacements or alternatives to AGP.

Probiotics are live beneficial microorganisms (bacteria) that may exert a beneficial effect on the host (Hou et al. 2015) and have been suggested as an alternative as some bacteria

http://dx.doi.org/10.19103/AS.2017.0013.28

can aid in the establishment or composition of the GIT microbiota, making it more diverse, competitively inhibiting the establishment of pathogenic bacteria and stimulating the immune system (Callaway et al. 2008). On these grounds, supplementing probiotics at birth and weaning may benefit the pigs, helping them overcome the challenges in early life and around the time of weaning. Prebiotics, on the other hand, are non-digestible food ingredients, and are thought to benefit the host by providing substrates for the beneficial microbiota, giving them a competitive advantage. The term synbiotics is used when the combination of the probiotics and prebiotics is used and they are purported to exert a beneficial effect on the host (Gaggìa, Mattarelli and Biavati 2010), and may be more beneficial compared to supplementing with either alone. Despite considerable research on these feed additives, the variety in different strains/types, doses and duration of supplementation to pigs makes it difficult to devise firm and generalised conclusions towards their overall effectiveness and suitability under different conditions of pork production. There is a need for a better understanding on how the microbiota is composed and how it functions, in addition to how probiotics, prebiotics and synbiotics work and the complex interactions that occur in the GIT ecosystem.

In this chapter, we will briefly review the microbiota of the GIT of the young pig and the important role it plays in the early stages of life. We will define probiotics, prebiotics and synbiotics, in addition to reviewing the conducted research on the topic, including a case study where a probiotic product (Peribios™) has been used commercially to modulate the production around parturition. We will also cover some possible future trends and examine where to look for more information on the topic.

2 Microbiota and their importance to the pig

The microbiota facilitate a key number of vital processes in the host (i.e. the pig), where bacteria are the predominant microorganisms that colonise the GIT of pigs. The relationship between the microbiota and the host is dynamic, providing the microbiota with a habitat, nutrition and the opportunity to reproduce, whilst the host benefits from a stable functional GIT ecosystem. In quite general terms, if the microbiota is healthy and well 'balanced', it enables effective digestion and absorption of nutrients, in addition to an increase in the resistance to infectious diseases (Ducatelle et al. 2015; Bäckhed et al. 2005; Sears 2005; Kyriakis et al. 1999).

The microbiota is present along the entire length of the GIT with the largest population and greatest diversity distally in the caecum and the colon. Due to a rapid transit time and more variable pH in the stomach and upper parts of the small intestine, bacterial adhesion and proliferation can be difficult; hence, the absolute numbers of bacteria and their diversity are lower. As the transit time slows down in the distal small intestine, an increase in the proliferation and colonisation is possible. There are trillions of bacteria present in the GIT of the pig; however and somewhat surprisingly, they tend to group into a relatively small number of phyla, with the most common bacteria generally being *Lactobacillus*, *Streptococcus*, *Peptococcus*, *Eubacterium*, *Clostridium*, *Bifidobacterium* and *Bacteroides*. Microbial communities have been found to be relatively stable from days 1 to 21 of life, with Bacteroidaceae, Clostridiaceae, Lachnospiraceae, Lactobacillaceae and Enterobacteriaceae composing the majority of the microbiota throughout this period. After weaning, the populations of *Bacteroides* and Enterobacteriaceae, generally decline, whilst the populations of Lactobacillaceae, Prevotellaceae, Ruminococcaceae and Veilonellaceae generally increase. This change in populations is largely due to the rapid change in diet, from a milk-based diet to a more plant-based adult feed (Frese et al. 2015; Heo et al. 2009, 2012; Kenny et al. 2011; Halas et al. 2007; De Angelis et al. 2006; Guarner and Malagelada 2003).

The formation of the normal GIT microbiota therefore starts at birth, as the piglet is exposed to microorganism in their immediate environment. Various bacteria are introduced into the GIT via a passage through the vagina and ingestion of faeces from the sow. This microbiota shows a gradual and natural succession, except for perturbations from when there are major nutritional changes such as after weaning. The formation of the GIT microbiota is influenced by factors such as intestinal pH, availability of substrate, mucus secretion, redox reactions, stress, peristalsis and transit time along the GIT. In a fully mature GIT environment, all niches are occupied, and all nutrients are nearly utilised. This typically prevents pathogenic bacteria from achieving a position in the ecosystem, and is critical to overall health and efficient production (Heo et al. 2012; Callaway et al. 2008; Hopwood and Hampson 2003; Demecková et al. 2002).

Weaning under natural conditions is a gradual process, which allows the piglet to be introduced to solid feed and the new fermentative capacity that is increasingly established over a period of approximately ten weeks. The commercial process, however, occurs much earlier (e.g. anywhere between 18 and 35 days of age) and is abrupt, and the piglet has to adapt to the myriad of changes in a very short period of time. It takes two to three weeks after weaning for the GIT microbiota to adapt and entirely cultivate their fermentative capacity, and hence to reach a high level of stability. Therefore, supplementing strategic feed additives around this time (to be discussed later) may help reinforce the composition of the microbiota and navigate the composition in a direction that may mitigate some of the post-weaning challenges. This abrupt change at weaning is challenging for the piglet and may lead to colibacillosis (post-weaning diarrhoea; PWD), stunted growth rates, decreased welfare and increased morbidity and/or mortality (Giang et al. 2011; Castillo et al. 2007; De Angelis et al. 2006).

A healthy microbiota is generally regarded as being characterised by high levels of diversity, where the diversity is thought to allow for a more rapid adaptation to environmental changes. The microbiota, and in particular the beneficial bacteria, is important for the GIT 'health' of young pigs as it stimulates the immune system and inhibits pathogens, and hence impacts on welfare and production. The microbiota coats the GIT preventing the adhesion of pathogens, produces natural antibiotics and antifungals, detoxifies food components, maintains barrier functions, provides vitamin and cofactor production and encourages anti-inflammatory responses. This healthy microbiota is particularly important during stressful periods in life, such as weaning, as the immune system is still immature and the piglet still relies on the sows' milk for the immune function to prevent pathogenic growth and disease (Ducatelle et al. 2015; Kenny et al. 2011; Castillo et al. 2007; Marchesi and Shanahan 2007).

In this respect, the microbiota plays a critical role in supporting the development and function of the gut-associated immune system. Animals raised in sterile environments show a profound retardation in the development of GIT morphology and immune function. When comparing conventional and germ-free pigs, for instance, findings suggest reduced intestinal motility and a weaker/underdeveloped immune system in germ-free pigs compared to conventionally reared pigs. However, the supplementation of a microbiota similar to that found in conventionally reared piglets reverses these characteristics. There is a possibility that piglets born in commercial conditions, where the housing is regularly cleaned and/or sterilised between batches of farrowing sows, obtain a considerably different microbiota compared to pigs born in, for example, a natural outdoor farrowing system. Inman et al. (2010) and Mulder et al. (2009) demonstrated that in addition to significantly affecting the composition of the microbiota, being exposed to an appropriate microbiota at an early age also affected the elements of the innate immune function in both neonates and adult animals. As the microbiota play a critical part in the development of the immune system and the functional adult structure, early life is an important time in the establishment of a healthy, diverse and beneficial microbiota (Cho, Zhao and Kim 2011; Boirivant, Amendola and Butera 2008; Wagner 2008; Calder et al. 2006; Williams et al. 2006; McFall-Ngai 2002).

The changes associated with weaning cause the GIT microbiota to restructure, and a decrease in milk-utilising bacteria, such as lactobacilli, and an increase in potentially pathogenic bacteria can be observed. The increase in potentially pathogenic bacteria, coupled with the decrease in beneficial bacteria, may lead to enteric diseases such as PWD caused by enterotoxigenic or enteropathogenic strains of *Escherichia coli*, negatively impacting welfare and production (Gareau, Wine and Sherman 2009; Heo et al. 2009; Boudry et al. 2004; Pluske et al. 2002). Traditionally, in-feed antimicrobials such as antibiotics, ZnO and/or $CuSO_4$ have been supplemented in diets and/or in the water supply to aid with the prevention of PWD, and enhance the production performance and welfare and reduce morbidity and mortality. However, due to concerns related to the development of antimicrobial resistance with antibiotics of human relevance, environmental accumulation of some metals, as well as bans of certain antimicrobials, there is a worldwide recognition that other (viable) in-feed strategies need to be explored. A discussion on probiotics and prebiotics, especially their role in the development of the beneficial microbial population in the GIT, will now be described.

3 Probiotics

Over the years, many definitions have been proposed for probiotics; however, the most accepted definition is that of 'live microorganisms, which when administered in adequate amounts confer a health benefit on the host' (FAO/WHO 2001). There are several conditions that microorganisms have to meet to be considered as a probiotic: the microbe (bacterium) has to be able to resist low pH and be capable of surviving contact with bile (as it needs to transition through the stomach where the pH is between 2.0 and 2.5), freely adhering to intestinal surfaces, stabilising intestinal microbiota, be non-pathogenic to the host, exert one or more beneficial functions for the host, have the ability to proliferate and colonise the GIT permanently or temporarily, as well as having an array of mechanisms to surpass pathogenic bacteria (de Lange et al. 2010; Halas et al. 2007; Tomasik and Tomasik 2003).

Published benefits from the supplementation of probiotic/probiotics in pigs date back many years and include improved absorption and digestion, stimulation of GIT immunity and increased resistance to infectious diseases of the GIT, which all together may improve the production performance. The mode of action usually varies somewhat between different probiotics; however, in general, it is suggested that it involves a competitive exclusion through mucosal adhesion, competition of nutrients in the GIT, decrease in luminal pH, production of bacteriocins/defensin and/or modulation of the mucosal and systematic immune responses and strengthening of intestinal barrier (Corcionivoschi et al. 2010; Gaggìa, Mattarelli and Biavati 2010; Jacela et al. 2010; Ng et al. 2009).

In this regard, probiotics likely compete with pathogenic bacteria for epithelial binding sites and hence prevent their colonisation. It is generally agreed that bacteria need to adhere to the intestinal epithelium to colonise and thus cause disease. It has been suggested that the competitive exclusion is based on the binding to the same glycoconjugate receptor sites on the epithelial surface (Cho, Zhao and Kim 2011; Roselli et al. 2005; Li, Jiang and Ma 2003; Walker 2000; Umesaki et al. 1997).

Probiotics also produce substances that exert an inhibitory effect on pathogenic bacteria, such as organic acids, antioxidants and bacteriocins. They are inhibitory to both Gram-positive and Gram-negative bacteria and may also affect metabolism and toxin production. The organic acids lower the pH, inhibiting the growth of bacteria that are sensitive to acidic conditions, in addition to permeating the outer membrane of Gram-negative bacteria. Acetic acid has the strongest inhibitory effect, affecting both bacteria and yeast. Lactic acids produce antibacterial peptides; however, they only have a narrow range and therefore only act on closely related bacteria. The majority of bacteriocins act on pathogenic bacteria by interfering with essential enzymes or permeabilising their membrane (Cho, Zhao and Kim 2011; Corcionivoschi et al. 2010; Wohlgemuth, Loh and Blaut 2010; Li, Jiang and Ma 2003).

Probiotics may modulate the innate immune system through the stimulation of toll-like receptors (TLRs), which modulate the regulation of pro- and anti-inflammatory cytokines from enterocytes and intestinal immune cells that are recruited to sites of inflammation. The TLRs are considered to be one of the GIT's primary ways of detecting and instigating responses to microbial molecular markers. Some probiotics, such as *Bacillus cereus* var. toyoi, have also been shown to modulate the adaptive immune system through the alteration of immunoglobulin production (Wohlgemuth, Loh and Blaut 2010; Roselli et al. 2007; Sauerwein, Schmitz and Hiss 2007; Scharek et al. 2007b; Bontempo et al. 2006).

Probiotic effects are strain-specific and therefore different results may be obtained if studies are carried out using different strains of the same probiotic species. For pigs, appropriate strains are typically selected based on pig origin, production of antimicrobial substances, tolerance to acid and bile, efficacy and safety, adhesion ability to the intestinal mucosa, antibiotic resistance pattern and stability under relevant commercial processes. The most commonly used probiotic organisms are bacillus, yeast and lactic-acid-producing bacteria (LAB), where *Lactobacillus*, *Bifidobacterium* and *Enterococcus* genera are the most common (Hou et al. 2015; Butel 2014; de Lange et al. 2010; Stein and Kil 2006). For a comprehensive review of the effects of probiotics see Llao and Nyachoti 2017, Heo et al. 2012; Corcionivoschi et al. 2010; de Lange et al. 2010; Gaggia et al. 2010 and Kenny et al. 2010, amongst some.

Lactobacillus spp. is a Gram-positive, non-spore-forming, microaerophilic rod, which is prevalent during the early stages of development of the microbiota. It has shown to lower the pH in the stomach and intestines, limit the growth of pathogens, colonise the intestinal epithelium impeding pathogenic bacteria to adhere, aid in digestion and absorption and modulate immune regulatory actions including increased macrophage activity and enhanced production of immunoglobulins. The major species of *Lactobacillus* found in the GIT of pigs are *L. amylovorus*, *L. salivarius* and *L. reuteri*, and in addition, *L. acidophilus*, *L. crispatus*, *L. johnsonii*, *L. agilis*, *L. plantarum*, *L. fermentum*, *L. brevis* and *L. mucosae* have been occasionally recovered (Ohashi and Ushida 2009; Myers 2007; Dworkin et al. 2006).

The bacteria *L. reuteri* 15007 (previously called *L. fermentum* 15007) has exhibited beneficial effects on the performance, the deterrence of diarrhoea, decreases signs of stress and modulates GIT microbiota and the immune system in pigs. Several studies have found an increase in average daily gain (ADG), average daily feed intake (ADFI) and feed conversion ratio (FCR) in weaned pigs when supplemented with *L. reuteri* (Wang et al. 2009b, 2011; Yu et al. 2008; Chang et al. 2001). In addition, improved performance, a decrease in PWD, as well as the enhancement of T-cell differentiation and inducement of ileum cytokine expression have been observed, suggesting the modulation of the immune functions. Wang et al. (2009a, 2013) found an improved antioxidant status in weaned and growing pigs, respectively, when fed a probiotic. Furthermore, improvements in the performance and a decrease in the incidence of PWD and shedding of enterotoxigenic *E. coli* (ETEC) have also been observed in studies using a combination of *L. reuteri* and *L. plantarum* (Zhao and Kim 2015; De Angelis et al. 2007). *L. plantarum* is tolerant against bile salts and low pH and has an antagonistic potential against the potential intestinal pathogens (De Vries et al. 2006). Research has found similar results as with *L. reuteri*, and Suo et al. (2012) also discovered an improvement in pork meat quality when supplemented with *L. plantarum*. A more extensive overview of research executed on different *Lactobacillus* spp. is shown in Table 1.

Enterococcus spp. represents Gram-positive, facultative anaerobic cocci of the Streptococcaceae family, and is commonly found amongst the intestinal microbiota of pigs where the most prevalent species is *E. faecium*. This bacterium is resistant to bile and low pH, is able to adhere to the intestinal mucus and produce antimicrobial substances with a broad inhibitory spectrum. *E. faecium* NCIMB 10415 (also known as SF68) has been found to reduce the pathogenic bacterial load of healthy piglets, and shows a trend for increased ADG and improved FCR, and an increase in absorption, as well as improvement of the immune system in weaned and finisher pigs (Gaggìa, Mattarelli and Biavati 2010; Myers 2007; Strompfova et al. 2006).

Increased ADG was observed in nursery piglets supplemented daily with *E. faecium* DSM 10663 NCIMB 10415 (EcF), in addition to reduced frequency of PWD (Zeyner and Boldt 2006). Taras et al. (2006) investigated the effectiveness of a long-term application of the probiotic strain *E. faecium* NCIMB 10415 on the performance and health characteristics of sows and their offspring, and found a trend for an improvement in FCR in week 8 of life. In addition, they found a trend for a decrease in the loss of piglets in the first three days of life compared to non-supplemented piglets. Similarly, Chen et al. (2006) found a trend for an increase in ADG for weeks 4–8, compared with a negative control, when investigating the effects of *E. faecium* SF68 on growth performance, nutrient digestibility, blood characteristics and faecal noxious gas content in finisher pigs. Furthermore, in a study where suckling piglets were supplemented with *E. faecium* EK13, increased leukocyte count and phagocytic activity index of leukocytes were detected. Broom et al. (2006) found a trend of decreasing levels of total IgG concentration, whilst Scharek et al. (2005) found no increase at weaning. They did, however, find increased levels of IgG at day 56, in addition to the reduction of cytotoxic T-cell population. With regard to increased absorption, Lodemann et al. (2006) observed an increase in glucose transport, which was suggested to contribute to an increase in the absorption of nutrients. Strompfova et al. (2006) also made observations, increase in total protein and urea, supporting the fact that the supplementation of *E. faecium*

Table 1 Summary of literature demonstrating the use of *Lactobacillus* spp. as a probiotic in pigs

Strain	Dose	Stage of growth	Results	Reference
L. reuteri I5007	6×10^9 CFU/g	Newborn piglets	Increased ADG (day 8–14 and 1–14), butyric, isobutyric and isovaleric acid contents and villous height in the jejunum on day 14, and decreased incidence of diarrhoea compared with the control group. Increased the colonic microbial population diversity and particularly reduced *Clostridium* spp. No significant effect on body weight, villous height or crypt depth on day 7 or crypt depth on day 14 compared with the control. No statistically significant effect was seen on lactase, maltase and sucrose levels on day 7 or 14, or branched chain fatty acids on day 7	Liu et al. 2014
L. reuteri I5007	10^8 CFU/mL	Weaned pigs	Increased ADG and FCR, decreased diarrhoea on day 10, stimulated T-cell differentiation and increased ileum interferon-γ and TNF-α expression. No statistically significant effects on ADFI	Wang et al. 2009b
L. reuteri I5007	10^8 CFU/mL	Weaned pigs	Increased ADG, ADFI and improved antioxidant status of oxidatively stressed pigs compared with non-supplemented group. No significant effect was observed on FCR or levels of cortisol and adrenalin	Wang et al. 2013
L. reuteri I5007	5.8×10^7 CFU/g	Weaned pigs	Increased ADG, ADFI and apparent crude protein digestibility. Increased serum-specific anti-OVA IgG levels. No significant effect on FCR	Yu et al. 2008
Lactobacilli complex	10^5 CFU/g	Weaned pigs	Increased ADG and ADFI in the first 14 d after weaning, and improved apparent digestibility of CP and phosphorus compared with carbadox-fed pigs. Decreased diarrhoea and *E. coli* and aerobic bacteria counts in jejunum, ileum, caecum and colon mucosa. Increased lactobacilli and anaerobic bacteria counts in jejunum, ileum, caecum and colon mucosa; however, counts in the stomach mucosa were not affected by the supplementation of the complex. Similarly, the aerobic and anaerobic counts in the duodenum mucosa were unaffected. No significant effect on FCR or VFA levels	Huang et al. 2004
L. reuteri BSA131	2×10^8, 2×10^8 CFU/g	Weaned pigs	Improved ADG, FCR and total count of faecal lactobacilli and decreased total faecal enterobacteria count compared with the negative control. No significant effect was observed for total weight gain or ADFI	Chang et al. 2001
L. reuteri X-1	10^8 CFU/g	Weaned pigs	Improved final live weight, ADG and FCR compared with the control group. Decreased serum IgG concentrations and increased serum DAO and D-lactate concentration compared with the control. In addition the supplementation increased serum cysteine levels and decreased serum levels of methionine, tyrosine and histidine. No significant effect was seen on ADFI, IL-1β, TNF-α, IL-6 or ET-1 or serum levels of threonine, glycine and lysine for pigs supplemented with lactobacilli	Wang et al. 2011

(Continued)

Table 1 (*Continued*)

Strain	Dose	Stage of growth	Results	Reference
L. reuteri I5007	1.02×10^7 CFU/g	Growing pigs	At the end of the growing phase (50 kg), there was an increase in total antioxidant capacity in serum compared with the negative control and no difference between treatments regarding the concentration of malondialdehyde, and activities of superoxide dismutase, catalase, glutathione peroxidase and copper-zinc-superoxide dismutase. At the end of the finishing phase (90 kg), activities of superoxide dismutase and glutathione peroxidase were increased and levels of malondialdehyde were decreased compared with negative and positive controls. Increased catalase activity in the liver compared with the negative control, but not the positive control. Increased activity of superoxide dismutase and copper-zinc-superoxide dismutase in muscle was increased compared with the negative control, but not the positive control. Malondialdehyde content was reduced compared with both positive and negative controls	Wang et al. 2009a
L. reuteri I5007	10^8 CFU/g	Weaned pigs	Improved ADG and final body weight at day 28, reduced apoptosis of enterocytes and attenuated oxidative stress in small intestine compared with the negative control group. No statistical difference was seen on ADFI	Wang et al. 2012
L. reuteri and *L. plantarum* complex	10^6 CFU/g	Weaned pigs	Improved ADG for days 0–28 compared with the negative control, but not differing from the positive control. However, no statistical difference for ADG for days 0–14 and days 15–28 compared with neither positive nor negative control. No statistical difference for ADFI or FCR compared with neither positive nor negative control. Improved N and GE digestibility on day 28 compared with the negative control. No statistical difference was observed for DM digestibility. Improved *E. coli* and *Lactobacillus* shedding, and diarrhoea score compared with the negative control. Improved faecal noxious gas emission compared with the negative control group, but not the positive control	Zhao and Kim, 2015
L. reuteri 3S7 and *L. plantarum* 4.1	10^{10} CFU/g	Sows and piglets	Decreased faecal excretion of enterobacteria and decreased β-glucuronidase activity of all pigs. No increase in faecal *Lactobacillus* was observed in sows; however, an increase was observed in piglets supplemented with *Lactobacillus* compared with the negative control. The endogenous microbiota did not differ between sows and piglets	De Angelis et al. 2007
L. plantarum ZJ316	1×10^9 CFU/d	Weaned pigs	Improved ADG, FCR and pork meat quality, and decreased ADFI and diarrhoea compared with pigs fed the antibiotic mequindox. No difference in mortality was observed. The villous height was greater in the ileum, jejunum and duodenum, compared with antibiotic-supplemented group. Crypt depth was decreased in the ileum of *Lactobacillus*-supplemented pigs; however, no statistical difference was observed in crypt depth of the jejunum and duodenum. Villous height:crypt depth ratio was increased in the ileum and jejunum compared with antibiotics supplemented group. SCFA levels and the overall bacterial community were not altered by the supplementation of lactobacilli	Suo et al. 2012

Strain	Dose	Stage of growth	Results	Reference
L. plantarum Lq80	10^{10} cells	Weaned pigs	Increased the lactobacilli count and population diversity compared with the control group	Takahashi et al. 2007
L. plantarum (PC-81-11-02, Alltech®)	5×10^9, 5×10^{10}, 3×10^9, 3×10^{10} CFU/mL	Weaned pigs	Increased microbiota diversity in colon and ETEC shedding on day 39, and decreased incidence of diarrhoea compared with the negative control. No statistically significant effect was observed on ADFI, ADG or FCR	Pieper et al. 2010
L. plantarum (PC-81-11-02, Alltech®)	6–7 × 10^{10} CFU/mL liquid feed	Sows and piglets	Reduced coliform count in sows faeces collected 7 days after parturition compared with control. Sows' colostrum mitogenic activity increased compared with the control. Increase in total faecal concentration of acetic and propionic acids in sows three weeks after the start of supplementation compared with the negative control	Demecková et al. 2002
L. plantarum (DSM13367), *L. acidophilus* (DSM13241), *L. casei* (ATCC55544), *L. pentosus* (DSM14025), *Bifidobacteriumanimalis* (DSM15954)	10^9 CFU/g	Neonatal piglets	Decreased atrophy of small intestine mucosa, and decreased gut dysfunction, pathogen numbers, gastric octanoic acid and gastric lactic acid contents compared with the negative control. Increased brush border enzyme activity compared with the negative control. Decreased levels of lactase and maltase compared with porcine colostrum-fed piglets. No statistical difference was observed for acetic + butyric acid levels	Siggers et al. 2008
L. acidophilus (C3), *plantarum* (1K8 and 3K2), *Enterococcusfaecium* (6H2), *Pediococcus pentosaceus* (D7)	(4 × 106), (2 × 106 and 7 × 106), (3 × 108), (3 × 106) CFU/g	Weaned pigs	Improved ADG, ADFI and FCR for the first two weeks after weaning compared with the control group. Increased levels of lactic and acetic acids in ileum and colon on days 14 and 35, and propionic acid in the ileum on day 35 compared with the control. No probiotic effect on performance during three to five weeks after weaning. There was no difference in ADG, ADFI and FCR between the different probiotic-fed piglets	Giang et al. 2010a
L. acidophilus (C3), *plantarum* (1K8 and 3K2), *Enterococcusfaecium* (6H2), *Pediococcus pentosaceus* (D7)	(4×10^6), (2×10^6 and 7×10^6), (3×10^8), (3×10^6) CFU/g	Weaned pigs	Improved ADG, ADFI and FCR in the first two weeks after weaning, and ADG and FCR for the overall period compared with the control. Improved incidences of diarrhoea and faecal score in week 2 and overall compared with the control. Improved digestibility for the first two weeks; however, no difference was observed for the period 3–5 weeks after weaning. LAB counts were increased in the stomach and ileum on days 14 and 35, as well as in the colon and rectum on day 35 compared with the control. *E. coli* counts were only decreased in the ileum on day 35 compared with the control. On day 14, an increase in OA, VFA, lactic and acetic acids was observed in the ileum, and the colon had an increase in OA, VFA and acetic acid when compared with the control. On day 35, an increase in OA and VFA could be observed in the ileum and colon, whilst an increase in acetic and propionic acids was only observed in the ileum	Giang et al. 2010b

(Continued)

Table 1 (*Continued*)

Strain	Dose	Stage of growth	Results	Reference
L. sobrius DSM 16698	10^{10} CFU/g	Weaned pigs	Increase in ADG; however, no increase in ADFI compared with the control. Reduction in ileal ETEC prevalence, but no difference in faecal score and ETEC shedding. An increased number of days with faecal score were observed compared with the control. An increase in IgA levels in saliva and blood serum was observed; however, IgA content secreted from jejunum was not affected. No effect on pH, villous height and crypt depth was observed	Konstantinov et al. 2008
L. murinus (DPC6002 and DPC6003), *L. pentosus* (DPC6004), *L. salivarius* (DPC6005), *Pediococcus pentosaceus* (DPC6006)	4×10^{10} (fermentate) or 4×10^{9} (suspension) CFU/d	Weaned pigs	Trend for increased ADG for fermented and suspension groups compared with the control. Decreased risk of diarrhoea in both groups compared with the control. Reduced shedding of *Salmonella* on day 15 post-infection for both groups compared with the control	Casey et al. 2007
L. lactis (NZ9000)	1.8 ug/mL	Weaned pigs	Final weight, overall ADG and ADFI were improved compared with the control and similar compared with antibiotic control. No effect observed on FCR. Enhanced intestinal development by increased villi height compared with the control group. Sucrase activities in duodenum and jejunum were similar compared with the antibiotic-treated group, but improved compared with the control. Sucrase levels in ileum were greater than both control and antibiotic-treated groups. The APA, APN and DPP-IV activities in the duodenum were increased compared with both control and antibiotic-treated groups. No effects on crypt depth and brush border lactase activity in the duodenum, jejunum and ileum were observed. Numbers of *E. coli* in ileum were reduced compared with the control, and similar compared with antibiotic-treated group. Both probiotic treatment and antibiotic-treated group decreased levels of *Enterococcus* compared with the control. No significant difference was observed for *Bifidobacterium* in the ileum	Wang et al. 2014

CFU = colony forming units; ADG = average daily gain; BCFA = branched chain fatty acids, FCR = feed conversion ratio; TNF-α = tumour necrosis factor alpha; ADFI = average daily feed intake; antiOVA IgG = anti-ovalbumin immunoglobulin G; CP = crude protein; VFA = volatile fatty acid; DAO = diamine oxidase; IL-1β = interleukin 1 β; IL-6 = interleukin-6; ET-1 = endothelin-1; DM = dry matter; SCFA = short chain fatty acids; ETEC = enterotoxigenic *E. coli*; LAB = lactic acid bacteria; FLF: fermented liquid feed; OA = organic acids; IgA = immunoglobulin A; APA = amino peptidase A; APN = amino peptidase N; DPP – IV = dipeptidase IV.

Table 2 Summary of literature demonstrating the use of *Enterococcus* spp. as a probiotic in pigs

Strain	Dose	Stage of growth	Results	Reference
E. faecium NCIMB 10415 (SF68)	1.6×10^6, 1.2×10^6, 1.7×10^5, 2.0×10^5 viable cells/g feed	Pregnant sows, lactating sows, nursed piglets and weaned pigs	Increased glucose transport	Lodemann et al. 2006
E. faecium EK13	10^9 CFU/mL	Newborn piglets	Decreased faecal count of CPS and CNS on day 1, *E. coli* on day 7 and increase in *Enterococcus* count on day 7 compared with the control. No statistically faecal bacterial differences were observed on day 14. Increase in bacterial count of *Enterococcus* spp. in the jejunum, ileum, colon and caecum at the end of the experiment. No other statistical differences in bacterial count were observed in the GIT. Decreased pH in duodenum on day 7; however, no differences were observed in other segments of the GIT compared with the control. Increased levels of lactic and propionic acids were observed in the colon compared with the control. No other significant changes in OA were observed in the jejunum, ileum or colon. Leukocyte, haemoglobin and haematocrit values were significantly higher at day 14, and also for phagocytic activity index of leukocytes and blood glutathione peroxidase concentrations at day 7 compared with the control. No effect was observed on body weight and ADG	Strompfova et al. 2006
E. faecium NCIMB10415 (SF68)	1.6×10^6, 1.2×10^6, 1.7×10^5, 2.0×10^5 viable cell/g of feed	Gestating sows, lactating sows, newborn piglets, weaned pigs	Pigs born per litter dead as well as nursed piglets at day 3, 7, 14 and 28 did not differ between treatments. Piglet losses during the first three days tended to be less in the probiotic group. A trend for a lower body weight compared with the control group was observed at days 42 and 49. The only significant improvement in the performance was a trend for a lower FCR during week 8 in the probiotic-fed group. Overall incidence of liquid faeces from days 28 to 56 was decreased in the probiotic group compared with the control; however, the total days with liquid faeces or diarrhoea did not differ in affected animals between treatments	Taras et al. 2006
E. faecium NCIMB10415 (SF68)	1.4×10^9 CFU/kg feed	Weaned piglets	Supplementation of *E. faecium* had no effect on performance data compared with the control, whilst ZnO tended to have an increase in ADG and ADFI on days 7–13. A statistical trend for an interaction between *E. faecium* and ZnO was observed for ADG and FCR on days 7–13. No effect on bacteriological data was observed for *E. faecium*, whereas ZnO tended to reduce the numbers of anaerobes and LAB count on day 20 after weaning. An interaction between *E. faecium* and ZnO was observed on day 6 for small intestinal tissue LAB. A trend to decrease serum IgG and increase intestinal IgA on day 20, when supplementing with *E. faecium*	Broom et al. 2006

(Continued)

Table 2 (*Continued*)

Strain	Dose	Stage of growth	Results	Reference
E. faecium NCIMB10415 (SF68)	1.6×10^9, 1.2×10^9, 1.7×10^8, 2.0×10^8 CFU/kg feed	Gestating sows, Lactating sows, suckling piglets, weaned pigs	No increase in total IgG levels at weaning; however, higher serum levels of IgG levels were observed at day 56 compared with the control. There were no statistical differences in IgA levels observed between the two treatment groups of piglets. No statistical difference in IgG or IgA levels was observed in sows. A lower percentage of $CD8^+$ (suppressor/cytotoxic) cells in upper jejunum was observed on day 14; however, there was no significant difference in the T-cell populations in the lymphoid tissues, subpopulations of helper T cells (CD4+) or relative amounts of $CD8^+$ T cells. Neither villous height nor crypt depth was affected by the probiotic treatment. A reduction in the frequency of β-haemolytic and enteropathogenic O141 *E. coli* was seen compared with the control	Scharek et al. 2005
E. faecium NCIMB 10415 (SF68)	100, 25, 30 ppm	Sows, suckling piglets, weaned pigs	Higher shedding of ***Salmonella*** compared to the control group. Increased levels of IgM in the first week after infection, and increased levels of IgA on days 7, 14, 21 and 28 compared with the control group. There was no significant difference between the two treatment groups in IgG levels. No difference was observed for body weight, body temperature and faecal consistency	Szabó et al. 2009
E. faecium NCIMB 10415 (SF68)	1.75×10^{11}	Finishing pigs	Trend for increase in ADG in weeks 4–8 compared with the control. Tendency to have a higher ADG. No significant differences were observed in ADFI and FCR. Increased digestibility of DM and N and decreased faecal NH_3-N and H_2S concentrations compared with the antibiotic-fed group. There were no effects on WBC, RBC and lymphocyte compared with the control	Chen et al. 2006
E. faecium DSM 10663 NCIMB 10415 (SF68)	2.8×10^9, 1.26×10^9, 2.9×10^8, 5.8×10^8 CFU/g	Nursing piglets	Decreased diarrhoea score and increased ADG compared with the negative control. There was no observed difference in viable born, stillborn and weaned piglets and their age at weaning between the two treatment groups	Zeyner and Boldt 2006

CPS = coagulase-positive staphylococci; CNS = coagulase-negative staphylococci; GIT = gastrointestinal tract; OA = organic acids; ADG = average daily gain; ADFI = average daily feed intake; FCR = feed conversion ratio; MLA = mesenteric lymph node; LAB = lactic acid bacteria; IgG = immunoglobulin G; IgA = immunoglobulin A; $CD4^+$ = cluster of differentiation 4; $CD8^+$ = cluster of differentiation 8; IgM = immunoglobulin M; DM = dry matter; N = nitrogen; WBC = whole blood cell count; RBC = red blood cell count.

has a positive effect on the digestibility and absorption. A more extensive overview of research on different *Enterococcus* spp. is shown in Table 2.

Bifidobacterium spp. represents Gram-positive bacteria, anaerobic rods that can produce acetic and lactic acids. Similar to *Lactobacillus*, they lower the intestinal pH and competitively inhibit the colonisation of pathogens (Myers 2007). Shu, Qu and Gill (2001) conducted an experiment supplementing weaned pigs with *B. lactis* HN019 and found a decrease in diarrhoea for the first two days after weaning as well as a decrease in rotavirus and *E. coli* in the faeces. They also observed an increase in FCR and an improved immune response compared with the control group.

Bacillus spp. represent Gram-positive, facultatively anaerobic, spore-forming organisms and are mostly soil organisms and are not usually a common constituent of the GIT microbiota. They are preferred by some probiotic manufacturers due to the high resistance of their spores to harsh environments and long-term storage at ambient temperatures. *Bacillus* spp. may grow in the GIT and consume oxygen to maintain the anaerobic environment for the prevention or therapy of GIT disorders. They are generally used alone or in combination with LAB or yeast (Hu et al. 2014; Tsukahara et al. 2013; Cho, Zhao and Kim 2011; Giang et al. 2011; Gaggìa, Mattarelli and Biavati 2010).

Bacillus spp. has been found to improve ADG and FCR, the diarrhoea score and mortality, respectively, and enhance the intestinal immune system. Improvement in the performance when supplementing weaned pigs with *B. cereus* var. toyoi, as well as a decrease in PWD, has been observed by Papatsiros et al. (2011) and Taras et al. (2005). Similar results have been found when supplemented with *B. subtilis* (Hu et al. 2014; Alexopoulos et al. 2004). Tsukahara et al. (2013), however, found no effects on the performance. Scharek et al. (2007a) found the enhancement of intestinal epithelial CD8/CD3 double positive cell populations, and CD25 + cell counts in the lamina propria mucosa, indicating an improved health status of the weaned pigs. A more extensive overview of research performed on different *Bacillus* spp. is shown in Table 3.

In contrast to bacteria that have been discussed to date, *Saccharomyces* spp. represents anaerobically fermenting yeast; however, it is also a part of the microbiota of the GIT. *Saccharomyces cerevisiae* and the strain *S. boulardii*, a biotype of *S. cerevisiae*, are available for commercial use in pig nutrition. Yeast cultures are rich sources of nutritional metabolites. Mannan oligosaccharides and β-glucans may prevent pathogen adhesion as well as strengthen the immune system. They are highly resistant to degeneration during their course through the GIT. It has also been proposed that yeast has the ability to stimulate digestion, and hence improved growth performance, in addition to aiding in the maintenance of the microbial equilibrium in the gut of young pigs (Gaggìa, Mattarelli and Biavati 2010; Price et al. 2010; Bontempo et al. 2006; Van Heugten, Funderburke and Dorton 2003). In this regard, several experiments have found the inclusion of *S. cerevisiae* to improve performance. Van der Peet-Schwering et al. (2007) found an increase in FCR for the first two weeks after weaning, in addition to a trend in improved ADG and live weight for the overall period, whereas Bontempo et al. (2006) observed an increase in the final live weight as well as ADG in weaned pigs supplemented with *S. cerevisiae* spp. *boulardii*. However, in contrast, Gagnon et al. (2007) detected a decrease in ADG in the three weeks after weaning, whereas Price et al. (2010) found no improvement in growth performance after weaning. A more extensive overview of research performed on different *Saccharomyces* spp. is shown in Table 4.

Table 3 Summary of literature demonstrating the use of *Bacillus* spp. as a probiotic in pigs

Strain	Dose	Stage of growth	Significant results	Reference
B. cereus var. toyoi	1×10^9 spores/kg feed	Weaned pigs	Decreased morbidity rates between days 0–7 and the overall period (days 0–35) compared with the negative control. No statistical difference in mortality rates between treatments. Significant decrease in diarrhoea score and ETEC count in faeces on day 14 was observed compared with the control. Average BW increased significantly at days 14, 21, 28 and 35 compared with the control. ADG improved on all days except for the period days 29–35, and FCR was improved at days 0–7, 8–14, 22–28 and the overall period compared with the control. No statistical difference in ADFI was observed compared with the control	Papatsiros et al. 2011
B. cereus var. toyoi	2.6×10^5, 4.0×10^5, 1.3×10^6, 1.4×10^6 CFU/g dry feed	Gestating sows, lactating sows, nursed piglets, weaned pigs	Enhanced intestinal immune system for piglets at the time of weaning and shortly thereafter, and decreased frequency of pathogen-associated *E. coli* serogroups compared with the control. No or limited influence on morphology of the mucous membrane with regard to villus height and crypt depth was observed. Number of goblet cells was not affected	Scharek et al., 2007a
B. cereus var. toyoi	2.6×10^5, 4.0×10^5, 1.3×10^6, 1.4×10^6 CFU/g DM	Gestating sows, lactating sows, nursed piglets, weaned pigs	Number of piglets per litter born alive and dead, as well as nursed piglets at days 3, 7, 14 and 28 of life, did not differ significantly. Number of piglets on day 7 tended to be higher compared with the control. More piglets were nursed for a longer time in the probiotic group, resulting in a significant higher sum of nursing piglets within one litter in the probiotic group. The control sow group tended to support a higher BWG per nursing day sum than the probiotic sow group. The control piglets were significantly heavier from birth to day 49, with the exception of day 42, compared with the probiotic fed pig. Nevertheless, ADG was only significantly higher in the control pigs in week 7 compared with the probiotic pigs, and was accompanied by a trend for decreased ADFI for the probiotic pigs. Contradictorily, probiotic-fed pigs had a higher ADG in week 6 and 8 compared with the control pigs. An increased FCR was observed for weeks 6, 8 and the post-weaning period compared with control pigs. Incidences of scours from after weaning reduced compared with control pigs	Taras et al. 2005

Strain	Dose	Stage of growth	Significant results	Reference
B. cereus var. toyoi	1.35×10^6 CFU/g dry feed	Gestating sows, lactating sows, nursed piglets, weaned pigs	Decrease in CD8 + T cells, and no difference in the percentage of CD4+, TcR1+, CD21+ and CD14 + compared with the control. However, the CD4+:CD8 + ratio was increased compared with the control. Increased production in PBMCs of IL-4 and IFN-γ compared with the control group. No difference in the proliferative response of peripheral immune cells compared with the control	Schierack et al. 2007
B. subtilis	1.2×10^9 CFU/kg	Weaned pigs	Decreased mortality and scouring compared to the control group. Growth performance not affected by dietary treatment. Similarly PUN, α-acid glycoprotein, TNF-α, digestive tract pH and ammonia or organic acid concentration were unaffected. Lactic acid bacterial concentrations in the colonic mucosa were increased when compared with the positive control. Probiotic-fed pigs had a more diverse microbial population and longer peaks compared with the negative control	Bhandari et al. 2008
B. subtilis *S. boulardii* LAB complex	3 mL/kg feed	Grower-finisher	No statistical difference was observed in performance data (ADG, ADFI and FRC) compared with the control. However, when supplementing with *Bacillus subtilis* and *Saccharomyces* (BS), or *B. subtilis*, *Saccharomyces* and a LAB complex (BSL), an improvement in ADG and FCR was observed in the growth period (28.8–54.2 kg) compared with the control and *B. subtilis* (B) group. In addition, an increase in the ADG for the overall period was observed. No statistical improvements in digestibility were observed compared with the control; however, BS had improved CF digestibility and BSL had improved digestibility for CP, CF, organic matter and nitrogen retention. BSL had an increase in lactic acid bacteria, and both BS and BSL had decreased counts of *E. coli* compared with control and B groups. No statistical differences were observed in the finisher period in either treatments	Giang et al. 2011
B. subtilis KN-42	2×10^9(L), 4×10^9(M) or 20×10^9(H) CFU/g	Weaned pigs	Increased FCR for treatment H in phase I compared with positive and negative controls, whereas M and L had an increased FCR compared with the negative control. In phase II, an increase in ADG was observed for M and H; in addition, M also had an increased FCR compared with the negative control. For the overall period, M and H had an increase in ADG and FCR compared with the negative control. Decreased incidence of diarrhoea for all treatments and all phases as well as the overall period for all treatments. H had increased levels of faecal *Lactobacillus* and decreased faecal levels of *E. coli* compared with the negative control. The levels of faecal *E. coli* were statistically similar for all probiotic groups compared with the positive control	Hu et al. 2014

(Continued)

Table 3 (*Continued*)

Strain	Dose	Stage of growth	Significant results	Reference
B. subtilis DB9011	5×10^8 CFU/g	Weaned pigs	No effect on the performance data (ADG, ADFI and FCR). Lower clinical oedema score, and oedema score in the ileum as well as a decrease in STEC numbers in the ileum compared with the control group	Tsukahara et al. 2013
B. subtilis CH201/ DSM5750 *B. licheniformis* CH200/ DSM5749	0.05, 0.1 or 0.2 g/kg feed with 3.2×10^9 viable spores/g	Growing pigs	ADG tended to increase in a linear fashion, whilst ADFI increased in a linear fashion. No effects observed for FRC and total tract apparent digestibility of DM and N. Decrease in NH_3 emission from slurry compared with the negative control throughout the entire experiment. Highest inclusion of probiotics decreased H_2S emission from slurry compared with the control. Mercaptan emission from slurry was not affected by dietary treatment. A decrease in pH at 72, 96 and 120 hours was observed for the two treatment groups with the highest inclusion rate of probiotics compared with the control	Wang et al. 2009c
B. subtilis *B. licheniformis*	1.28×10^6 viable spores/g of feed	Gestating sows, lactating sows, nursing piglets	Sows had higher body weight at weaning, and lost less weight during lactation compared with the control. The sows consumed more feed during the first 14 d postpartum, with their litters also consuming more creep feed during lactation compared with the control. More piglets weaned, lower pre-weaning mortality, increased weaning weight and decreased diarrhoea score compared with the control. No statistical difference in the number of piglets per litter born alive and dead, as well as piglet weight at birth and 12 days postpartum. Fat and protein content was higher in the sows milk on day 14 postpartum compared with the control. No differences in cholesterol (% of fat), lactose or total solids were observed between treatments. A lower proportion of sows returned to oestrus after service compared with the control; however, no statistical differences were observed in the weaning to oestrus interval as well as the proportion of sows with MMA syndrome	Alexopoulos et al. 2004

ETEC = enterotoxigenic *E. coli*; BW = body weight; ADG = average daily gain; FCR = feed conversion ratio; ADFI = average daily feed intake; CFU = colony forming unit; BWG = body weight gain; CD8+ + = cluster of differentiation 8; CD4+ = cluster of differentiation 4; TcRi+ = γδ T cells; CD21+ = mature B cells; CD14+ = monocytes; PBMC = peripheral blood mononuclear cell; IL-4 = interleukin-4; INF-γ = interferdon-γ; PUN = plasma urea nitrogen; TNF-α = tumour necrosis factor alpha; CF = crude fibre; CP = crude protein; DM = dry matter; STEC = Shiga toxin 2e-producing *E. coli*; MMA syndrome = mastitis, metritis and agalactia syndrome.

Table 4 Summary of literature demonstrating the use of *Saccharomyces* spp. as a probiotic in pigs

Strain	Dose	Stage of growth	Significant results	Reference
S. cerevisiae (XPC, Diamond V. Cedar Rapids, IA)	0.2% of diet	Weaned pigs	*Salmonella* infection increased RT, however, no statistical difference between dietary treatments. Tendency of decreased *Salmonella* shedding compared with the control. No effects on ADG, FCR or LW in PRE period; however a tendency of decreased ADFI compared with the control. In POST period, there was a diet × inoculation interaction, indicating that infected pigs supplemented with probiotics gained more BW and had a decreased RT compared with the control pigs. Overall, an interaction trend was observed for FCR and RT and a modest trend for faecal shedding of *Salmonella* during SCK for BW, ADFI, ADG and FCR. A diet × inoculation interaction for villi height, indicating probiotics increased villous height in non-infected pigs but had the opposite effect in *Salmonella*-challenged pigs. No differences in total species richness were observed between treatments; however, a shift in composition and apparent abundance were observed. Increased *Bacteroides* and *Lactobacillus* spp., and decreased Firmicutes compared with the control. Diet × inoculation interaction indicated increased *Bacteroides* in infected piglet-fed probiotics, and trend for an increase in total bacteria in infected pig-fed probiotics compared with infected control pigs. Increase in *Bacteroides* and *Lactobacillus* in probiotic-fed pigs during SCK was also observed	Price et al. 2010
S. cerevisiae	10^9 CFU, 10^9 CFU/kg feed	Suckling piglets, weaned pigs	No statistical difference in ADG and LW, compared with control, was observed for any treatment groups during trial. No diarrhoea was observed either. A treatment × weaning interaction was observed for folic acid plasma concentration, indicating a trend for the decrease in folic acid after weaning. No day or treatment effect on vitamin B_{12} plasma levels. No significant difference for SCB group in the leukocyte population in the ileum. A treatment × weaning interaction was observed for the percentage of macrophages in the MLN. No other treatment effect was observed for leukocyte populations in the MLN. There was a tendency for $CD4^+$ in the blood to be affected by treatment and weaning. Decreased counts of facultative anaerobic bacterial colonies compared with the control after ETEC challenge	Lessard et al. 2009

(Continued)

Table 4 (*Continued*)

Strain	Dose	Stage of growth	Significant results	Reference
S. cerevisiae spp. *boulardii* CNCM 1–1079	2 × 10^9 CFU/kg feed	Weaned pigs	Increased FCR, and a transient decrease in *E. coli* and coliform counts at four weeks after weaning compared with the control. Villous height, villous length, crypt depth, goblet cell number and thickness of the mucus layer in the small intestine were not affected by the probiotic treatment	Le Bon et al. 2010
S. cerevisiae SC47	8 × 10^9 live cells/g, 2.4 × 10^7 CFU/g	Weaned pigs	Exp. 1: no effect on growth performance or faecal *Lactobacillus* count. Reduced digestibility of DM, fat, and energy in the pre-starter phase compared with the negative control Exp. 2: no effect on growth performance in addition to reduced total bacteria and *Lactobacillus* counts compared with the negative control	Van Heugten et al. 2003
S. cerevisiae spp. *boulardii*	2 × 10^6 CFU/g feed	Weaned pigs	Increased final LW and ADG; however, no effects on ADFI and FCR. Greater villous height and crypt depth, and consequently the V:C ratio tended to be lower compared with the control. Increased numbers of epithelial cells with immunoreactive nuclei (proliferating cells), adherent mucous gel thickness and increased numbers of macrophages compared with the control	Bontempo et al. 2006
S. cerevisiae spp. *boulardii*	10^9 CFU, 10^9 CFU/kg feed	Nursing piglets, weaned pigs	No effect on ADG, pig weight, pH or concentration of acetic, propionic, butyric and lactic acids compared with the control. Interaction effect between treatment and period (before and after weaning), indicating increase in microbiota diversity in SCB after weaning compared with the control	Gagnon et al. 2007
S. cerevisiae CCTCC Y200007	4.3 × 10^9 CFU/g of LY, HKY or SFY	Weaned pigs	Increased FCR at days 15–21 and 0–21 when supplemented with LY compared with the control. No statistical difference observed for ADG and ADFI for any treatment group. Decreased levels of urea nitrogen on days 7 and 21 for LY, HKY and SFY compared with the control. On day 21, increased levels of T_3 and T_4 were observed for all three treatment groups, and in addition, LY and SFY had increased levels of GH and IGF-1 compared with the control. Increased villous height and V:C ratio on day 7 in duodenum and jejunum for LY and SFY groups compared with the control. An increase in the V:C ratio in the duodenum on day 21 for LY and SFY, and in the jejunum for LY compared with the control. Increased concentrations of IgA, IL-2 and IL-6 levels in the serum on days 7 and 21 for LY and SFY compared with the control. HKY had increased levels of IL-2 and IL-6 on day 21 compared with the control. Increased $CD4^+/CD8^+$ ratio of lymphocytes at day 7 for LY compared with the control. Both LY and SFY had increased LPR on day 7 compared with the control	Jiang et al., 2015

Strain	Dose	Stage of growth	Significant results	Reference
S. cerevisiae	0.2%	Weaned pigs	Decreased diarrhoea, and increased serum TNF-α pre- and post-ETEC challenge compared with the control. An increase in ileal colon thickness observed when compared with the control	Kiarie et al. 2012
S. cerevisiae	0.125%	Weaned pigs	Increased FCR for days 1–14, trend for increased ADG and FCR on days 15–35 compared with the control. For the overall period, a trend for increased ADG, and an increase LW and FCR observed compared with the control. No effect observed on ADFI. Dietary treatment had no effect on blood cell composition for most parameters, except for an increase in the numbers of platelets compared to the control and an increase in $CD4^+$ and $CD8^+$ compared with the positive control. Villous height and depth were not affected by dietary treatment	Van der Peet-Schwering et al. 2007
S. cerevisiae CNCM 1-4407	1010 CFU/g, 1 g, 5 g	Sows, suckling pigs, weaned pigs	Decreased daily diarrhoea score, ETEC shedding and LW after ETEC challenge compared with the control. No difference in total immunoglobulin concentrations in serum and colostrum among sows during farrowing. No difference in BW before and during ETEC challenge; however, an increased BW was observed after challenge compared with the control	Trckova et al. 2014

RT = rectal temperature; ADG = average daily gain; FCR = feed conversion ratio; LW = live weight; PRE period = days 0–14; ADFI = average daily feed intake; POST period = days 21–35; BW = body weight; SCK = days 14–21; CFU = colony forming unit; SCB = *Saccharomyces cerevisiae* spp.; INF-γ = interferon-γ; MLN = mesenteric lymph nodes; $CD4^+$ = cluster of differentiation 4; ETEC = enterotoxigenic *E. coli*; DM = dry matter; LW = live weight; V:C ratio = villous height:crypt depth ratio; LY = live yeast; HKY = heat killed yeast; SFY = superfine powder of yeast; T_3 = triiodothyronine; T_4 = tetraiodothyronine; GH = growth hormone; IGF-1 = inulin-like growth factors-1; IgA = immunoglobulin A; IL-2 = interleukin-2; IL-6 = interleukin-6; $CD4^+/CD8^+$ = cluster of differentiation 4/cluster of differentiation 8; LPR = T-lymphocyte proliferation rate; TNF-α = tumour necrosis factor alpha; $CD4^+$ = cluster of differentiation 4; $CD8^{++}$ = cluster of differentiation 8.

4 Prebiotics

A prebiotic is defined as a non-digestible food ingredient that benefits the host by stimulating the growth of beneficial bacteria in the colon (Gibson and Roberfroid 1995). Prebiotics achieve this function through being a readily available substrate for the beneficial bacteria to grow, and hence a prebiotic is expected to benefit the host in a similar matter to probiotics. Prebiotics that have been identified for use in pigs are generally carbohydrates containing a different molecular structure, namely dietary carbohydrates such as fibres and resistant starch, and non-digestible oligosaccharides (NDOs). Fructooligosaccharide (FOS), oligofructose (OF) and inulin, galactooligosaccharides, transgalacto-oligosaccharides (TOS) and lactulose are all NDOs and meet the criteria for being defined as a prebiotic. Prebiotics typically reach the large intestines without being hydrolysed or absorbed in the stomach or small intestines, they should be selective for the beneficial bacteria, and the fermentation from the substrate should have beneficial effects within the host. However, it is recognised that some prebiotics (e.g. resistant starch) are fermented to some extent the ileum. At present, only inulin and TOS fully meet these definitions. In general, main bacterial targets for manipulating populations of probiotics have been lactobacilli and *Bifidobacteria*. However, other strains have also been found to benefit from prebiotics (Cammarota et al. 2014; Slavin 2013; Gaggìa, Mattarelli and Biavati 2010; Jacela et al. 2010; Halas et al. 2007; Gibson and Roberfroid 1995).

The effects of prebiotics in pigs, therefore, have been suggested to exert their effect in two ways, namely (1) they may be fermented by beneficial bacteria such as lactobacilli and *Bifidobacteria*, which would give the beneficial bacteria a competitive advantage by giving them the opportunity to work more efficiently, and (2) they may interfere with the attachment of pathogenic bacteria to the GIT wall, which would alter the microbiota. Research has also indicated that some prebiotics may modulate the activity of the immune system directly. For a comprehensive review of the effects of prebiotics, see Heo et al. (2012), McDonald (2011), Roberfroid et al. (2010), Heo et al. (2012), McDonald (2011), Roberfroid et al. (2010), Halas et al. (2007) and Seifert and Watzl (2007).

In the weaned pig, Konstantinov et al. (2004) supplemented pigs with four dietary fermentable carbohydrates: inulin, lactulose, wheat starch and sugar beet pulp, and observed an increase in *Lactobacillus* in the ileum and the colon. Mikkelsen, Jakobsen and Jensen (2003) found little or no effect in the levels of *Lactobacillus* and *Enterococcus* in the faecal bacterial populations when supplemented with TOS or FOS; however, the faecal population of yeast was significantly higher in the pigs supplemented with TOS. No effect was observed in the *Lactobacillus* population when growing pigs (9–12 weeks of age) were supplemented with inulin; however, Loh et al. (2006) detected an increase in colonic *Bifidobacteria*. This coincides with Lynch et al. (2007), who also observed an increase in caecal *Bifidobacteria* with inulin supplementation.

The supplementation of prebiotics may aid in the establishment and diversity of the GIT microbiota, as well as improving the functionality of small and large intestines (Bauer et al. 2006; Williams, Verstegen and Tamminga 2001). Konstantinov et al. (2003) supplemented weaned pigs with sugar beet pulp and FOS, and observed an increase in the diversity as well as a more rapid stabilisation of the microbiota. In addition, Pierce et al. (2006) observed an improvement in the intestinal architecture when supplementing with inulin. Liu et al. (2008) found increased villous height and villous:crypt ratio in the ileum and jejunum when supplementing weaned pigs with chitooligosaccharide (COS).

From a production viewpoint, Houdijk et al. (1998) found no effect on the overall growth rate of weanling pigs when supplementing with FOS and TOS, although a temporary depressed feed intake was observed in the first three weeks, and subsequently a lower ADG was recorded for the first three weeks also. As opposed to this, Liu et al. (2008) found an improvement in overall ADG, ADFI and FCR when adding COS to the diet of weanling pigs. An increase in apparent total tract digestibilities of dry matter, gross energy (GE), crude protein (CP), crude fat, Ca and P compared to control pigs was also observed. They also observed a decrease in the incidence of PWD, a lower diarrhoea score, as well as lower faecal *E. coli* count at day 21. Furthermore, Awati et al. (2006) and Shim, Williams and Verstegen (2005b) found a decrease in ammonia content in the caecum, proximal colon and faeces when supplementing with a blend (unmolassed sugar beet pulp, native wheat starch, lactulose and inulin) and FOS, respectively. Shim, Williams and Verstegen (2005b) also observed a stimulation of fermentation increasing the total VFA, resulting in a lowered pH for the supplemented group; however, Smiricky-Tjardes et al. (2003) found a decrease in the fermentation and VFA production. Houdijk et al. (1998) found no change in faecal pH when supplementing with FOS and TOS. Similarly, Lynch et al. (2007) found no effect on intestinal pH with the supplementation of inulin. However, they did observe an increase in faecal N secretion and a decrease in the urine N:faecal N ratio, but no effect was observed on the N retention.

5 Synbiotics

A synbiotic is defined as the combination of a prebiotic and a probiotic (or prebiotics and probiotics). It has been suggested that the use of synbiotics would be a more beneficial strategy for preventing and controlling diseases during early life and weaning, compared with supplementing with either one of it. The prebiotics would provide a readily available substrate for the probiotics, allowing enhanced growth and enhancing their survival rate, colonisation and effects.

Some research has shown promising results for the use of synbiotics in pigs (Gaggìa, Mattarelli and Biavati 2010; Bomba et al. 2002; Gibson and Roberfroid 1995). Nemcová et al. (2007) supplemented suckling piglets with a combination of maltodextrins, FOS and *L. plantarum*, and found a decrease in adhesion of *E. coli* 08:K88 in the jejunum and colon. Shim et al. (2005a) offered a diet containing OF and a probiotic mixture containing *L. acidophilus*, *L. bulgaricus*, *B. subtilis* and *S. cerevisiae*. Piglets that were supplemented with the OF and probiotic mixture grew faster than the control group. In addition, they were found to have a significantly lower total number of coliform bacteria in the colon; an increase in *Bifidobacteria* in the ileum and the colon was also observed. Another study supplemented weaned pigs with a combination of mannan oligosaccharide, lactose, sodium acetate and ammonium nitrate as the prebiotic, and a probiotic from either bacteria, mould, yeast or a combination of the three (Lee et al. 2009). When compared to a positive control containing antibiotics, there was no significant difference in growth performance, total faecal bacteria or *E. coli* populations between any of the treatment groups. However, the Shigella counts were lower in the symbiotic-treated pigs, in addition to increased faecal acetic acid concentrations. When supplementing nursery pigs with a synbiotic consisting of probiotics (non-virulent *E. coli*) and raw potato starch, an increase in ADG was observed after infection with ETEC compared with both a negative control and the probiotic-fed group (Krause et al. 2010). Further, an increase in microbial diversity in the GIT as well as a decrease in PWD was observed. An increase in short-chain fatty acid levels in the colon was also observed compared with the control and the probiotic-fed pigs.

In summary, and as has been shown with the material presented in this chapter to date, an abundant amount of research, both under research conditions and commercially (albeit that data for the latter are often not presented), has been conducted on the production and intestinal health effects of probiotics and prebiotics, or their combination, and especially in the post-weaning period. On balance, it is difficult, however, to arrive at clear conclusions regarding their efficacy, reliability and reproducibility due to the wide-ranging variety of organisms, strains, strain origin, types of products, doses, viability of the strain in the feed and through the GIT and duration of feeding. There are also great differences in environments, production practices and feed composition and form between studies, all of which adds to the challenge of drawing firm and clear conclusions from the available data. Fundamentally, there is still a lack of detailed knowledge on the influence of the complex interactions inside the GIT ecosystem as well as the definite mode(s) of action of each probiotic strain or prebiotic material.

6 Case study: using probiotics to modulate production around parturition

A ban on the use of AGP in pig and poultry production occurred in the European Union on 1 January 2006, following voluntary withdrawals previously in Sweden and Denmark. In conjunction with limitations on the therapeutic use of ZnO in some European countries, as well as the recent FEEDAP's (Panel on Additives and Products or Substances used in Animal Feed) proposal to reduce the amount of copper from 170 mg/kg of feed to 25 mg/kg of piglet feed, the need to find an alternative nutritional means to maintain (or improve) GIT structure and function (gut health) is paramount.

One way to overcome the challenges associated with weaning piglets is the modulation of the microbiota of the GIT by adding probiotics and/or prebiotics to the feed, as already discussed in this chapter. In this respect, favourably colonising the GIT of the suckling piglet in the farrowing house with beneficial bacteria (a probiotic) to allow for a better transition after weaning is a practical example being adopted on some farms in Europe, and is the basis of this case study. Such a practice might enable a better colonisation than after weaning, where feed intake (and hence intake of the feed additive) is typically low and variable, thereby causing differential GIT colonisation and hence production outcomes.

In this example, a patented concept of Peribios™ (from DSM) is explained and an on-farm case study from Nevele, Belgium, is presented to demonstrate the hypothesis. Peribios™ is part of DSM's farm colonisation concept for improving the 'gut health' of sows and piglets around parturition, and comprises a powder with CYLACTIN® (*Enterococcus* faecium NCIMB 10415), a lactic acid fermenting probiotic, as well as some clay minerals. CYLACTIN® has the potential to be efficacious in sows and suckling piglets when supplemented to both sows and suckling piglets at 7.0×10^8 and 1×10^9 CFU/kg feed, respectively. CYLACTIN® also has the potential to be efficacious in weaned piglets when supplemented at the minimum recommended dose of 3.5×10^8 CFU/kg feed (EFSA 2015).

Table 5 Case study data during lactation following implementation of the farm colonisation and Peribios™ concept in 2014. Data were obtained from an 850 sow (Danbred) farrow-to-finish farm in Nevele, Belgium (called 'Test Farm' in the table). Piglets were weaned at approximately 21–23 days of age

Measurement	2012	2013	2014: Implementation of Peribios™ concept on the Test Farm
Live-born piglets per sow	15.1	15.4	–
Live-born piglets per sow: Test Farm	15.4	15.6	16.3
% mortality	13.6	13.7	–
% mortality: Test Farm	11.9	11.7	11.1
Weaned piglets per litter	13.1	13.3	–
Weaned piglets per litter: Test Farm	13.6	13.8	14.5
Weaned piglets per sow per year	29.6	30.2	–
Weaned piglets per sow per year: Test Farm	34.3	34.7	36.6

In this case study (Table 5), CYLACTIN® was included in the sow feed (8.0×10^8 CFU/kg feed) during the farrowing transition period and in lactation, and Peribios™ was sprinkled in the farrowing area twice per day on the suckling piglets during the first three days of life, at 50–70 g per litter per day (one 10 kg bag per 50 sows). This practice has the prospect of delivering healthier piglets, with less neonatal diarrhoea and a concomitant lower use of antibiotics, as well as providing a favourable microbiota as piglets leave the farrowing house. Data in Table 5 show the farrowing house performance in 2014, during which the concept was introduced, relative to the lactation performance on the same farm in 2012 and 2013.

A statistical analysis of the data could not be undertaken because of the practical and applied nature of the study; however, the data indicate that the intervention using the Peribios™ concept including the probiotic was beneficial for an increase in the number of live-born piglets, the number of weaned piglets per litter and an increase in approximately two piglets weaned per sow per year compared to when the concept was not applied (2012 and 2013). This improvement is larger than the genetic improvement achieved between 2012 and 2013.

7 Future trends and conclusion

Looking towards the future, a more complete understanding of the GIT microbiota and how it relates to the GIT (gut health), as well as performance, is the key. In this regard, a trend for the future – although it is underway to some extent currently – is to define 'gut health' far more succinctly than at present, where the definition (and hence the application to practise through the use of various feed additives such as probiotics and prebiotics, for example) is relatively loose. The myriad of interactions occurring in the GIT among diet, environment (external) influences and the microbiota makes the prediction of production and/or 'gut health' outcomes of using feed additives such as probiotics and prebiotics hard to be carried out. Nevertheless, rapid advances in more efficient and cheaper sequencing technologies to identify both the composition and functions of the microbiota in the GIT, as well as in the host (e.g. in the epithelium), in response to such interventions, hold promise to define in greater clarity what 'gut health' is, and what it means practically when feed additives such as probiotics and prebiotics are provided to pigs. These technologies will only become cheaper, making potentially for a more personalised (even 'on-farm') approach to the 'gut health' status of a particular farm or farms in a system in response to dietary additives.

In this regard, and given that the sizeable body of research conducted in this field has, overall, yielded generally variable results (for the reasons already provided), yet knowing that additional pork-producing countries will be under pressure to reduce the use of antimicrobials, thereby opening up the international feed additives' market even more; we believe it is inevitable that more focussed/targeted research will occur in the future. As alluded to above, more and more controlled studies are likely to be conducted on-farm, or under more commercial conditions, rather than in established research facilities because modern-day producers will require greater evidence than in the past for the efficacy of products if they are to use them in their production systems. Concomitantly, this will define under what conditions products such as probiotics and prebiotics (or a synbiotic approach) might 'work', that is, improve production, reduce PWD and reduce pig removals after weaning. Such an approach does not discount the importance of research facilities for exploring fundamental, underlying principles associated with the use of such products, because clearly, this is also very important; we are suggesting a more complementary approach to unravelling issues associated with this subject.

Finally, we believe that a synbiotic approach will likely become more prevalent than a single-use approach of probiotics and prebiotics in the future. This is because no single product can replace an antimicrobial such as AGP or $ZnO/CuSO_4$; hence, a combined approach – and likely with other products – will be necessary for the producers to adopt to overcome the complete loss of antimicrobials currently available for use in the post-weaning period. This may necessitate higher-cost diets; however, the challenge will always be there for cost-efficient feed additives to be on the market.

8 Where to look for further information

The primary sources of information related to this topic for readers should be peer-reviewed information such as that from papers, scientific articles and scientific presentations. In recent years, there has been a growth in online journals and content related to the microbiota and the GIT system, with journals such as *Beneficial Microbes*, *Emerging Microbes and Infections* and *PLoS ONE* regularly publishing relevant articles on this topic. However, it is understood that not all people have access to such information, therefore and for a good general overview of the topic, books such as *Probiotics: The Scientific Basis* (Fuller 1992), *Handbook of Prebiotics* (Gibson and Roberfroid 2008) and *Direct-Fed Microbials and Prebiotics for Animals: Science and Mechanisms of Action* (Callaway and Ricke 2011) are advised. More recently, the Food and Agricultural Organisation (2016) published a document (available online) titled, *Probiotics in animal nutrition – Production, impact and regulation* (Bajagai Y.S., Klieve A.V., Dart P.J. and Bryden W.L. 2016, *Probiotics in animal nutrition – Production, impact and regulation* (Makkar, HPS). The Food and Agricultural Organisation. Available from: http://www.fao.org/3/a-i5933e.pdf), which 'presents a unique and exhaustive review of the state-of-knowledge on the use of probiotics in various animal production systems, and their impact on animal productivity. It focuses specifically on definitions, production, mechanisms of action, applications in diverse animal production systems, effects, safety and potential public health risks of probiotics. Also covered are the labelling of probiotic products and global regulatory status of probiotics in animal feed'.

For general information and especially from a commercial or industry perspective (that can sometimes be 'sponsored content') related to probiotics and prebiotics (mostly the former), websites/subscription sites such as Pig Progress, pig333.com, Feed eNews, Pig International, Feed International and National Hog Farmer are useful sources. These sites typically provide more up to date information than that appearing through peer-reviewed sources such as journal articles, where it can take up to 12 months (or even longer) for research to be published.

As might be anticipated, this field of research has attracted considerable research effort, especially in Europe. A relatively recent COST (European Cooperation in Science and Technology) project (COST is the longest-running European framework supporting transnational cooperation among researchers, engineers and scholars across Europe) titled 'European network on the factors affecting the gastro-intestinal microbial balance and the impact on the health status of pigs (PiGutNet)' will aim to define both environmental and host genetic factors affecting the GIT microbiota and the complex interactions between microbiota and gut maturation, to maintain a healthy gut throughout life. The network will coordinate databases and unravel innovative tools to define the status of intestinal eubiosis in pigs. The most important outcomes will be genome/metabolome-wide association studies and the provision of a road map to increase pig resistance against GIT infections. This will have an important translational potential, being the foundation for European companies to develop strategies in the areas of feed additives and animal husbandry, resulting in improved animal health and welfare, consumer protection and competitive advantage for the European agriculture.

9 References

Alexopoulos, C., Georgoulakis, I., Tzivara, A., Kritas, S., Siochu, A. and Kyriakis, S. 2004, 'Field evaluation of the efficacy of a probiotic containing Bacillus licheniformis and Bacillus subtilis spores, on the health status and performance of sows and their litters', *Journal of Animal Physiology and Animal Nutrition*, vol. 88, no. 11–12, pp. 381–92.

Awati, A., Williams, B., Bosch, M., Gerrits, W. and Verstegen, M. 2006, 'Effect of inclusion of fermentable carbohydrates in the diet on fermentation end-product profile in feces of weanling piglets', *Journal of Animal Science*, vol. 84, no. 8, pp. 2133–40.

Bäckhed, F., Ley, R. E., Sonnenburg, J. L., Peterson, D. A. and Gordon, J. I. 2005, 'Host-bacterial mutualism in the human intestine', *Science*, vol. 307, no. 5717, pp. 1915–20.

Bauer, E., Williams, B. A., Smidt, H., Mosenthin, R. and Verstegen, M. W. 2006, 'Influence of dietary components on development of the microbiota in single-stomached species', *Nutrition Research Reviews*, vol. 19, no. 01, pp. 63–78.

Bhandari, S. K., Xu, B., Nyachoti, C. M., Giesting, D. M. and Krause, D. O. 2008, 'Evaluation of alternatives to antibiotics using a Escherichia coli K88 model of piglet diarrhoea: Effects on gut microbial ecology', *Journal of Animal Science*, vol. 86, pp. 836–47.

Boirivant, M., Amendola, A. and Butera, A. 2008, 'Intestinal microflora and immunoregulation', *Mucosal Immunology*, vol. 1, pp. S47–S49.

Bomba, A., Nemcova, R., Gancarcikova, S., Herich, R., Guba, P. and Mudronova, D. 2002, 'Improvement of the probiotic effect of micro-organisms by their combination with maltodextrins, fructo-oligosaccharides and polyunsaturated fatty acids', *British Journal of Nutrition*, vol. 88, no. S1, pp. S95–9.

Bontempo, V., Di Giancamillo, A., Savoini, G., Dell'Orto, V. and Domeneghini, C. 2006, 'Live yeast dietary supplementation acts upon intestinal morpho-functional aspects and growth in weanling piglets', *Animal Feed Science and Technology*, vol. 129, no. 3, pp. 224–36.

Boudry, G., Péron, V., Le Huërou-Luron, I., Lallès, J. P. and Sève, B. 2004, 'Weaning induces both transient and long-lasting modifications of absorptive, secretory, and barrier properties of piglet intestine', *The Journal of Nutrition*, vol. 134, no. 9, pp. 2256–62.

Broom, L., Miller, H., Kerr, K. and Knapp, J. 2006, 'Effects of zinc oxide and Enterococcus faecium SF68 dietary supplementation on the performance, intestinal microbiota and immune status of weaned piglets', *Research in Veterinary Science*, vol. 80, no. 1, pp. 45–54.

Butel, M.-J. 2014, 'Probiotics, gut microbiota and health', *Médecine et Maladies Infectieuses*, vol. 44, no. 1, pp. 1–8.

Calder, P. C., Krauss-Etschmann, S., de Jong, E. C., Dupont, C., Frick, J-S., Frokiaer, H., Heinrich, J., Garn, H., Koletzko, S., et al. 2006, 'Early nutrition and immunity–progress and perspectives', *British Journal of Nutrition*, vol. 96, no. 04, pp. 774–90.

Callaway, T., Edrington, T., Anderson, R., Harvey, R., Genovese, K., Kennedy, C., Venn, D. and Nisbet, D. 2008, 'Probiotics, prebiotics and competitive exclusion for prophylaxis against bacterial disease', *Animal Health Research Reviews*, vol. 9, no. 02, pp. 217–25.

Callaway, T. R. and Ricke S. C. (Eds) 2011, Direct-fed microbials and prebiotics for animals: science and mechanisms of action, *Springer-Vederlag*, New York.

Cammarota, G., Ianiro, G., Bibbo, S. and Gasbarrini, A. 2014, 'Gut microbiota modulation: Probiotics, antibiotics or fecal microbiota transplantation?', *Internal and Emergency Medicine*, vol. 9, no. 4, pp. 365–73.

Casey, P. G., Gardiner, G. E., Casey, G., Bradshaw, B., Lawlor, P. G., Lynch, P. B., Leonard, F. C., Stanton, C., Ross, R. P., et al. 2007, 'A five-strain probiotic combination reduces pathogen shedding and alleviates disease signs in pigs challenged with Salmonella enterica serovar Typhimurium', *Applied and Environmental Microbiology*, vol. 73, no. 6, pp. 1858–63.

Castillo, M., Martín-Orúe, S. M., Nofrarías, M., Manzanilla, E. G. and Gasa, J. 2007, 'Changes in caecal microbiota and mucosal morphology of weaned pigs', *Veterinary Microbiology*, vol. 124, no. 3, pp. 239–47.

Chang, Y-H., Kim, J.-K., Kim, H.-J., Kim, W.-Y., Kim, Y.-B. and Park, Y.-H. 2001, 'Selection of a potential probiotic Lactobacillus strain and subsequent in vivo studies', *Antonie Van Leeuwenhoek*, vol. 80, no. 2, pp. 193–9.

Chen, Y., Min, B., Cho, J., Kwon, O., Son, K., Kim, I. and Kim, S. 2006, 'Effects of dietary enterococcus faecium SF68 on growth performance, nutrient digestibility, blood characteristics and-faecal noxious gas content in finishing pigs', *Asian Australasian Journal of Animal Sciences*, vol. 19, no. 3, p. 406.

Cho, J. H., Zhao, P. Y. and Kim, I. H. 2011, 'Probiotics as a dietary additive for pigs: A review', *Journal of Animal and Veterinary Advances*, vol. 10, no. 16, pp. 2127–34.

Corcionivoschi, N., Drinceanu, D., Pop, I. M., Stack, D., Ştef, L., Julean, C. and Bourke, B. 2010, 'The effect of probiotics on animal health', *Scientific Papers Animal Science and Biotechnologies*, vol. 43, no. 1, pp. 35–41.

De Angelis, M., Siragusa, S., Berloco, M., Caputo, L., Settanni, L., Alfonsi, G., Amerio, M., Grandi, A., Ragni, A., et al. 2006, 'Selection of potential probiotic lactobacilli from pig feces to be used as additives in pelleted feeding', *Research in Microbiology*, vol. 157, no. 8, pp. 792–801.

De Angelis, M., Siragusa, S., Caputo, L., Ragni, A., Burzigotti, R. and Gobbetti, M. 2007, 'Survival and persistence of Lactobacillus plantarum 4.1 and Lactobacillus reuteri 3S7 in the gastrointestinal tract of pigs', *Veterinary Microbiology*, vol. 123, no. 1, pp. 133–44.

de Lange, C., Pluske, J., Gong, J. and Nyachoti, C. 2010, 'Strategic use of feed ingredients and feed additives to stimulate gut health and development in young pigs', *Livestock Science*, vol. 134, no. 1, pp. 124–34.

De Vries, M. C., Vaughan, E. E., Kleerebezem, M. and de Vos, W. M. 2006, 'Lactobacillus plantarum—survival, functional and potential probiotic properties in the human intestinal tract', *International Dairy Journal*, vol. 16, no. 9, pp. 1018–28.

Demecková, V., Kelly, D., Coutts, A., Brooks, P. and Campbell, A. 2002, 'The effect of fermented liquid feeding on the faecal microbiology and colostrum quality of farrowing sows', *International Journal of Food Microbiology*, vol. 79, no. 1, pp. 85–97.

Ducatelle, R., Eeckhaut, V., Haesebrouck, F. and Van Immerseel, F. 2015, 'A review on prebiotics and probiotics for the control of dysbiosis: Present status and future perspectives', *Animal*, vol. 9, no. 01, pp. 43–8.

Dworkin, M., Falkow, S., Rosenberg, E., Schleifer, K-H. and Stackebrandt, E. 2006, *The Prokaryotes: Vol. 4: Bacteria: Firmicutes*, Cyanobacteria, Springer Science and Business Media.

EFSA. 2015, 'Scientific opinion on the safety and efficacy of Cylactin® (Enterococcus faecium NCIMB 10415) as a feed additive for pigs for fattening, piglets and sows', *EFSA Journal*, vol. 13, no. 7, pp. 4158–76.

FAO/WHO 2001, 'Health and nutritional properties of probiotics in food including powder milk with live lactic acid bacteria', *Prevention*, vol. 5, no. 1.

Frese, S. A., Parker, K, Calvert, C. C. and Mills, D. A. 2015, 'Diet shapes the gut microbiome of pigs during nursing and weaning', *Microbiome*, vol. 3, no. 1, p. 1.

Fuller, R. (Ed) 1992, *Probiotics: The Scientific basis*, Springer, Netherlands.

Gaggìa, F, Mattarelli, P. and Biavati, B. 2010, 'Probiotics and prebiotics in animal feeding for safe food production', *International Journal of Food Microbiology*, vol. 141, pp. S15–S28.

Gagnon, N., Talbot, G., Ward, P., Roy, D., Dupuis, M., Farnworth, E., Tompkins, T. and Lessard, M. 2007, 'Evaluation of bacterial diversity in the gut of piglets supplemented with probiotics using ribosomal intergenic spacer analysis', *Canadian Journal of Animal Science*, vol. 87, no. 2, pp. 207–19.

Gareau, M. G., Wine, E and Sherman, P. M. 2009, 'Early life stress induces both acute and chronic colonic barrier dysfunction', *NeoReviews*, vol. 10, no. 4, pp. e191–7.

Giang, H. H., Viet, T. Q., Ogle, B. and Lindberg, J. E. 2010a, 'Effects of different probiotic complexes of lactic acid bacteria on growth performance and gut environment of weaned piglets', *Livestock Science*, vol. 133, no. 1, pp. 182–4.

Giang, H. H., Viet, T. Q., Ogle, B and Lindberg, J. E. 2010b, 'Growth performance, digestibility, gut environment and health status in weaned piglets fed a diet supplemented with potentially probiotic complexes of lactic acid bacteria', *Livestock Science*, vol. 129, no. 1, pp. 95–103.

Giang, H. H., Viet, T. Q., Ogle, B and Lindberg, J. E. 2011, 'Effects of supplementation of probiotics on the performance, nutrient digestibility and faecal microflora in growing-finishing pigs', *Asian-Australasian Journal of Animal Sciences*, vol. 24, no. 5, pp. 655–61.

Gibson, G. and Roberfroid, M. 1995, 'Dietary modulation of the human colonic microbiota: Introducing the concept of prebiotics', *Journal of Nutrition*, vol. 125, pp. 1401–12.

Gibson, G.R. and Roberfroid M. (Eds) 2008, *Handbook of prebiotics*, CRC Press, USA.

Guarner, F. and Malagelada, J.-R. 2003, 'Gut flora in health and disease', *The Lancet*, vol. 361, no. 9356, pp. 512–19.

Halas, D., Heo, J., Hansen, C. F., Kim, J., Hampson, D., Mullan, B. and Pluske, J. 2007, 'Organic acids, prebiotics and protein level as dietary tools to control the weaning transition and reduce post-weaning diarrhoea in piglets', *CAB Reviews: Perspectives in Agriculture, Veterinary Science, Nutrition and Natural Resources* . Vol 2 (no. 079), p. 13.

Heo, J. M., Kim, J. C., Mullan, B. P., Hampson, D. J. and Pluske, J. R. 2009, 'Feeding a diet with decreased protein content reduces indices of protein fermentation and the indices of

postweaning diarrhoea in weaned pigs challenged with an enterotoxigenic strain of Escherichia coli', *Journal of Animal Science*, vol. 87, pp. 2833–43.
Heo, J. M., Opapeju, F. O., Pluske, J. R., Kim, J. C., Hampson, D. J. and Nyachoti, C. M. 2012, 'Gastrointestinal health and function in weaned pigs: A review of feeding strategies to control post-weaning diarrhoea without using in-feed antimicrobial compounds', *Journal of Animal Physiology and Animal Nutrition*, vol. 97, pp. 201–37.
Hopwood, D. E. and Hampson, D. J. 2003, 'Interactions between the intestinal microflora, diet and diarrhoea, and their influences on piglet health in the immidiate post-weaning period'. In: J. R. Pluske, J. Le Dividich and M. W. A. Verstegen (Eds), *Weaning the Pig: Concepts and Consequences*, pp. 199–218. Nottingham University Press, United Kingdom.
Hou, C., Zeng, X., Yang, F., Liu, H. and Qiao, S. 2015, 'Study and use of the probiotic Lactobacillus reuteri in pigs: A review', *Journal of Animal Science and Biotechnology*, vol. 6, no. 1, p. 1.
Houdijk, J., Bosch, M., Verstegen, M. and Berenpas, H. 1998, 'Effects of dietary oligosaccharides on the growth performance and faecal characteristics of young growing pigs', *Animal Feed Science and Technology*, vol. 71, no. 1, pp. 35–48.
Hu, Y., Dun, Y., Li, S., Zhao, S., Peng, N. and Liang, Y. 2014, 'Effects of Bacillus subtilis KN-42 on growth performance, diarrhea and faecal bacterial flora of weaned piglets', *Asian-Australasian Journal of Animal Sciences*, vol. 27, no. 8, pp. 1131–40.
Huang, C., Qiao, S., Li, D., Piao, X. and Ren, J. 2004, 'Effects of lactobacilli on the performance, diarrhea incidence, VFA concentration and gastrointestinal microbial flora of weaning pigs', *Asian Australasian Journal of Animal Sciences*, vol. 17, no. 3, pp. 401–9.
Inman, C., Haverson, K., Konstantinov, S., Jones, P., Harris, C., Smidt, H., Miller, B., Bailey, M. and Stokes, C. 2010, 'Rearing environment affects development of the immune system in neonates', *Clinical and Experimental Immunology*, vol. 160, no. 3, pp. 431–9.
Jacela, J. Y., DeRouchey, J. M., Tokach, M. D., Goodband, R. D., Nelssen, J. L., Renter, D. G. and Dritz, S. S. 2010, 'Feed additives for swine: Fact sheets–prebiotics and probiotics, and phytogenics', *Journal of Swine Health Production*, vol. 18, no. 3, pp. 132–6.
Jiang, Z., Wei, S., Wang, Z., Zhu, C., Hu, S., Zheng, C., Chen, Z., Hu, Y., Wang, L., et al. 2015, 'Effects of different forms of yeast Saccharomyces cerevisiae on growth performance, intestinal development, and systemic immunity in early-weaned piglets', *Journal of Animal Science and Biotechnology*, vol. 6, no. 1, p. 1.
Kenny, M., Smidt, H., Mengheri, E. and Miller, B. 2011, 'Probiotics–do they have a role in the pig industry?', *Animal*, vol. 5, no. 03, pp. 462–70.
Kiarie, E., Scott, M., Krause, D., Khazanehei, H., Khafipour, E. and Nyachoti, C. 2012, 'Interactions of fermentation product and in-feed antibiotic on gastrointestinal and immunological responses in piglets challenged with K88', *Journal of Animal Science*, vol. 90, no. Suppl. 4, pp. 1–3.
Konstantinov, S. R., Awati, A., Smidt, H., Williams, B. A., Akkermans, A. D. and De Vos, W. M. 2004, 'Specific response of a novel and abundant Lactobacillus amylovorus-like phylotype to dietary prebiotics in the guts of weaning piglets', *Applied and Environmental Microbiology*, vol. 70, no. 7, pp. 3821–30.
Konstantinov, S. R., Smidt, H., Akkermans, A. D., Casini, L., Trevisi, P., Mazzoni, M., De Filippi, S., Bosi, P. and De Vos, W. M. 2008, 'Feeding of Lactobacillus sobrius reduces Escherichia coli F4 levels in the gut and promotes growth of infected piglets', *FEMS Microbiology Ecology*, vol. 66, no. 3, pp. 599–607.
Konstantinov, S. R., Zhu, W-Y., Williams, B. A., Tamminga, S, de Vos, W. M. and Akkermans, A. D. 2003, 'Effect of fermentable carbohydrates on piglet faecal bacterial communities as revealed by denaturing gradient gel electrophoresis analysis of 16S ribosomal DNA', *FEMS Microbiology Ecology*, vol. 43, no. 2, pp. 225–35.
Krause, D, Bhandari, S., House, J. and Nyachoti, C. 2010, 'Response of nursery pigs to a synbiotic preparation of starch and an anti-Escherichia coli K88 probiotic', *Applied and Environmental Microbiology*, vol. 76, no. 24, pp. 8192–200.
Kyriakis, S., Tsiloyiannis, V., Vlemmas, J., Sarris, K., Tsinas, A., Alexopoulos, C. and Jansegers, L. 1999, 'The effect of probiotic LSP 122 on the control of post-weaning diarrhoea syndrome of piglets', *Research in Veterinary Science*, vol. 67, no. 3, pp. 223–8.

Le Bon, M, Davies, H. E., Glynn, C., Thompson, C., Madden, M., Wiseman, J., Dodd, C. E. R., Hurdidge, L., Payne, G., et al. 2010, 'Influence of probiotis on gut health in the weaned pig', *Livestock Science*, vol. 133, pp. 179–81.

Lee, S. J., Shin, N. H., Ok, J. U., Jung, H. S., Chu, G. M., Kim, J. D., Kim, I. H. and Lee, S. S. 2009, 'Effects of dietary synbiotics from anaerobic microflora on growth performance, noxious gas emission and fecal pathogenic bacteria population in weaning pigs', *Asian Australasian Journal of Animal Sciences*, vol. 22, pp. 1202–8.

Lessard, M., Dupuis, M., Gagnon, N., Nadeau, E., Matte, J., Goulet, J. and Fairbrother, J. 2009, 'Administration of Pediococcus acidilactici or Saccharomyces cerevisiae boulardii modulates development of porcine mucosal immunity and reduces intestinal bacterial translocation after Escherichia coli challenge1, 2', *Journal of Animal Science*, vol. 87, no. 3, p. 922.

Li, D. F., Jiang, J. Y. and Ma, Y. X. 2003, 'Early weaning diets and feed additives'. In R. J. Xu and P. Cranwell, (Eds), *The Neonatal Pig: Gastrointestinal Physiology and Nutrition*, pp. 247–74. Nothingham University Press, United Kingdom.

Liao, S. F. and Nyachoti, M. 2017, 'Using probiotics to improve swine gut health and nutrient utilization', *Animal Nutrition*.

Liu, H., Zhang, J., Zhang, S., Yang, F., Thacker, P. A., Zhang, G., Qiao, S. and Ma, X. 2014, 'Oral administration of Lactobacillus fermentum I5007 favors intestinal development and alters the intestinal microbiota in formula-fed piglets', *Journal of Agricultural and Food Chemistry*, vol. 62, no. 4, pp. 860–6.

Liu, P., Piao, X., Kim, S., Wang, L., Shen, Y., Lee, H. and Li, S. 2008, 'Effects of chito-oligosaccharide supplementation on the growth performance, nutrient digestibility, intestinal morphology, and fecal shedding of and in weaning pigs', *Journal of Animal Science*, vol. 86, no. 10, pp. 2609–18.

Lodemann, U., Hübener, K., Jansen, N. and Martens, H. 2006, 'Effects of Enterococcus faecium NCIMB 10415 as probiotic supplement on intestinal transport and barrier function of piglets', *Archives of Animal Nutrition*, vol. 60, no. 1, pp. 35–48.

Loh, G., Eberhard, M., Brunner, R. M., Hennig, U., Kuhla, S., Kleessen, B. and Metges, C. C. 2006, 'Inulin alters the intestinal microbiota and short-chain fatty acid concentrations in growing pigs regardless of their basal diet', *The Journal of Nutrition*, vol. 136, no. 5, pp. 1198–202.

Lynch, M., Sweeney, T., Callan, J., Flynn, B. and O'Doherty, J. 2007, 'The effect of high and low dietary crude protein and inulin supplementation on nutrient digestibility, nitrogen excretion, intestinal microflora and manure ammonia emissions from finisher pigs', *Animal*, vol. 1, no. 8, pp. 1112–21.

Marchesi, J. and Shanahan, F. 2007, 'The normal intestinal microbiota', *Current Opinion in Infectious Diseases*, vol. 20, no. 5, pp. 508–13.

McDonald, P. 2011, *Animal Nutrition*, Pearson education. 7th edition, Harlow, England.

McFall-Ngai, M. J. 2002, 'Unseen forces: The influence of bacteria on animal development', *Developmental Biology*, vol. 242, no. 1, pp. 1–14.

Mikkelsen, L. L., Jakobsen, M. and Jensen, B. B. 2003, 'Effects of dietary oligosaccharides on microbial diversity and fructo-oligosaccharide degrading bacteria in faeces of piglets post-weaning', *Animal Feed Science and Technology*, vol. 109, no. 1, pp. 133–50.

Mulder, I. E., Schmidt, B., Stokes, C. R., Lewis, M., Bailey, M., Aminov, R. I., Prosser, J. I., Gill, B. P., Pluske, J. R., et al. 2009, 'Environmentally-acquired bacteria influence microbial diversity and natural innate immune responses at gut surfaces', *BMC Biology*, vol. 7, no. 1, p. 1.

Myers, D. 2007, 'Probiotics', *Journal of Exotic Pet Medicine*, vol. 16, no. 3, pp. 195–7.

Nemcová, R., Bomba, A., Gancarčíková, S., Reiffova, K., Guba, P., Koščová, J., Jonecova, Z., Scirankova, L. and Bugarský, A. 2007, 'Effects of the administration of lactobacilli, maltodextrins and fructooligosaccharides upon the adhesion of E. coli O8: K88 to the intestinal mucosa and organic acid levels in the gut contents of piglets', *Veterinary Research Communications*, vol. 31, no. 7, pp. 791–800.

Ng, S., Hart, A., Kamm, M., Stagg, A. and Knight, S. 2009, 'Mechanisms of action of probiotics: Recent advances', *Inflammatory Bowel Diseases*, vol. 15, no. 2, pp. 300–10.

Ohashi, Y. and Ushida, K. 2009, 'Health-beneficial effects of probiotics: Its mode of action', *Animal Science Journal*, vol. 80, no. 4, pp. 361–71.

Papatsiros, V., Tassis, P., Tzika, E., Papaioannou, D., Petridou, E., Alexopoulos, C. and Kyriakis, S. 2011, 'Effect of benzoic acid and combination of benzoic acid with a probiotic containing Bacillus cereus var. toyoi in weaned pig nutrition', *Polish Journal of Veterinary Sciences*, vol. 14, no. 1, pp. 117–25.
Pieper, R., Janczyk, P., Urubschurov, V., Hou, Z., Korn, U., Pieper, B. and Souffrant, W. B. 2010, 'Effect of Lactobacillus plantarum on intestinal microbial community composition and response to enterotoxigenic Escherichia coli challenge in weaning piglets', *Livestock Science*, vol. 133, no. 1, pp. 98–100.
Pierce, K., Sweeney, T., Brophy, P., Callan, J., Fitzpatrick, E., McCarthy, P. and O'Doherty, J. 2006, 'The effect of lactose and inulin on intestinal morphology, selected microbial populations and volatile fatty acid concentrations in the gastro-intestinal tract of the weanling pig', *Animal Science*, vol. 82, no. 03, pp. 311–18.
Pluske, J. R., Pethick, D. W., Hopwood, D. E. and Hampson, D. J. 2002, 'Nutritional influences on some major enteric bacterial diseases of pigs', *Nutrition Research Reviews*, vol. 15, no. 02, pp. 333–71.
Price, K., Totty, H., Lee, H., Utt, M., Fitzner, G., Yoon, I., Ponder, M. and Escobar, J. 2010, 'Use of fermentation product on growth performance and microbiota of weaned pigs during infection', *Journal of Animal Science*, vol. 88, no. 12, pp. 3896–908.
Roberfroid, M., Gibson, G. R., Hoyles, L., McCartney, A. L., Rastall, R., Rowland, I., Wolvers, D., Watzl, B., Szajewska, H., et al. 2010, 'Prebiotic effects: Metabolic and health benefits', *British Journal of Nutrition*, vol. 104, no. S2, pp. S1–S63.
Roselli, M., Finamore, A., Britti, M. S., Bosi, P., Oswald, I. and Mengheri, E. 2005, 'Alternatives to in-feed antibiotics in pigs: Evaluation of probiotics, zinc or organic acids as protective agents for the intestinal mucosa. A comparison of in vitro and in vivo results', *Animal Research*, vol. 54, no. 3, pp. 203–18.
Roselli, M., Finamore, A., Britti, M. S., Konstantinov, S. R., Smidt, H., de Vos, W. M. and Mengheri, E. 2007, 'The novel porcine Lactobacillus sobrius strain protects intestinal cells from enterotoxigenic Escherichia coli K88 infection and prevents membrane barrier damage', *The Journal of Nutrition*, vol. 137, no. 12, pp. 2709–16.
Sauerwein, H., Schmitz, S. and Hiss, S. 2007, 'Effects of a dietary application of a yeast cell wall extract on innate and acquired immunity, on oxidative status and growth performance in weanling piglets and on the ileal epithelium in fattened pigs', *Journal of Animal Physiology and Animal Nutrition*, vol. 91, no. 9–10, pp. 369–80.
Scharek, L., Altherr, B., Tölke, C. and Schmidt, M. 2007a, 'Influence of the probiotic Bacillus cereus var. toyoi on the intestinal immunity of piglets', *Veterinary Immunology and Immunopathology*, vol. 120, no. 3, pp. 136–47.
Scharek, L., Guth, J., Filter, M. and Schmidt, M. F. 2007b, 'Impact of the probiotic bacteria Enterococcus faecium NCIMB 10415 (SF68) and Bacillus cereus var. toyoi NCIMB 40112 on the development of serum IgG and faecal IgA of sows and their piglets', *Archives of Animal Nutrition*, vol. 61, no. 4, pp. 223–34.
Scharek, L., Guth, J., Reiter, K., Weyrauch, K., Taras, D., Schwerk, P., Schierack, P., Schmidt, M., Wieler, L., et al. 2005, 'Influence of a probiotic Enterococcus faecium strain on development of the immune system of sows and piglets', *Veterinary Immunology and Immunopathology*, vol. 105, no. 1, pp. 151–61.
Schierack, P., Wieler, L. H., Taras, D., Herwig, V., Tachu, B., Hlinak, A., Schmidt, M. F. and Scharek, L. 2007, 'Bacillus cereus var. toyoi enhanced systemic immune response in piglets', *Veterinary Immunology and Immunopathology*, vol. 118, no. 1, pp. 1–11.
Sears, C. L. 2005, 'A dynamic partnership: Celebrating our gut flora', *Anaerobe*, vol. 11, no. 5, pp. 247–51.
Seifert, S. and Watzl, B. 2007, 'Inulin and oligofructose: Review of experimental data on immune modulation', *The Journal of Nutrition*, vol. 137, no. 11, pp. 2563S–67S.
Shim, S., Verstegen, M., Kim, I., Kwon, O. and Verdonk, J. 2005a, 'Effects of feeding antibiotic-free creep feed supplemented with oligofructose, probiotics or synbiotics to suckling piglets

increases the preweaning weight gain and composition of intestinal microbiota', *Archives of Animal Nutrition*, vol. 59, no. 6, pp. 419–27.

Shim, S., Williams, I. and Verstegen, M. 2005b, 'Effects of dietary fructo-oligosaccharide on villous height and disaccharidase activity of the small intestine, pH, VFA and ammonia concentrations in the large intestine of weaned pigs', *Acta Agriculturae Scandinavica, Section A-Animal Science*, vol. 55, no. 2–3, pp. 91–7.

Shu, Q., Qu, F. and Gill, H. S. 2001, 'Probiotic treatment using Bifidobacterium lactis HN019 reduces weanling diarrhea associated with rotavirus and Escherichia coli infection in a piglet model', *Journal of Pediatric Gastroenterology and Nutrition*, vol. 33, no. 2, pp. 171–7.

Siggers, R. H., Siggers, J., Boye, M., Thymann, T., Mølbak, L., Leser, T., Jensen, B. B. and Sangild, P. T. 2008, 'Early administration of probiotics alters bacterial colonization and limits diet-induced gut dysfunction and severity of necrotizing enterocolitis in preterm pigs', *The Journal of Nutrition*, vol. 138, no. 8, pp. 1437–44.

Slavin, J. 2013, 'Fiber and prebiotics: Mechanisms and health benefits', *Nutrients*, vol. 5, no. 4, pp. 1417–35.

Smiricky-Tjardes, M., Grieshop, C., Flickinger, E., Bauer, L. and Fahey, G. 2003, 'Dietary galactooligosaccharides affect ileal and total-tract nutrient digestibility, ileal and fecal bacterial concentrations, and ileal fermentative characteristics of growing pigs', *Journal of Animal Science*, vol. 81, no. 10, pp. 2535–45.

Stein, H. H. and Kil, D. Y. 2006, 'Reduced use of antibiotic growth promoters in diets fed to weanling pigs: Dietary tools, part 2', *Animal Biotechnology*, vol. 17, no. 2, pp. 217–31.

Strompfova, V., Marciňáková, M., Simonová, M., Gancarčíková, S., Jonecová, Z., Sciranková., Koščová, J., Buleca, V., Čobanová, K., et al. 2006, 'Enterococcus faecium EK13—an enterocin a-producing strain with probiotic character and its effect in piglets', *Anaerobe*, vol. 12, no. 5, pp. 242–8.

Suo, C., Yin, Y., Wang, X., Lou, X., Song, D., Wang, X. and Gu, Q. 2012, 'Effects of lactobacillus plantarum ZJ316 on pig growth and pork quality', *BMC Veterinary Research*, vol. 8, no. 1, p. 1.

Szabó, I., Wieler, L. H., Tedin, K., Scharek-Tedin, L., Taras, D., Hensel, A., Appel, B. and Nöckler, K. 2009, 'Influence of a probiotic strain of Enterococcus faecium on Salmonella enterica serovar Typhimurium D. T.104 infection in a porcine animal infection model', *Applied and Environmental Microbiology*, vol. 75, no. 9, pp. 2621–8.

Takahashi, S., Egawa, Y., Simojo, N., Tsukahara, T. and Ushida, K. 2007, 'Oral administration of Lactobacillus plantarum strain Lq80 to weaning piglets stimulates the growth of indigenous lactobacilli to modify the lactobacillal population', *The Journal of General and Applied Microbiology*, vol. 53, no. 6, pp. 325–32.

Taras, D., Vahjen, W., Macha, M. and Simon, O. 2005, 'Response of performance characteristics and fecal consistency to long-lasting dietary supplementation with the probiotic strain Bacillus cereus var. toyoi to sows and piglets', *Archives of Animal Nutrition*, vol. 59, no. 6, pp. 405–17.

Taras, D., Vahjen, W., Macha, M. and Simon, O. 2006, 'Performance, diarrhea incidence, and occurrence of virulence genes during long-term administration of a probiotic strain to sows and piglets', *Journal of Animal Science*, vol. 84, no. 3, pp. 608–17.

Tomasik, P. J. and Tomasik, P. 2003, 'Probiotics and prebiotics', *Cereal Chemistry*, vol. 80, no. 2, p. 113.

Trckova, M., Faldyna, M., Alexa, P., Zajacova, Z. S., Gopfert, E., Kumprechtova, D., Auclair, E. and D'Inca, R. 2014, 'The effects of live yeast on postweaning diarrhea, immune response, and growth performance in weaned piglets', *Journal of Animal Science*, vol. 92, no. 2, pp. 767–74.

Tsukahara, T., Tsuruta, T., Nakanishi, N., Hikita, C., Mochizuki, M. and Nakayama, K. 2013, 'The preventive effect of Bacillus subtilus strain DB9011 against experimental infection with enterotoxcemic Escherichia coli in weaning piglets', *Animal Science Journal*, vol. 84, no. 4, pp. 316–21.

Umesaki, Y., Okada, Y., Imaoka, A., Setoyama, H. and Matsumoto, S. 1997, 'Interactions between epithelial cells and bacteria, normal and pathogenic', *Science*, vol. 276, no. 5314, pp. 964–5.

Van der Peet-Schwering, C., Jansman, A., Smidt, H. and Yoon, I. 2007, 'Effects of yeast culture on performance, gut integrity, and blood cell composition of weanling pigs', *Journal of Animal Science*, vol. 85, no. 11, pp. 3099–109.

Van Heugten, E., Funderburke, D. and Dorton, K. 2003, 'Growth performance, nutrient digestibility, and fecal microflora in weanling pigs fed live yeast', *Journal of Animal Science*, vol. 81, no. 4, pp. 1004–12.

Wagner, R. D. 2008, 'Effects of microbiota on GI health: Gnotobiotic research'. In: Huffnagle, G. B., Noverr, M. C. (Eds) *GI Microbiota and Regulation of the Immune System*. Landes Bioscience and Springer Science+Business Media, Springer, pp. 41–56.

Walker, W. A. 2000, 'Role of nutrients and bacterial colonization in the development of intestinal host defense', *Journal of Pediatric Gastroenterology and Nutrition*, vol. 30, pp. S2–S7.

Wang, A., Cai, C., Zeng, X., Zhang, F., Zhang, G., Thacker, P., Wang, J. and Qiao, S. 2013, 'Dietary supplementation with Lactobacillus fermentum I5007 improves the anti-oxidative activity of weanling piglets challenged with diquat', *Journal of Applied Microbiology*, vol. 114, no. 6, pp. 1582–91.

Wang, A., Yi, X., Yu, H., Dong, B. and Qiao, S. 2009a, 'Free radical scavenging activity of Lactobacillus fermentum in vitro and its antioxidative effect on growing–finishing pigs', *Journal of Applied Microbiology*, vol. 107, no. 4, pp. 1140–8.

Wang, A., Yu, H., Gao, X., Li, X. and Qiao, S. 2009b, 'Influence of Lactobacillus fermentum I5007 on the intestinal and systemic immune responses of healthy and E. coli challenged piglets', *Antonie Van Leeuwenhoek*, vol. 96, no. 1, pp. 89–98.

Wang, D., Xu, S., Lin, Y., Fang, Z., Che, L., Xue, B. and Wu, D. 2014, 'Recombinant porcine epidermal growth factor-secreting Lactococcus lactis promotes the growth performance of early-weaned piglets', *BMC Veterinary Research*, vol. 10, no. 1, p. 1.

Wang, S.-P., Yang, L., Tang, X.-S., Cai, L.-C., Liu, G., Kong, X.-F., Blachier, F. and Yin, Y.-L. 2011, 'Dietary supplementation with high-dose Bacillus subtilis or Lactobacillus reuteri modulates cellular and humoral immunities and improves performance in weaned piglets', *Journal of Food Agriculture Environment*, vol. 9, no. 2, pp. 181–7.

Wang, X., Yang, F., Liu, C., Zhou, H., Wu, G., Qiao, S., Li, D. and Wang, J. 2012, 'Dietary supplementation with the probiotic Lactobacillus fermentum I5007 and the antibiotic aureomycin differentially affects the small intestinal proteomes of weanling piglets', *The Journal of Nutrition*, vol. 142, no. 1, pp. 7–13.

Wang, Y., Cho, J., Chen, Y., Yoo, J., Huang, Y., Kim, H. and Kim, I. 2009c, 'The effect of probiotic BioPlus 2B® on growth performance, dry matter and nitrogen digestibility and slurry noxious gas emission in growing pigs', *Livestock Science*, vol. 120, no. 1, pp. 35–42.

Williams, A. M., Probert, C. S., Stepankova, R, Tlaskalova-Hogenova, H., Phillips, A. and Bland, P. W. 2006, 'Effects of microflora on the neonatal development of gut mucosal T cells and myeloid cells in the mouse', *Immunology*, vol. 119, no. 4, pp. 470–8.

Williams, B. A., Verstegen, M. W. and Tamminga, S 2001, 'Fermentation in the large intestine of single-stomached animals and its relationship to animal health', *Nutrition Research Reviews*, vol. 14, no. 02, pp. 207–28.

Wohlgemuth, S., Loh, G. and Blaut, M. 2010, 'Recent developments and perspectives in the investigation of probiotic effects', *International Journal of Medical Microbiology*, vol. 300, no. 1, pp. 3–10.

Yu, H., Wang, A., Li, X. and Qiao, S. 2008, 'Effect of viable Lactobacillus fermentum on the growth performance, nutrient digestibility and immunity of weaned pigs', *Journal of Animal and Feed Sciences*, vol. 17, no. 1, pp. 61–9.

Zeyner, A. and Boldt, E. 2006, 'Effects of a probiotic Enterococcus faecium strain supplemented from birth to weaning on diarrhoea patterns and performance of piglets', *Journal of Animal Physiology and Animal Nutrition*, vol. 90, no. 1–2, pp. 25–31.

Zhao, P. and Kim, I. 2015, 'Effect of direct-fed microbial on growth performance, nutrient digestibility, fecal noxious gas emission, fecal microbial flora and diarrhea score in weanling pigs', *Animal Feed Science and Technology*, vol. 200, pp. 86–92.

Chapter 14

Meeting individual nutrient requirements to improve nutrient efficiency and the sustainability of growing pig production systems

Candido Pomar, Agriculture and Agri-Food Canada (AAFC), Canada; Ines Andretta, Universidade Federal do Rio Grande do Sul, Brazil; and Luciano Hauschild, Universidade Estadual Paulista, Brazil

1 Introduction

Farm animals are raised to produce commodities such as food (e.g. meat and dairy products), fibre and labour. The nutrient losses associated with the conversion of feed nutrients into animal products contribute significantly to production costs and to the excretion of excess nutrients and nutrient constituents (e.g. nitrogen, phosphorus, trace minerals, carbon and methane). Feed costs may account for over 60% of the overall production costs for various livestock such as pigs, poultry and cattle. However, the efficiency with which farm animals transform the energy and nutrients extracted from feed into animal products is generally low. For instance, protein, which is one of the most limiting and costly nutrients in livestock feed, is retained by growing pigs at an efficiency rate ranging from 15% (Flachowsky and Kamphues, 2012) to 33% (Dourmad et al., 1999; Poulsen et al., 1999). Dietary nutrients such as protein and phosphorus, which are not retained in the animal's body, are excreted via the faeces and urine, and can contribute to environmental problems such as nitrate pollution of ground and surface water. Improving feed and nutrient efficiency is essential

http://dx.doi.org/10.19103/AS.2017.0013.31

because of the challenges associated with the expected increase in the human population, the limited arable land and the environmental problems frequently linked with farm animal production (Niemann et al., 2011).

Precision livestock farming is an innovative production approach that improves the management of livestock by using the principles and technologies of process engineering (Wathes et al., 2008). Precision animal feeding is part of the precision livestock farming approach and involves the use of feeding techniques that deliver the right amount of feed with the right composition at the right time to a group of animals (Parsons et al., 2007; Niemi et al., 2010; Cangar et al., 2008; Pomar et al., 2014) or to individual animals within the group (Andretta et al., 2014, 2016b). The on-farm application of precision feeding requires the design and development of measuring devices (e.g. to measure the animals' feed intake and body weight [BW]), computational methods (e.g. to estimate voluntary feed intake and nutrient requirements in a timely manner based on the animals' actual growth) and feeding systems capable of providing the required amount of feed with a composition that will optimize animal performance while minimizing the use of farm resources (Pomar et al., 2009). Meeting individual pig nutrient requirements on a daily basis using precision feeding techniques is the most efficient method of improving nutrient efficiency while ensuring the sustainability of growing pig production systems. This chapter describes the principal sources of variations in nutrient inefficiency, the methods used to estimate the amount of nutrients required by animals and the impact of using precision feeding techniques in growing-finishing production systems.

2 Sources of nutrient inefficiency

Only some of the nutrients ingested by livestock contribute to the animal's metabolism and are retained in animal products. In growing animals, nutrient efficiency can be expressed as the ratio between the amount of nutrients ingested and the amount actually retained in the animal's body. Nutrient loss is the difference between the amount of nutrients ingested and retained, and occurs at different stages of the nutrient use process. Some of the ingested nutrients are lost in the digestive tract through cellular intestinal desquamation and secretion of enzymes and other substances during digestion. After absorption, another portion of the dietary nutrients is used for maintenance, for which the main determinants are basal endogenous intestinal losses, skin and hair losses, and the minimum nutrient catabolism associated with basal turnover of body tissues and the irreversible synthesis of essential body compounds (Moughan, 1999). Nutrients are also lost during the development of animal products (e.g. body lean, milk and eggs). In growing animals, losses associated with the use of the first-limiting amino acid for body protein deposition (PD) can largely be attributed to its inevitable catabolism (Mohn et al., 2000a). These inevitable amino acid losses should be differentiated from other metabolic losses related to the preferential amino acid catabolism, which results from the catabolism of excess amino acids, from the excretion of chemically unavailable absorbed amino acids (e.g. heat-damaged proteins) (van Barneveld et al., 1994) and from the use of amino acids for the synthesis of non-protein body compounds (Moughan, 1989). In growing pigs fed with cereal-soybean meal-based diets, the sum of the undigested protein and the losses associated with digestion, maintenance functions and body PD may represent 40% of the total amount of protein ingested, and similar values have been obtained for dietary phosphorus (Dourmad et al., 1999). These sources of nutrient inefficiency are difficult to minimize because they occur during digestion and essential metabolic processes.

In addition to the inevitable nutrient losses associated with the digestion and metabolism of nutrients, farm animals may receive more nutrients than they need, and all excess nutrients are excreted and contribute to the overall nutrient inefficiency. To minimize the supply of excess nutrients and thus maximize the efficiency with which they are used by the animal, it is essential to a) precisely estimate the amount of nutrients that will be available for the animals' metabolism in the feed ingredients provided to the animals, b) estimate the amount of nutrients required by each animal throughout the growing period, c) formulate balanced diets that limit excess nutrients and d) concomitantly adjust the dietary supply of nutrients to match the animals' estimated requirements (Pomar et al., 2009).

3 Current methods of estimating nutrient requirements

Energy and nutrients such as amino acids, minerals, vitamins and water have to be provided to animals in adequate amounts for maintenance, growth and production. These nutrients are provided by the feed ingredients, whose nutritional value is estimated based on their composition, digestibility, availability and metabolic fate, taking into consideration the addition of enzymes, feed physical treatments and other factors (Noblet and van Milgen, 2004; National Research Council [NRC], 2012; Stein et al., 2007). The methods used to precisely estimate available nutrients is not covered in this chapter, but this step should be considered essential for improving nutrient efficiency and successfully applying precision feed formulation techniques and precision livestock farming in general.

Once the nutritional value of the feed ingredients is determined, the amount of nutrients that should be provided to the animals to match their dietary needs has to be carefully estimated. To estimate the optimal nutrient supply, it is essential to identify the animals that will receive the feed, the length of time during which the animals will receive the feed, the farmer's production objectives and other factors (e.g. farm environment) that may affect the animals' response to the nutrient supply. The concept of nutrient requirements therefore needs to be clearly defined, taking into consideration the production objectives (e.g. feed efficiency) and feeding method (e.g. number and length of feeding phases in group feeding). The feeding method is especially important, because one animal and a heterogeneous group of animals do not respond the same way to a given nutrient intake (Pomar et al., 2003). Nutrient requirements can be defined as the minimum amount of nutrients needed by an animal to satisfy the biological processes associated with tissue maintenance and repair, growth and all other productive processes, thus preventing signs of deficiency and allowing the animal to perform its necessary functions normally. When feeding one animal for a short period, these biological processes can be associated with body maintenance, growth and production, and the animal's nutrient requirements can be estimated by adding the amounts of nutrients needed for each of these processes. However, when feeding a population of animals, the concept of maintenance, growth and production requirements may not be appropriate. Individual animals (e.g. pigs) within a given population differ significantly in terms of BW, growth potential, health and other characteristics, and therefore differ in terms of the amount of nutrients they each need at a given point in time (Knap, 2000; Knap et al., 2003). Using the average pig to characterize average population responses or to estimate the nutrient requirements of the population can have important limitations, given that the response of a population to varying levels of nutrients differs in magnitude and form from that of an individual animal. For example, the linear-plateau response used to represent the response to amino acid supply in growing pigs is compatible with the curvilinear response that is observed under experimental and production conditions (Pomar et al., 2003; St-Pierre and Weiss, 2012). Therefore, when feeding populations of animals, nutrient requirements should be seen as the amount of nutrients needed for specified production purposes, such as optimal growth rate, PD and feed efficiency (Hauschild et al., 2010), or for the optimal balance between the proportions of pigs that will be overfed and underfed (Brossard et al., 2009; Pomar et al., 2003). Furthermore, because factors such as population heterogeneity, production environment, diseases and other stressors may change the amount of certain nutrients needed for optimal performance, nutrient requirements should not be seen as a fixed amount, but rather as a value that evolves over time and is specific to each animal population and production environment (St-Pierre and Weiss, 2012).

In commercial animal production systems, nutrient requirements are commonly estimated using the empirical and factorial methods (Patience et al., 1995). In the empirical method, nutrient requirements are estimated by feeding groups of pigs with increasing levels of the nutrient under evaluation and measuring one or more sets of response criteria (e.g. growth rate) within a given time or weight interval. The nutrient level at which the optimal population response is observed is identified as the population requirement for this nutrient and growing interval. The nature of this population response may be related to biological, technical, economic and/or environmental aspects (Jean dit Bailleul et al., 2000), but different response criteria may also suggest different nutrient requirement estimates (Baker, 1986; Hauschild et al., 2010). In the factorial method, daily nutrient requirements are estimated by adding together the requirements for maintenance and production (Fuller and Chamberlain, 1982; NRC, 2012). These requirements are estimated for each nutrient, taking into account the efficiency with which each nutrient is used for each metabolic function (van Milgen and Noblet, 2003). Ultimately, both methods of estimating nutrient requirements are based on experimental results derived from studying the relationship between nutrient intake and animal responses. However, these methods address the needs of different animals and different production situations. The empirical method estimates optimal nutrient requirements from a population perspective during a relatively long period, whereas the factorial method estimates the requirements of one animal during a short period, normally one day (Pomar, 2007). As indicated earlier, care must be taken when using factorial methods to estimate population requirements, because the relationship between the empirical and factorial methods is difficult to establish and is strongly affected by factors related to the animal, growing period and population heterogeneity (Pomar et al., 2003; Hauschild et al., 2010).

Factorial estimates of nutrient requirements are often extrapolated to populations on the assumption that the animal used for the factorial estimate is representative of the population. However, because of between-animal variations and the dynamic nature of animal responses, it is difficult to determine which animal in a population should be used to estimate the population requirements. It is common practice to use the average pig to represent the population. However, care has to be taken when making this assumption, because using the average pig to feed the population implies that half of the population will be overfed and the other half will be underfed (Hauschild et al., 2010; Brossard et al., 2009), leading to a reduction in population performance. For instance, Hauschild et al.

(2010) found that the maximum average daily gain (ADG) in pigs between 25 and 50 kg BW was reached with a lysine (Lys)/NE ratio 12% higher than the requirement estimated with the factorial method for the average pig. The maximum population ADG response in this BW interval corresponded closely to that of the pig in the 82nd percentile of the population. This value is in agreement with other *in vivo* (Brossard et al., 2014) and *in silico* results (Brossard et al., 2009). The latter authors suggested that oversupplying the average pig in the population by 15% at the beginning of the feeding phase maximizes population performance. For the feed conversion, however, the empirical estimates corresponded to those for a pig in the 58th percentile of the population. The relationship between the nutrient requirement estimates obtained with the empirical and factorial methods therefore has to be taken into account by either using the factorial method to estimate population requirements or using empirical values to calibrate factorial methods.

Several mathematical models based on factorial methods are proposed today as a way to estimate nutrient requirements in growing animals. This is because mathematical models can partially represent the complexity of animal responses to nutrient intake and the numerous modulating factors within the model. These models have been developed to simulate the growth of a single pig (NRC, 2012; Birkett and de Lange, 2001, Black et al., 1986; Moughan et al., 1987; Pomar et al., 1991; van Milgen et al., 2008; Whittemore and Fawcett, 1976) or the growth of pig populations (Brossard et al., 2009; Ferguson et al., 1997; Ferguson, 2008; Vautier et al., 2013; Knap, 1999; Wellock et al., 2004; Pomar et al., 2003). As indicated previously, caution is required when using single-animal models to estimate population responses to a given nutrient supply. Some single-animal models, however, have been adjusted to approximate population requirements using the average pig as the population representative. For instance, daily NRC (2012) Lys requirements are 13% higher than the average pig's daily requirements (Remus et al., 2015). This model was calibrated for maximum population responses by implicitly considering between-animal variability in adjusting different estimates (e.g. post-absorptive efficiency of nutrient use) from values that had been established in individual animals (e.g. Pomar et al., 2003, cited by NRC, 2012).

Before these models can be used in practical situations, they have to be properly calibrated using data collected from historical reference populations. Model users are therefore confronted with the difficulty of identifying the right non-living reference population to represent the animal responses of a future population, as well as the fact that the present-day animals may follow different feed intake and growth patterns than those observed in the chosen reference population. Model users have to be careful to identify any differences that may exist between the reference and target populations, as well as any changes in the evolution of this target population during growth.

4 Real-time estimation of individual pig nutrient requirements

Like broiler chickens and many other livestock species, pigs are typically raised and fed in groups, usually with the same feed being given to all animals in the group during a given period. As previously stated, nutrient requirements vary widely between animals in a population (Brossard et al., 2009; Pomar et al., 2003) and evolve over time according to individual patterns (Andretta et al., 2014; Hauschild et al., 2010). There are two important sources of variation that can be controlled to improve animal production efficiency in conventional livestock production systems: 1) variation between animals within the group receiving the same feed and 2) changes in individual or group nutrient requirements over time. Given that, for most nutrients, underfed animals will exhibit reduced performance while overfed animals will exhibit near-optimal performance, nutrients have to be provided in sufficient amounts to satisfy the requirements of the most demanding animals in the group in order to obtain maximum herd production performance (i.e. growth) (Hauschild et al., 2010). In this situation, almost all animals receive more nutrients than they actually need. Providing animals with high levels of nutrients to maximize herd performance is common practice in conventional commercial livestock operations, even though maximum growth does not ensure maximum economic efficiency (Hauschild et al., 2010; Niemi et al., 2010). Furthermore, to account for the variability not only between animals but also between feed ingredients and other uncontrolled factors (e.g. environment, health status), nutritionists include safety margins when formulating diets for maximum population responses. The need for these safety margins can be seen as an admission of our inability to precisely estimate the nutritional value of feed ingredients and the nutrient requirements of groups of animals (Patience, 1996).

Precision nutrition may play an important role in animal production because innovative monitoring approaches are making it easier to determine nutrient requirements in real time. Precision feeding involves the use of feeding techniques that provide animals with diets tailored to the production objectives (e.g. maximum or controlled production rates), including environmental and animal welfare standards. The development of practical precision feeding systems requires the integration of three types of activities: 1) automatic collection of data, 2) processing of the collected data and 3) actions to control the system (Aerts et al., 2003; Banhazi et al., 2012; Berckmans, 2004). The use of mechanistic models, described earlier in this chapter, in the development of precision feeding systems has been criticized because these models are overly complex and the information required by the model to simulate practical conditions is not always available (Aerts et al., 2003; Wathes et al., 2008). New mathematical model approaches are therefore required to estimate real-time nutrient requirements for precision feeding in order to feed individual pigs according to their own patterns of feed intake and growth. The automatic collection of individual feed intake and weight data, together with the computational power and reliability of modern information technologies, enable advanced recursive technologies to be used in the development of precision feeding systems (Wathes et al., 2008; Pomar et al., 2009).

In an effort to develop sustainable precision farming systems for swine, a mathematical model was proposed for estimating daily amino acid requirements and optimal dietary concentrations for each individual growing-finishing animal in a herd according to its actual BW and actual growth and feed intake patterns (Hauschild et al. 2012). This model follows the 'grey box model' approach (Roush, 2006), which is a combination of 'black box' (empirical) and 'knowledge-based' (mechanistic) components (Fig. 1).

The empirical component of the model is used to estimate the starting-day ($t + 1$) daily feed intake (DFI), BW and daily gain (DG) based on measurements taken in each individual animal up to the current day (t). The objective of this approach is to follow the dynamic DFI and BW trajectories of each pig in the herd in real time. Short-term variations over time and temporary declines in performance are not taken into account in this model. Forecasting methods using exponential smoothing techniques, which weigh recent and past observations in different ways, are appropriate for this model because they reduce fluctuations from irregular observations in the observed time series (Claycombe and

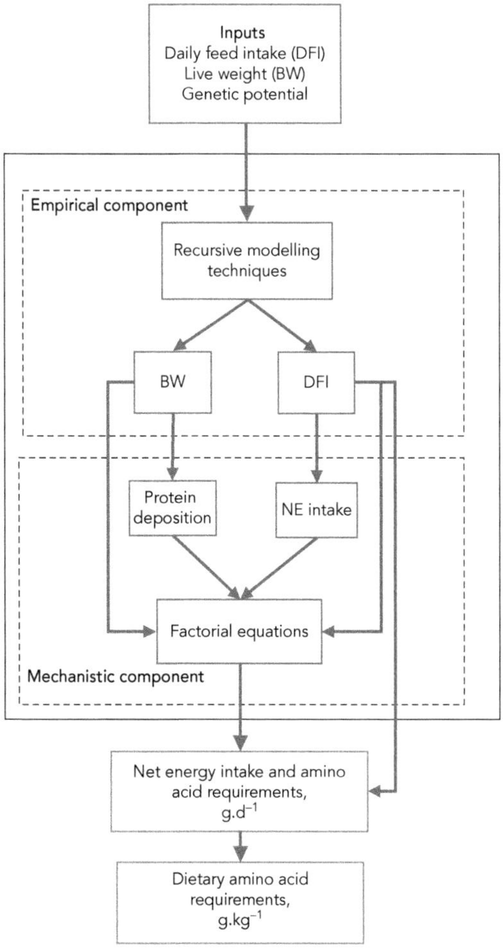

Figure 1 General outline of the model with the empirical and mechanistic model components used to estimate daily nutrient requirements for each individual in one pig population, according to their actual growth and feed intake patterns (from Hauschild et al., 2012).

Sullivan, 1977). The double-exponential smoothing forecasting time series method was chosen for this model because the DFI and BW of pig populations show clear long-run trends and because this method can work with a limited number of observations. The double-exponential forecasting equation, which produces an *i*-period-ahead forecast at time *t*, was evaluated (Hauschild et al., 2012) for individuals using individual DFI and BW data collected from nine performance tests (involving 2406 pigs) at the CDPQ Experimental Station in Deschambault (Quebec, Canada) and calibrated (Rivest et al., 2012).

The mechanistic component of the model was used to estimate starting-day nutrient requirements and optimal dietary amino acid concentrations for each pig in the herd based on the starting-day expected DFI (or NE intake), BW and DG information previously estimated by the empirical model component. Daily PD was assumed to be a given proportion of DG (PD/DG), which would evolve over time according to a function calculated from trials held at the Lennoxville Experimental Farm (Rivest et al., 2013, unpublished results). In this mechanistic model component, BW, DFI and DG are the driving variables used in a factorial procedure that estimates the standardized ileal digestible (SID) Lys (g/day) of each individual animal. Maintenance requirements are estimated as suggested by van Milgen et al. (2008b). The SID Lys growth requirements are calculated assuming 7% Lys in protein gain (Mahan and Shields, 1998) and 72% SID Lys retention efficiency (Mohn et al., 2000b). Protein and other amino acid (AA) requirements are estimated in relation to SID Lys, according to an ideal AA profile. The optimal concentration of SID Lys (g/NE) that should be provided in each pig's individual daily-adjusted diet is then calculated by dividing the sum of the maintenance and growth requirements by the expected NE intake. Other nutrient requirements, including minerals or vitamins, are not yet explicitly estimated in the model version described.

Thus, the empirical model component uses up-to-date data on each pig to estimate the starting-day expected DFI, BW and DG values, and the mechanistic model component then uses these forecasted values to estimate the SID Lys and other AA requirements, as well as the optimal concentration of these nutrients in the feed for the starting day. An initial evaluation of the model (Hauschild et al., 2012) was performed with data from a study that investigated the effect of feeding pigs with a three-phase or daily multiphase system. The empirical component results indicated that the DFI and BW trajectories of individual pigs fed *ad libitum* could be predicted 1 day (DFI) or 7 days (BW) ahead with small errors. The mechanistic model component estimated the optimal SID Lys/NE ratio with reasonable variation between animals (average CV = 7%) and over time (average CV = 14%). Thus, the AA requirements estimated by the model are animal- and time-dependent and follow the individual DFI and BW growth patterns in real time. The empirical approach used in the proposed model to estimate current-day DFI, BW and DG does not provide information on the biological mechanisms involved in the changes in these variables. However, it was concluded that the proposed model can accurately follow the DFI and BW trajectories of each individual pig in real time (Hauschild et al., 2012). Based on these trajectories and classical factorial equations, the model dynamically estimates the AA requirements of each individual animal. The factorial method has been calibrated in two animal trials (Zhang et al., 2012; Cloutier et al., 2015), and the overall approach of estimating real-time AA requirements has been challenged in three validation trials (Andretta et al., 2014, 2016b; Pomar et al., 2014). This model has recently been updated (Rivest et al., 2012) and will soon be evaluated under commercial conditions.

5 Case study

All calibration and validation trials were performed at Agriculture and Agri-Food Canada's Sherbrooke Research and Development Centre using similar protocols. Pigs were maintained under thermo-neutral conditions in all trials. Water was provided through low-pressure nipple drinkers, and feed was provided individually through electronic feeding stations. The method of operation of these feeders has been described previously (Pomar et al., 2011, 2015). Briefly, the feeding stations identify each pig when its head enters the feeder and then deliver a blend of feeds containing the estimated concentration of nutrients required by the pig upon animal request and according to the assigned experimental treatment. A serving refers to the amount of feed delivered upon each effective serving request. A time delay is imposed to ensure that pigs eat each serving before requesting a new one. The serving size was progressively increased during the experiments and ranged from 15 to 25 g. A meal includes all of the servings delivered during each visit to the feeder. Pigs leave the feeder hopper empty or leave very small amounts of feed after each visit, thus ensuring that each pig receives the assigned amount of blended feed. These feeds were independently formulated in each of the trials and differed from each other in terms of nutrient density. The same data collection methodology was used in all trials. Feed intake was recorded automatically for each visit by a computerized device. Pigs were weighed weekly, and their body composition was measured at the beginning of each feeding phase and at the end of the trials by dual-energy X-ray densitometry (DXA, GE Lunar Prodigy Advance, GE Healthcare).

In the first calibration trial, growing (25–55 kg BW, n = 60) and finishing (70–100 kg BW, n = 60) pigs were used to calibrate the mechanistic component of the mathematical model. Four feeds (A1, A2, B1 and B2) were blended to provide each animal with a diet containing 60%, 70%, 80%, 90%, 100% or 110% of the estimated Lys requirements. Feeds A1 and B1 were formulated to supply 110% of the assumed SID AA requirements of a 20-kg and a 110-kg BW pig with high PD potential, respectively (NRC, 1998). Feeds A2 and B2 had the same nutrient composition as A1 and B1, respectively, with the exception of their SID Lys components, which were formulated to supply 60% of the assumed SID Lys requirements of a 20-kg and a 110-kg BW pig. Dietary P and Ca requirements were estimated in this trial and the subsequent calibration and validation trials according to the method of Jondreville and Dourmad (2005). The detailed results of these trials were described by Zhang et al. (2012). To summarize, the maximum average DFI (ADFI) and average daily PD (APD) were observed in the animals that were fed according to their estimated requirements (100%) during the growing phase but that received diets that provided 110% of their estimated Lys requirements during the finishing phase. Although the model seemed to accurately estimate the Lys requirements of pigs from 25 to 55 kg BW, it appeared to underestimate the requirements of heavier animals.

A second calibration project was developed after the model was updated to reflect the previous results. In this second trial, the performance of barrows from two genetic lines (terminal cross, n = 75; maternal cross, n = 72) was studied during the growing and finishing periods. Four Lys supply levels were tested, respectively, representing 70%, 85%, 100% and 115% of the estimated requirements. Four feeds were independently formulated, with their nutrient concentrations determined in the same manner as in the previous trial. The detailed results of this trial were presented by Cloutier et al. (2015). In this project, the model slightly underestimated the pigs' Lys requirements during the growing period, given that optimum APD and ADG were achieved at 115% of Lys requirements. However, the method adequately estimated the finishing pigs' requirements, given that maximal performance was achieved at 100% of Lys requirements.

After further adjustments, the model was challenged in three validation trials. These projects were designed to evaluate the impact that switching from conventional to precision feeding systems in growing-finishing pig operations would have on productive performance, nutrient utilization, body composition and environmental costs.

In the first validation trial, described by Pomar et al. (2014), 96 pigs were used to study the effect of feeding them in a three-phase-feeding system (3PF) or in a precision feeding system in which they received diets adjusted daily by group (PFG). Feeds for the 3PF and PFG treatments were obtained by mixing two feeds, one with a high nutrient concentration and the other with a low nutrient concentration. The pigs assigned to the PFG treatment consumed 7.3% less protein ($P < 0.01$), tended to gain more weight (+2.4%, $P = 0.08$) and excreted 12% less nitrogen ($P < 0.01$) than the 3PF pigs.

The second validation project, described by Andretta et al. (2014), was developed using 60 barrows (41.2 ± 3.9 kg initial BW) to compare conventional 3PF feeding programmes to precision feeding programmes by adjusting the blended proportions of feeds daily to meet the estimated nutritional requirements of the group (equivalent to the PFG treatment tested in the previously described trial) or of each pig individually (PFI). Two experimental feeds (named A and B) were independently formulated for this trial on the basis of NE and SID AA, using the same ingredient composition database. Feed A had high nutrient density levels and was formulated to satisfy the requirements of the most demanding pigs at the beginning of the first growing period, whereas feed B had a low nutrient density level and was formulated to satisfy the requirements of the least demanding pigs at the end of the last growing period, as in previous experiments. The PFG and PFI programmes showed ADFI, ADG, feed efficiency, protein and phosphorus retention, final BW, and carcass and primal cut weight values that were similar to those obtained for the 3PF programme. However, applying the PFI feeding programme reduced SID Lys intake by 27%, estimated nitrogen excretion by 22% and estimated phosphorus excretion by 27% ($P < 0.05$) compared with the conventional programme.

A similar methodology was used in the third validation trial, conducted by Andretta et al. (2016b), to assess the performance of 70 pigs (30.4 ± 2.2 kg initial BW) receiving a 3PF programme or daily individual phase-feeding programmes in which the blending proportions of the feeds were updated daily to meet 110%, 100%, 90% or 80% of the estimated Lys requirements. It is important to note that the precision feeding programme corresponding to 100% of the nutrient requirements was equivalent to the PFI treatment described in the previous project. According to the results, ADFI, feed conversion and body fat mass were similar among treatments. Feeding pigs using a daily-adjusted programme providing 110%, 100% or 90% of the estimated individual Lys requirements also did not influence ADG, protein retention, BW or body protein mass in comparison with the animals in the 3PF programme. However, feeding pigs individually with diets tailored to match 100% of nutrient requirements made it possible to reduce ($P < 0.05$) digestible Lys intake by 26%, estimated nitrogen excretion by 30% and feeding costs by 10% relative to the group phase-feeding programme. Feeding behaviour was also evaluated in this project, and the studied responses (number of meals, duration of meals, intervals between meals, feed consumed per meal and feed consumption rates) were not affected by the feeding programmes (Andretta et al., 2016a).

Recently, the potential environmental impact of applying precision feeding during the growing-finishing phase was estimated through life cycle assessment (Andretta et al., 2017, unpublished results). Data collected in the validation projects described above were used in a simulation model developed for the scenario of Brazilian pig production. The results showed that applying the PFG programme may mitigate the potential climate change impact by up to 4% compared with the 3PF programme. Better results were achieved with the PFI programme. Making a further adjustment by feeding pigs according to their individual nutrient requirements may mitigate the potential climate change impact by up to 6% and the potential eutrophication and acidification impact by up to 5% compared with the 3PF programme. The results obtained in all validation projects and in the life cycle assessment study clearly indicate that precision feeding is an effective approach for improving the sustainability of pig production by reducing nutrient intake, nutrient excretion and feeding costs.

6 Summary

In current pig production systems, pigs are raised in groups and the same feed is provided to all pigs during long periods. However, nutrient requirements vary widely among animals and evolve over time according to individual patterns. Precision feeding is proposed as a means of controlling these two important sources of variation. Precision feeding involves feeding animals with diets tailored to the production objectives (e.g. maximum or controlled production rates), including environmental and animal welfare standards. The development of practical precision feeding systems requires the integration of three types of activities: 1) automatic data collection, 2) data processing and 3) control actions. Real-time estimation of nutrient requirements is the most important aspect of data processing in precision feeding systems. To this end, a new method of estimating nutrient requirements that integrates current knowledge on the pig metabolism has been developed and incorporated into a mathematical model. This model includes empirical and mechanistic model components. The empirical component estimates DFI, BW and DG based on individual pig information collected in real time. Based on the DFI, BW and DG estimates, the mechanistic model component uses classical factorial methods to estimate the optimal concentration of AAs that must be offered to each pig to meet its requirements. The mechanistic model component was calibrated in two animal trials, and the overall approach of estimating real-time AA requirements was challenged in three validation trials. The recent research projects described in this chapter clearly indicated that feeding pigs in accordance with their individual requirements is essential to maximize nutrient efficiency and ensure the sustainability of the pig industry by reducing nutrient excretion and feeding costs. The new nutritional approach represents a paradigm shift in pig feeding because the optimal dietary nutrient level is no longer considered a static population attribute, but rather a dynamic process that evolves independently for each animal. Precision feeding is a highly promising avenue for improving resource use efficiency (e.g. nutrients) in comparison with conventional group phase-feeding programmes and is expected to receive increasing attention in the coming years from both the research and the industry sectors.

7 Future trends

Meeting individual pig nutrient requirements has had very promising results, and the adoption of the precision feeding approach described in this chapter may represent a major advance for modern pig farming by improving nutrient efficiency and the sustainability of the overall production system. However, further development of this feeding approach is required. The empirical component of the model provides reasonable estimates of feed intake and growth patterns in unstressed pigs (typical individuals). If the animals face health, environmental or other stressors at any point in their life, the current recursive approach may not be appropriate (Hauschild et al., 2012). Furthermore, this empirical model component takes into account only long-term trends in feeding behaviour, but recent studies indicate that short-term feeding patterns (e.g. diurnal feeding) may also affect nutrient use. It is also known that a wide range of pathogens, social and environmental conditions, and the degree of on-farm hygiene induce a series of responses in the animal, such as reduced feed intake and increased energy expenditure, body protein synthesis and catabolism, and body temperature. The extent and duration of the animal's responses to these stressors depend on the type and intensity of its exposure to pathogens and on its ability to stimulate its immune response. Nonetheless, of all these animal responses, feed intake is the one that can most easily be observed in commercial facilities. Characterization of short- and long-term feeding patterns in individual pigs and identification of any disruptions of these trends are being included in research projects today with the objective of developing automatic tools that would allow early detection of health issues. Pigs seem to have a limited ability to modify their feed intake to compensate for other nutrient deficiencies or excesses. However, pigs seem to be able to offset some nutrient deficiencies by increasing their nutrient efficiency rather than modifying feed intake. This metabolic conditioning has been observed for phosphorus, lysine, methionine and threonine, but it has not yet been incorporated into the current model. Given that pigs are frequently raised in tropical and subtropical conditions, animal responses to environmental stressors such as heat stress, health issues and other stressors also need to be addressed in future projects. Furthermore, ongoing projects have been planned in order to perform the required animal studies to optimize the formulation of premixes with a view to updating and calibrating the model currently used for real-time prediction of AA and phosphorus requirements and evaluating the technical, economic and environmental impact of precision feeding on commercial farms. In addition to refinements to the feeders and concepts used for individual precision feeding, the technology needs to be implemented and validated in commercial facilities.

8 References

Aerts, J.-M., Wathes, C. M. and Berckmans, D. (2003), Dynamic data-based modelling of heat production and growth of broiler chickens: development of an integrated management system, *Biosyst. Eng.*, 84, 257–66.

Andretta, I., Pomar, C., Kipper, M., Hauschild, L. and Rivest, J. (2016a), Feeding behavior of growing-finishing pigs reared under precision feeding strategies, *J. Anim. Sci.*, 94, 3042–50.

Andretta, I., Pomar, C., Rivest, J., Pomar, J., Lovatto, P. A. and Neto, J. R. (2014), The impact of feeding growing-finishing pigs with daily tailored diets using precision feeding techniques on animal performance, nutrient utilization, and body and carcass composition, *J. Anim. Sci.*, 92, 3925–36.

Andretta, I., Pomar, C., Rivest, J., Pomar, J. and Radunz, J. (2016b), Precision feeding can significantly reduce lysine intake and nitrogen excretion without compromising the performance of growing pigs, *Animal*, 10, 1137–47.

Baker, D. H. (1986), Problems and pitfalls in animal experiments designed to establish dietary requirements for essential nutrients, *J. Nutr.*, 116, 2339–49.

Banhazi, T. M., Lehr, H., Black, J. L., Crabtree, H., Schofield, P., Tscharke, M. and Berckmans, D. (2012), Precision livestock farming: An international review of scientific and commercial aspects, *Int. J. Agric. Biol. Eng.*, 5, 1–9.

Berckmans, D. (2004), Automatic on-line monitoring of animals by precision livestock farming, ISAH Conference on Animal Production in Europe: The Way Forward in a Changing World, 11–13 October 2004, Saint-Malo, France, 27–31.

Birkett, S. and De Lange, C. F. M. (2001), A computational framework for a nutrient flow representation of energy utilization by growing monogastric animals, *Br. J. Nutr.*, 86, 661–74.

Black, J. L., Campbell, R. G., Williams, I. H., James, K. J. and Davies, G. T. (1986), Simulation of energy and amino acid utilisation in the pig, *Res. Dev. Agric.*, 3, 121–45.

Brossard, L., Dourmad, J.-Y., Rivest, J. and Van Milgen, J. (2009), Modelling the variation in performance of a population of growing pig as affected by lysine supply and feeding strategy, *Animal*, 3, 1114–23.

Brossard, L., Vautier, B., Van Milgen, J., Salaun, Y. and Quiniou, N. (2014), Comparison of in vivo and in silico growth performance and variability in pigs when applying a feeding strategy designed by simulation to control the variability of slaughter weight, *Anim. Prod. Sci.*, 54, 1939–45.

Cangar, Ö., Aerts, J.-M., Vranken, E. and Berckmans, D. (2008). Effects of different target trajectories on the broiler performance in growth control, *Poult. Sci.*, 87, 2196–207.

Claycombe, W. and Sullivan, W. (1977), *Fundamentals of forecasting*. Reston Publishing Company, Inc., Reston, Virginia, USA.

Cloutier, L., Pomar, C., Letourneau Montminy, M. P., Bernier, J. F. and Pomar, J. (2015), Evaluation of a method estimating real-time individual lysine requirements in two lines of growing-finishing pigs, *Animal*, 9, 561–8.

Dourmad, J. Y., Seve, B., Latimier, P., Boisen, S., Fernandez, J., Van Der Peet-Schwering, C. and Jongbloed, A. W. (1999), Nitrogen consumption, utilisation and losses in pig production in France, The Netherlands and Denmark, *Livest. Prod. Sci.*, 58, 261–4.

Ferguson, N. S. (2008), Application of an integrated management system to improve performance and profitability in finishing pigs, Banff Pork Seminar, Banff.

Ferguson, N. S., Gous, R. M. and Emmans, G. C. (1997) Predicting the effects of animal variation on growth and food intake in growing pigs using simulation modelling, *Animal Sci.*, 64, 513–22.

Flachowsky, G. and Kamphues, J. (2012), Carbon footprints for food of animal origin: what are the most preferable criteria to measure animal yields?, *Animals*, 2, 108.

Fuller, M. F. and Chamberlain, A. G. (1982), Protein requirements of pigs. In Haresign, W. (Ed.), *Recent Advances in Animal Nutrition*. Butterworths, London, UK, pp. 175–86.

Hauschild, L., Lovatto, P. A., Pomar, J. and Pomar, C. (2012), Development of sustainable precision farming systems for swine: Estimating real-time individual amino acid requirements in growing-finishing pigs, *J. Anim. Sci.*, 90, 2255–63.

Hauschild, L., Pomar, C. and Lovatto, P. A. (2010), Systematic comparison of the empirical and factorial methods used to estimate the nutrient requirements of growing pigs, *Animal*, 4, 714–23.

Jean dit Bailleul, P., Bernier, J. F., Van Milgen, J., Sauvant, D. and Pomar, C. (2000), The utilization of prediction models to optimize farm animal production systems: the case of a growing pig model. In McNamara, J. P., France, J. and Beever, D. (Eds), *Modelling Nutrient Utilization in Farm Animals*. CABI International, Wallingford, Oxon, UK, pp. 379–92.

Jondreville, C. and Dourmad, J. Y. (2005), Le phosphore dans la nutrition des porcs, *INRA Prod. Anim.*, 18, 183–92.

Knap, P. W. (1999), Simulation of growth in pigs: evaluation of a model to relate thermoregulation to body protein and lipid content and deposition, *Anim. Sci.*, 68, 655–79.

Knap, P. W. (2000), Time trends of Gompertz growth parameters in 'meat-type' pigs, *Anim. Sci.*, 70, 39–49.

Knap, P. W., Roehe, R., Kolstad, K., Pomar, C. and Luiting, P. (2003), Characterization of pig genotypes, *J. Anim. Sci.*, 81(E. Suppl. 2), E187–95.

Mahan, D. C. and Shields, R. G., JR. (1998), Essential and nonessential amino acid composition of pigs from birth to 145 kilograms of body weight, and comparison to other studies, *J. Anim Sci.*, 76, 513–21.

Mohn, S., Gillis, A. M., Moughan, P. J. and de Lange, C. F. (2000a), Influence of dietary lysine and energy intakes on body protein deposition and lysine utilization in the growing pig, *J. Anim. Sci.*, 78, 1510–19.

Mohn, S., Gillis, A. M., Moughan, P. J. and de Lange, C. F. (2000b), Influence of dietary lysine and energy intakes on body protein deposition and lysine utilization in the growing pig, *J. Anim Sci.*, 78, 1510–19.
Moughan, P. J. (1989), Simulation of the daily partitioning of lysine in the 50 kg liveweight pig – A factorial approach to estimating amino acid requirements for the growth andaintenance, *Res. Dev. Agric.*, 6, 7–14.
Moughan, P. J. (1999), Protein metabolism in the growing pig. In Kyriazakis, I. (Ed.), *A Quantitative Biology of the Pig*. CAB International, Wallingford, Oxon, UK., pp. 299–331.
Moughan, P. J., Smith, W. C. and Pearson, G. (1987), Description and validation of a model simulating growth in the pig (20–90 kg liveweight), *NZ J. Agr. Res.*, 30, 481–90.
National Research Council (NRC) (1998). *Nutrient Requirements of Swine*, National Academy Press, Washington, DC., USA.
National Research Council (NRC) (2012). *Nutrient Requirements of Swine*, National Academy Press, Washington, DC., USA.
Niemann, H., Kuhla, B. and Flachowsky, G. (2011), Perspectives for feed-efficient animal production, *J. Anim. Sci.*, 89, 4344–63.
Niemi, J. K., Sevón-Aimonen, M.-L., Pietola, K. and Stalder, K. J. (2010), The value of precision feeding technologies for grow-finish swine, *Livest. Sci.*, 129, 13–23.
Noblet, J. and van Milgen, J. (2004), Energy value of pig feeds: Effect of pig body weight and energy evaluation system, *J. Anim. Sci.*, 82, E229–38.
Parsons, D. J., Green, D. M., Schofield, C. P. and WhittemoRE, C. T. (2007), Real-time control of pig growth through an integrated management system, *Biosyst. Eng.*, 96, 257–66.
Patience, J. F. (1996), Precision in swine feeding programs: an integrated approach, *Anim. Feed Sci. Tech.*, 59, 137–45.
Patience, J. F., Thacker, P. A. and De Lange, C. F. M. (1995), *Swine Nutrition Guide*. Prairie Swine Center, Saskatoon, Saskatchewan Canada.
Pomar, C. (2007), Predicting responses and nutrient requirements in growing animal populations: the case of the growing-finishing pig. In Hanigan, M. D., Novotny, J. A. and Marstaller, C. L. (Eds), *Mathematical modeling in nutrition and agriculture*. Virginia Polytechnic and State University.
Pomar, C., Harris, D. L. and Minvielle, F. (1991), Computer-simulation model of swine production systems: I. Modeling the growth of young pigs, *J. Anim. Sci.*, 69, 1468–88.
Pomar, C., Hauschild, L., Zhang, G. H., Pomar, J. and Lovatto, P. A. (2009), Applying precision feeding techniques in growing-finishing pig operations, *Rev. Bras. Zootecn.*, 38(Supl. special), 226–37.
Pomar, C., Kyriazakis, I., Emmans, G. C. and Knap, P. W. (2003), Modeling stochasticity: dealing with populations rather than individual pigs, *J. Anim. Sci.*, 81(E. Suppl. 2), E178–86.
Pomar, C., Pomar, J., Dubeau, F., Joannopoulos, E. and Dussault, J.-P. (2014), The impact of daily multiphase feeding on animal performance, body composition, nitrogen and phosphorus excretions, and feed costs in growing–finishing pigs, *Animal*, 8, 704–13.
Pomar, C., Pomar, J., Rivest, J., Cloutier, L., Letourneau-Montminy, M.-P., Andretta, I. and Hauschild, L. (2015), Estimating real-time individual amino acid requirements in growing-finishing pigs: towards a new definition of nutrient requirements? In Sakomura, N. K., Gous, R., Kyriazakis, I. and Hauschild, L. (Eds), *Nutritional modelling for pigs and poultry*. CAB International, Wallingford, UK, pp. 157–74.
Pomar, J., López, V. and Pomar, C. (2011), Agent-based simulation framework for virtual prototyping of advanced livestock precision feeding systems, *Comput. Electron. Agr.*, 78, 88–97.
Poulsen, H. D., Jongbloed, A. W., Latimier, P. and Fernandez, J. A. (1999), Phosphorus consumption, utilisation and losses in pig production in France, The Netherlands and Denmark, *Livest. Prod. Sci.*, 58, 251–9.
Remus, A., Pomar, C. and Hauschild, L. (2015), Amino acids requirements of growing pigs differ between different factorial methods, *J. Anim. Sci.*, 93(Supplement 2), 49 (Abstract).
Rivest, J., Pomar, C., Pomar, J. and Hauschild, L. (2012), Estimating optimal smoothing parameters for the daily feed intake and weight forecasts required to implement sustainable precision farming systems for swine, *Can. J. Anim. Sci.*, 91, 720 (Abstract).

Roush, W. B. (2006), Advancements in empirical models for predictions and prescription. In Fisher, C., Gous, R. and Morris, T. (Eds), *Mechanistic Modelling in Pig and Poultry Production*. CAB International, Wallingford, Oxfordshire, UK, pp. 97–116.

St-Pierre, N. R. and Weiss, W. P. (2012), Nutrient requirements, responses, and feed efficiency, Proceedings of the Four-State Dairy Nutrition and Management Conference, June 13–14 2012, Dubuque, IA. 1–7.

Stein, H. H., Seve, B., Fuller, M. F., Moughan, P. J. and de Lange, C. F. M. (2007), Invited review: amino acid bioavailability and digestibility in pig feed ingredients: terminology and application, *J. Anim. Sci.*, 85, 172–80.

Van Barneveld, R. J., Batterham, E. S. and Norton, B. W. (1994), The effect of heat on amino acids for growing pigs 3. The availability of lysine from heat-treated field peas *(Pisum sativum* cultivar Dundale) determined using the slope-ratio assay, *Br.J. Nutr.*, 72, 257–75.

Van Milgen, J. and Noblet, J. (2003), Partitioning of energy intake to heat, protein, and fat in growing pigs, *J. Anim. Sci.*, 81(E. Suppl. 2), E86–E93.

Van Milgen, J., Valancogne, A., Dubois, S., Dourmad, J.-Y., Sève, B. and Noblet, J. (2008), InraPorc: a model and decision support tool for the nutrition of growing pigs. *Anim. Feed Sci. Tech.*, 143, 387–405.

Vautier, B., Quiniou, N., Van Milgen, J. and Brossard, L. (2013), Accounting for variability among individual pigs in deterministic growth models, *Animal*, 7, 1265–73.

Wathes, C. M., Kristensen, H. H., Aerts, J.-M. and Berckmans, D. (2008), Is precision livestock farming an engineer's daydream or nightmare, an animal's friend or foe, and a farmer's panacea or pitfall?, *Comput. Electron. Agr.*, 64, 2–10.

Wellock, I. J., Emmans, G. C. and Kyriazakis, I. (2004), Modeling the effects of stressors on the performance of populations of pigs, *J. Anim. Sci.*, 82, 2442–50.

Whittemore, C. T. and Fawcett, R. H. (1976), Theoretical aspects of a flexible model to stimulate protein and lipid growth in pigs, *Anim. Prod.*, 22, 87–96.

Zhang, G. H., Pomar, C., Pomar, J. and Del Castillo, J. R. E. (2012), L'alimentation de précision chez le porc charcutier: Estimation des niveaux dynamiques de lysine digestible nécessaires à la maximisation du gain de poids, *Journées de la Recherche Porcine*, 44, 171–6.

Index